the Book
of the Cosmos

Imagining the Universe
from Heraclitus to Hawking

A Helix Anthology

edited by

Dennis Richard Danielson

§

HELIX BOOKS

PERSEUS PUBLISHING
Cambridge, Massachusetts

Many of the designations used by manufacturers and sellers to distinguish their products are claimed as trademarks. Where those designations appear in this book and Perseus Publishing was aware of a trademark claim, the designations have been printed in initial capital letters.

A CIP record for this book is available from the Library of Congress.
ISBN: 0–7382–0247-9

Perseus Publishing is a member of the Perseus Books Group

Text design by Jeff Williams
Set in 11-point Sabon by the Perseus Books Group

 2 3 4 5 6 7 8 9 10—03 02 01 00
First printing, June 2000

Perseus Publishing books are available at special discounts for bulk purchases in the U.S. by corporations, institutions, and other organizations. For more information, please contact the Special Markets Department at HarperCollins Publishers, 10 East 53rd Street, New York, NY 10022, or call 1–212–207–7528.

Find Helix Books on the World Wide Web at http://www.perseuspublishing.com

For Nora, Eva, Jessa, and John
in honour of their maternal grandmother
Eleanor Henshaw

The Greeks have borrowed a name for the universe from ornament, on account of the variety of the elements and the beauty of the stars. For it is called among them *kosmos*. . . . For with the eyes of the flesh we see nothing fairer than the universe.

–Isidore of Seville (d. 636)

The whole world is a shadow, a way, and a trace; a book with writing front and back.

–St. Bonaventure, 13th c.

Wisdom is to be read in the immense book of God, which is the world.

–Tommaso Campanella, 1616

Cosmology is a very useful science, and they are benefactors of mankind who help us to read this book of the visible heavens.

–Thomas Vivian, 1792

We especially need imagination in science. It is not all mathematics, nor all logic, but it is somewhat beauty and poetry.

–Maria Mitchell, 1871

What entity, short of God, could be nobler or worthier of [our] attention than the cosmos itself?

–Rudy Rucker, 1984

Cosmologists and artists have much in common.

–George Smoot, 1993

Contents

Part 1
Cosmological Origins

1 **We Have Seen But Few of His Works**   2

Torah, Sacred Poetry, Apocrypha, New Testament

Western cosmology arises, textually, from two different
traditions, the Greek and the Hebrew. It is the latter, the
"biblical" tradition, that gives us a world that is spoken
into being and bespeaks its Creator.

2 **Twice into the Same River?**   12

Heraclitus and Parmenides

Two of the first philosophers of nature embody an enduring
antithesis: the cosmology of change versus the cosmology of
permanence.

3 **The Things of the Universe Are Not Sliced Off
with a Hatchet**   18

Empedocles and Anaxagoras

Empedocles uncovers the roots of all physical things and
tells how Strife entangles them, but Anaxagoras declares
that Mind rules the cosmos.

Part 2
Ptolemy, Middle Earth, Middle Ages

PART 5
THE UNIVERSE RE-IMAGINED

PART 6
BEGINNINGS AND ENDS

Acknowledgments

This book, the making of which has been one of the greatest joys and adventures of my life, had two beginnings, a general and a specific. In the summer of 1994 my family and I were driving to a family reunion in Montana. My wife's mother, Eleanor Henshaw, had a few weeks earlier attended a lecture by Hugh Ross, a "popular" speaker whose specialty is cosmology. She had thought "Dennis would be interested in this," and had kindly brought along a cassette tape of the lecture for me to listen to in the car. Within weeks I was reading everything I could lay my hands on that had to do with cosmology, and was soon integrating cosmological perspectives into my literature courses whenever I could. In 1995–96, at the University of British Columbia, I had the good fortune to teach an honors seminar on Literature and Cosmology, to whose members I am still grateful for exploring with me the literary dimensions of cosmology and the cosmological dimensions of literature.

Then in the summer of 1997 I was in Oxford continuing my study of the literary history of cosmology. Like any normal academic I wanted to write a book about my interest. Like any normal academic, I was unsure about how to focus my efforts. And like any normal academic in Oxford, I wandered into Blackwell's Bookshop to imbibe that heady combination of exhilaration at seeing the world of books at its best, and despair at so much evidence seeming to suggest there's nothing left for one to contribute. In the Norrington Room, one book on the "Featured Titles" shelf caught my eye: *The Faber Book of Science*, edited by John Carey. For a moment I doubted. Could this be John Carey the Merton Professor of Literature, who twenty years earlier had been my own supervisor when I was a research student at Oxford? Indeed it was—and the book was a splendid anthology of short, outstanding examples of science writing from the sixteenth century to the present. I bought and devoured the book, and within about a day was hatching a plan to adopt and adapt Carey's format for a critical anthology of cosmological writing. I met with Carey to discuss the idea; and with his typically generous advice and encouragement, I embarked on the project whose completion you have before you.

Thus I begin my expressions of gratitude by offering deepest thanks to Eleanor Henshaw and to John Carey. I am also humbly grateful to many others who have encouraged me on my way with good fellowship and good example: to my wife, Janet Henshaw Danielson, a composer whose interest in the history of music has perfectly complemented mine in cosmology; my daughter and stylistic critic extraordinaire Nora Danielson; my colleague Lee M. Johnson, lover of poetry and the heavens; my running partners, Ron Donkersloot and Leo Pel, who early Sunday morning after early Sunday morning patiently endured my company as I expatiated on the moons of Jupiter, astronomic spectroscopy, spacetime curvature, and black holes. Thanks to Ken Kiers, now of Taylor University, who in the early days of my cosmological obsession gave much patient advice; to John Leonard and John Tanner, whose too-infrequent but valued conversation helped inspire me to grow upward and outward from Miltonic soil. Still others, some of them anonymous, have along the way provided me with excellent counsel. To Pamela Gossin I offer sincere thanks for an astute early reading of my proposal—even if in the end the book could not do everything her capacious mind wished it could—and to Philip Sampson, Susan Squier, Reiner Dienlin, and Joseph Jones for suggesting authors for inclusion. One joy that has accompanied the drudgery of acquiring permissions to use copyright material is the contact this has given me with living authors, and I would like to express my gratitude for the kindness of David Berlinski, Kitty Ferguson, Werner Gitt, John Archibald Wheeler, and Owen Gingerich. I'm grateful to Jeffrey Robbins for having faith in this project, Sean Abbot for suggesting many improvements to my text, and to Elizabeth Carduff and Marco Pavia of Perseus Publishing for seeing the book through to completion. Institutionally, I have been well supported by the University of British Columbia, in particular with a grant for interdisciplinary research from the Hampton Fund. The project would not have been possible without the rich resources of three very different libraries: the Bodleian in Oxford, the University of British Columbia Library, and the Bob Prittie (Public) Library in Burnaby, where I live. And finally, on a highly personal but relevant note (moving from cosmology to urology!) I want to thank Doctors Gidon Frame and John Warner, without whose perceptiveness and skill during the final three months of this book's composition I might not have lived to see its publication.

D.R.D.

Introduction:
Telescopes for the Mind

Canadian cosmologist Werner Israel tells of the time he was interviewed for a television show by someone who had carefully prepared a list of questions to ask him about lipstick, blusher, and mascara. Although the kinship between cosmology and cosmetology probably did little to advance the career of that interviewer, it actually helps me here to introduce an idea central to the purpose of this book. In lecturing about cosmology, I sometimes try to break the ice by asking how many members of the audience wear cosmetics—and then I take advantage of their candor by pointing out that our word *cosmetics* derives from the Greek verb meaning "to bring order out of chaos." My point, simply, is that cosmology, like its etymological cousin cosmetology, is indeed about order, and about beauty.

Are we not drawn to the heavens in the first place because they are beautiful and because they are awesome? Their grandeur humbles us, thrills us, calls forth our contemplation, and inspires a craving (as Alan Guth has put it) "that has been part of human consciousness from the writing of Genesis to the scientific era of relativity and quantum mechanics."[1] What *is* the cosmos? How did it come into being? How are we related to it, and what is our place in it? Furthermore, when we contemplate the universe, isn't what we see and experience molded by what others of our species have seen and thought elsewhere and before us? What *I* see is in large measure an amalgam of what *we* see and have seen—and it is a very long and complex *we*. From the beginning of human history, others have looked at and spoken and written about this cosmos that is the object of our awe and our contemplation. And, to echo Wordsworth, the world is rich and dear to us both for itself and for the sake of those others who have preceded us and shaped our vision.

To make available and audible the voices of some of "those others"—of exceptional minds across time who have spoken and written about the cosmos—is this book's principal aim. Although we most naturally talk about *looking* at the heavens, the essence of *The Book of the Cosmos* is more precisely the process of *thinking* that is mediated by writing and reading about the cosmos. Important as pictures are to our understanding of the universe, they can often virtually bypass our critical faculties and make us feel as if we have understood something, when actually the "vision" that moves and inspires us goes far beyond the pictorial. What I offer here, therefore, are cosmologists' voices as embodied in their writings, accompanied by only a small handful of pictures. Employing a capacious and nontechnical definition of cosmology—discourse concerned with the cosmos and with cosmic questions—I have selected these writings using a number of criteria both objective and subjective. But above all, I have chosen readings I think succeed in evoking that very mixture of the beautiful and the awesome that draws us to contemplate this great universe in the first place.

By *contemplate* I don't, however, imply passive observation. Part of the beauty of literature, including cosmological literature, is its capacity to join author and reader in *active* contemplation—in acts of imagination and acts of interpretation. It will be clear from Chapter 1 onward how persistent is the idea that we can hear the heavens speak, and that the cosmos is a book that we can read. The same profound analogy of verbal communication undergirds much cosmological writing and, as my title intimates, informs the overall conception of *The Book of the Cosmos* itself. Finally, this whole splendid dimension of the verbal—with its evocation of beauty, order, meaningfulness, and often ambiguity, as well as its engagement of human imagination—justifies the book's aesthetic agenda. Philosopher Charles Hartshorne has written that science is "a form of love or sympathy, sympathy for the ideas of others and love of reality as open to observational inquiry. It is the imaginative, socially critical, and observational feeling for nature."[2] I hope that readers will find much of such love, sympathy for ideas, observation, imagination, and criticism in *The Book of the Cosmos*, the more so for its attempt to display cosmology as an art as well as a science.

To indulge in one more brief fit of etymology, I'd like to add that *anthology* means, roughly, a gathering of flowers. In *The Book of the Cosmos* I have tried to gather a single big bunch of cosmological blossoms picked from a range of species. You will find here excerpts from poetry, philosophy, theology; from diaries, dinner speeches, and dialogues; from epics, essays, and epistles—as well as from the more standard garden variety of colorful scientific prose. This book also collects a wider chronological range of cosmological specimens, from the beginnings of the western tradition to the present, than has previously been pressed between two covers. For all its range, it does not pretend to be an encyclopedia or a comprehensive history of cosmology. Nor is it a science or astronomy textbook, even though it presents many texts that are genuinely scientific or astronomical. But if you were my Cockney grandfather, you still might want to call this the fattest bloomin' collection yet gathered from the fields, and from along the grand boulevards, of western cosmology.

As I have already hinted, *The Book of the Cosmos* is also an anthology that deliberately transgresses the boundaries that often separate academic "cultures" or disciplines. C. P. Snow's somewhat shopworn account of "two cultures"—scientists on one side and literary intellectuals on the other, each group speaking their own language and, in splendid isolation, fancying themselves the true bearers of wisdom—still unfortunately has some validity as a description of the academic scene today. This picture has been updated by John Brockman, according to whom there is now a "third culture" consisting of eloquent scientists who have usurped the role of "rendering visible the deeper meanings of our lives"; meanwhile, those in the humanities, particularly the literary types, whose culture "dismisses science," busy themselves with "comment on comments, the swelling spiral of commentary eventually reaching the point where the real world gets lost."[3] According to this account, it is not that the scientists have staked out a territory separate from that of the humanities but, rather, that they have extended their domain to take in the entire continent of true learning, while the "arts" people inhabit floating islands that drift ever farther from the scientific mainland. According to Murray Gell-Mann, one of Brockman's third-culture colleagues, "there are people in the arts and humanities . . . who are proud of knowing very little about science and technology, or about mathematics. The opposite phenomenon is very rare. You may occasionally find a scientist who is ignorant of Shakespeare, but you will never find a scientist who is *proud* of being ignorant of Shakespeare."[4]

Although I would reply by admitting a *measure* of justice in the third-culture scientists' complaint, I doubt that scorn or smugness directed at whatever culture helps to solve our problems. My main fields are English literature and intellectual history, and my knowledge of Shakespeare may well exceed my knowledge of mathematics. I'm not proud of my arrested mathematical adolescence. Yet I do believe that people whose education is primarily in the humanities can both learn much from and contribute significantly to the sciences. In fact, I'd like to see an end to the war among the cultures, with no victors nor losers declared, and to work instead for a grand coalition in which a small degree of humility regarding one's own particular discipline and a large degree of respect regarding others' disciplines is the rule.

Study of the cosmos provides an excellent realm in which to exercise such principles of academic disarmament and diplomacy. Accordingly, working from my home territory in literature and intellectual history, I have assembled here a firsthand (if necessarily abridged) *textual* history of cosmology, a history that begins long before cosmology's establishment as a specific scientific discipline. Nevertheless, while aiming in this way to enhance interest in the sciences by employing tools primarily from the humanities, I have also tried to avoid making *The Book of the Cosmos* an unduly academic volume. My preferred model is more that of Renaissance humanism than of postmodern academia with its generally laudable if not conspicuously successful promotion of "interdisciplinarity." Renaissance humanists were interested in all things principally because of their intuition that all things *are* interesting and their conviction that all things are connected. The philosopher Giovanni Pico's delight was precisely in declaring and exploring those connections, and part of his effort was directed toward encouraging the investigation of physical reality through the practice of a kind of "magic" stripped of its occult or demonic connotations—something we have come to call experimental science. And the poet Philip Sidney, while he certainly recognized the differences among the genres of history, philosophy, and poetry (which today we would simply call "fiction"), saw the various "kinds" of writing as engaging in the same moral and educational undertaking. All had their roles to play in the *scientiae* (literally, the "knowledges"), even if Sidney thought poetry did a somewhat superior job of teaching, moving, and delighting. A century later, Isaac Newton the physicist and mathematician was publishing his own theories in a journal called not physical or mathematical but *philosophical* transactions (that is, *Philosophical Transactions of the Royal Society*).

In keeping with this old-fashioned model of interconnectedness, I have, as indicated, chosen excerpts of writings on the nature of the universe from across a wide spectrum of "philosophical" writings—from poetry to history of science to physical theory—writings that in varying ways, I hope, may indeed teach, move, and delight. Most of these selections have in common the creative exercise of imagination, something not less vital to good science than to good literature. Moreover, on a practical level, a humanities approach to the human history of the cosmos (along with the constraint imposed by the curtailed extent of my own and many readers' mathematical education) offers the advantage that writings selected for *The Book of the Cosmos* may be accessible to both scientists and nonscientists. I don't pretend that every facet of the cosmic story can be understood by the nonspecialist, but I do try here to provide a lively, historically responsible textual foundation for further study and deeper appreciation of the narrative's connectedness and attractiveness. Again, one of the distinctive features of this book is that, in all but a few chapters, what you hear is the voices of the cosmologists themselves. The tour that it offers across the chronological sweep of written history affords a firsthand experience of the human passion, awe, and not-infrequent bewilderment, as well as the intellectual flair, displayed in literature about cosmic origins, about the structure of space and time, and about the purpose of the universe and of humankind within it.

An effort partially to bridge or transcend the academic disciplines harmonizes, I hope, with the aim I have already mentioned of presenting cosmology as an art as well as a science. Poets, dreamers, religious thinkers, philosophers, physicists—all may look at the cosmos and contemplate its mysteries, and each may write in a way that reflects its beauties. There really are more things in heaven and earth than can be contained in a single discipline. More and more scientists seem to share this recognition, although there are plenty of examples of books about the universe whose authors arbitrarily rule out views they deem to be influenced by religious or philosophical presuppositions—as if science (or any other discipline) were devoid of such things. This observation is not intended to kindle or promote any sort of polemic against science. On the contrary, *The Book of the Cosmos* is constructed upon a foundation of deep respect for science no less than for poetry, religion, and philosophy. And to repeat my point, serious attention to cosmological writing may contribute to closer and more fruitful relations among these areas of human endeavor. It may conduct us in the sharing of beautiful things.

The other "bringing together" that this book seeks to accomplish, as by now may be apparent, is that of past and present. Its organization is primarily chronological, its approach historical. A considerable number of books on theories of the universe begin with ten or fifteen pages of potted history, rounding up and identifying the usual suspects—Democritus, Aristotle, Ptolemy, Copernicus, Galileo, and possibly Cusanus, Bruno, Kepler, or Herschel. But these introductions often only perpetuate outworn clichés and, not infrequently, ignorance. (One study published by Oxford University Press in 1995 places Ptolemy in the second century B.C.!) Something else that such introductions often perpetuate is disrespect for the thinkers, poets, and scientists of the past. The purportedly scholarly or scientific summaries to which I refer can easily silence or swallow up earlier generations of cosmologists; and, unfortunately, such chronological snobbery has also naturally worked its way into Hollywood movies and the other media (including school textbooks) through which many of today's youth are learning their history of science. Of course, it is no sin to produce summaries based on reliable sources. But firsthand examination of those sources is even better, promoting as it often does the virtues of accuracy and respect. It is my hope that *The Book of the Cosmos*, by allowing so many cosmologists to speak for themselves, will provide some modest assistance to scientists and nonscientists alike in the exercise of such virtues.

One anxiety we understandably feel in reading about "old" ideas, however, is that they may be wrong, outdated, superseded. Leaving aside the neglected truism that today's up-to-date ideas may appear wrong, outdated, and superseded ten years from now, I think there are ways of approaching conceptions from the past without condescension and at the same time without disregard for the question of truth. Referring to the opinions of Aristotle, Aquinas, and Montaigne on the topic of consciousness, for example, one third-culture scientist complains that "these people have a vague hand-waving notion of what consciousness is about, with a religious tinge to it. Their work wouldn't fly at all in modern academics. Yet we're being told that if you haven't read them you aren't educated. Well, I'm reading them, but I'm not learning much from them."[5] The narrowness of the definition of learning implicit here is unfortunately part of what fosters disciplinary and chronological snobbery in the first place, as well as fueling a "now-centered," self-congratulatory tendency that I hope *The Book of the Cosmos* will help to subvert.

A much more useful approach to ideas of the past, or of the present for that matter, has been proposed by Daniel Dennett:

> If you look at the history of philosophy, you see that all the great and influential stuff has been technically full of holes but utterly memorable and vivid. They are what I call "intuition pumps"—lovely thought experiments. Like Plato's cave, and Descartes's evil demon, and Hobbes' vision of the state of nature and the social contract. . . . I don't know of any philosopher who thinks any one of those is a logically sound argument for anything. But they're wonderful imagination grabbers, jungle gyms for the imagination.[6]

Dennett's description of "intuition pumps" applies equally well to scientific thought experiments and to literary fictions or poetic "conceits." In accordance with this model, the eighty-five chapters of *The Book of the Cosmos* can function in a way that to varying degrees combines the scientific and the poetic—as exercisers of the imagination. In such discourse, truth is by no means irrelevant. Yet in modesty we must admit, even in up-to-date science, the circumscribed nature of our cognitive capacities: We may *approach* knowledge of the truth but cannot take undisputed possession of it. We do catch sight of it, but through a glass darkly. And so, in keeping with a technology central to the history of astronomy itself, I would like to propose a near equivalent of Dennett's device of the intuition pump and offer these eighty-five chapters as *telescopes for the mind.* For I hope that each of them, focusing on one aspect of the universe, be it the sun or the moon, Mars or a comet or the nebulae, the structure of the solar system or the shape of space—or on an aspect of how these things have appeared to human beings—may convey a meaningful glimmer of new instruction and delight as the mind's eye surveys the sweep, the richness, and the deep excitement of the human process of imagining the universe.

Before concluding my brief introduction, I must say a word or two more about this anthology's scope and limitations. As with any anthology, the book's selections reflect the tastes and interests of its compiler. There is no getting around this personal dimension, even if (as I hope) readers of *The Book of the Cosmos* might discern some objective grounds for the choices I have made in putting it together. Moreover, one of the hardest tasks of an anthologist is to draw the line and choose *not* to include some attractive selection. Such acts of exclusion, however, are aided by the constraints of space and economics, and in the end they are really neither avoidable nor regrettable—unless one wants an anthology too big to carry.

But there are two further, more specific limitations that I would like to mention. First, *The Book of the Cosmos* is a book that surveys the history of *western* cosmology. I initially thought I could also include stories about the universe and its origins from Africa, eastern Asia, North American aboriginal cultures, and so on. But such a collection would have been an encyclopedia rather than a book with closely connected historical joints and sinews. From quite early on I realized that simple demands of coherence would require a narrower scope than the one I had originally envisaged. Nevertheless, although it does make sense from a historical point of view to speak of *western* cosmology—as in the cases, for example, of those seventeenth-century European discoveries regarding laws of planetary motion or gravitational theory— the adjective *western* is increasingly inapplicable to the science and cosmology of the present. In spite of those who insist (with some justification) on the culturally "situated" nature of knowledge, there today seems little point in speaking of *western* optics or *western* gravitation, even though the *history* of current theories concerning them is predominantly western. Accordingly, I hope that the worldwide relevance of the history of western thought to present-day transcultural fascination with the cosmos will make this collection interesting and useful to readers in Tokyo, Beijing, New Delhi, and Nairobi as well as to those in London, New York, Vancouver, and Auckland.

The other limitation I will mention has been a more frustrating one to deal with, though for present practical purposes similarly unavoidable. An anthologist is dependent upon the availability of materials, and to my sincere regret there is still a paucity of published writings by cosmologists who are women. There are many historical reasons for this, some of them irremediable. However, the work of numerous women whose contributions to astronomy and cosmology are now acknowledged is either unavailable in English or else not yet adequately accessible in published form. For these reasons some women who perhaps ought to be in this anthology—such as Maria Cunitz from the seventeenth century and Jeanne Dumée and Louise du Pierry from the eighteenth—aren't. Neither is Antonia Maury, who in 1888, working for twenty-five cents an hour classifying stellar spectra at the Harvard Observatory, irritated the observatory's director, E. C. Pickering, by pointing out flaws in his system for classifying stellar spectra and employing a superior, independent system of her own.[7] Nor is Henrietta Swan Leavitt (although I summarize her main contribution in Chapter 64), simply because her most famous article, published in 1912 under Pickering's name, is simply too technical for the purposes of this anthology. Other distinguished as well as eloquent cosmological writers such as Mary Fairfax Somerville, Maria Mitchell, Agnes Mary Clerke, Annie Jump Cannon, Cecilia Payne-Gaposchkin, and more recently Kitty Ferguson and Vera Rubin, I am pleased to say have been included. Even before any of these, Aphra Behn appears as a brilliant translator—not a trivial role in a book of literary cosmology. Still, there would have been more women in evidence here had I been more successful in my attempts to lay hands on suitable materials. Given time and the efforts of scholars who even now are taking up the job of excavating and editing the work of women in this field, the present lack should prove partly remediable (see "Further Reading" at the end of the book). Maybe someday one of my three daughters, or I myself if I'm so fortunate, will be able to revise this collection with a fairer representation of cosmological writings by women.

Editorial Procedure

In the eighty-five short chapters that make up *The Book of the Cosmos* I have aimed at presenting readable (not critical) texts. By this I mean that I have done everything I could within the constraints of historical accuracy to present the readings in a form accessible to today's educated general reader. For example, when using materials published in English I have taken the liberty (a daring and dangerous thing for a literary scholar to do!) of regularizing spelling, punctuation, and usage of such things as capitals and italics, as well as expanding unfamiliar abbreviations. For the most part I have also silently substituted English translations in cases where authors have interspersed their writing with brief quotations from other languages, particularly Latin. Otherwise I have followed familiar conventions, using ellipses (...) where something is left out and square brackets ([ ]) where something is inserted. As for writings originally in languages other than English, I have presented (a) direct excerpts from published translations or (b) my own adaptation of published translations or (c) my own English translation. The "source" note at the end of each chapter indicates clearly which procedure has been followed.

NOTES

1. Alan H. Guth, *The Inflationary Universe: The Quest for a New Theory of Cosmic Origins* (Reading, Mass.: Addison-Wesley/Helix, 1997), p. xiv.

2. Charles Hartshorne, "Science as the Search for the Hidden Beauty of the World," in *The Aesthetic Dimension of Science*, ed. Deane W. Curtin (New York: Philosophical Library, 1980), pp. 95-96.

3. John Brockman, *The Third Culture* (New York: Simon & Schuster, 1995), p. 17-18.

4. Murray Gell-Mann, quoted in Brockman, *The Third Culture*, p. 22.

5. Roger Schank, quoted in Brockman, *The Third Culture*, p. 28.

6. Daniel Dennett, quoted in Brockman, *The Third Culture*, p. 182.

7. Dorrit Hoffleit, *Women in the History of Variable Star Astronomy* (Cambridge, Mass.: American Association of Variable Star Observers, 1993), pp. 2–3.

Part One

COSMOLOGICAL ORIGINS

We Have Seen
But Few of His Works

Torah, Sacred Poetry,
Apocrypha, New Testament

The two principal ancient legacies informing western cosmology are the Greek and the Hebrew. Although for centuries Christian philosophers and poets interwove and tried to harmonize Greek and Hebrew themes, Greek and Hebrew views of the world are strikingly contrary. Greek cosmology and cosmogony begin either with the world itself or with some form of primordial chaos that provides the stuff of the world, whereas biblical Hebrew teaching focuses on the world as a creation formed and governed by a transcendent creator.

The contrast can be expressed most simply as a contrast between models. The Greek model for the production of the world is agricultural or architectural. *In ancient Greek literature we read a great deal about elements, seeds, raw materials, geometrical shapes. If the gods are involved in the process of creation at all, they are like farmers who plant seeds and then amuse themselves elsewhere while the seeds sprout on their own. Or else they are like the mind as it seeks mastery over the moving parts of its own body; or like a craftsman who does the best he can with whatever raw materials are available.*

The Hebrew scriptures, however, present a contrast that begins with vocabulary itself. Cosmos *is a Greek word and a Greek concept, so to talk about "Hebrew cosmology" may already skew the discussion. The Hebrew expression* the sky and the earth *(or traditionally,* the heavens and the earth) *is more collective than is "cosmos," and it much less readily bespeaks a comprehensive unitary whole.*

In this sense the ancient Hebrew conception of the world is more contingent than the Greek, and more fragmentary. Its model of production is more poetic or narrative than agricultural or architectural. (Ancient Hebrew agriculture was often nomadic; and, although biblical writers do occasionally present God as a builder, there was no indigenous Hebrew architecture comparable to the Greek.) To use an only slightly anachronistic term, the Hebrew sky and earth are a literary production. Like the Torah or the Covenant, which provided the "framework agreement" governing the conditions for religious life from generation to generation, the world itself was spoken into existence by God, who is its author.

AND GOD SAID

In Genesis, the first book of the Bible, the creation story forms a framework narrative for the longer though more narrowly focused narratives that follow. It is a book not only of genesis but of genealogy: It tells its audience where they have come from, and to whom they are related. And the creation, resoundingly, is brought into being by one who speaks.

Chapter 1

In the beginning when God created the heavens and the earth, the earth was a formless void and darkness covered the face of the deep, while a wind from God swept over the face of the waters. Then God said, "Let there be light"; and there was light. And God saw that the light was good; and God separated the light from the darkness. God called the light Day, and the darkness he called Night. And there was evening and there was morning, the first day.

And God said, "Let there be a dome in the midst of the waters, and let it separate the waters from the waters." So God made the dome and separated the waters that were under the dome from the waters that were above the dome. And it was so. And God called the dome Sky. And there was evening and there was morning, the second day.

And God said, "Let the waters under the sky be gathered together into one place, and let the dry land appear." And it was so. God called the dry land Earth, and the waters that were gathered together he called Seas. And God saw that it was good. Then God said, "Let the earth put forth vegetation: plants yielding seed, and fruit trees of every kind on earth that bear fruit with the seed in it." And it was so. The earth brought forth vegetation: plants yielding seed of every kind, and trees of every kind bearing fruit with the seed in it. And God saw that it was good. And there was evening and there was morning, the third day.

And God said, "Let there be lights in the dome of the sky to separate the day from the night; and let them be for signs and for seasons and for days and years, and let them be lights in the dome of the sky to give light upon the earth." And it was so. And God made the two great lights—the greater light to rule the day and the lesser light to rule the night—and the stars. God set them in the dome of the sky to give light upon the earth, to rule over the day and over the night, and to separate the light from the darkness. And God saw that it was good. And there was evening and there was morning, the fourth day.

And God said, "Let the waters bring forth swarms of living creatures, and let birds fly above the earth across the dome of the sky." So God created the great sea monsters and every living creature that moves, of every kind, with which the waters swarm, and every winged bird of every kind. And God saw that it was good. God blessed them, saying, "Be fruitful and multiply and fill the waters in the seas, and let birds multiply on the earth." And there was evening and there was morning, the fifth day.

And God said, "Let the earth bring forth living creatures of every kind: cattle and creeping things and wild animals of the earth of every kind." And it was so. God made the wild animals of the earth of every kind, and the cattle of every kind, and everything that creeps upon the ground of every kind. And God saw that it was good.

For Democritus and other ancient Greeks, man is a little world, a microcosm. However, in the Hebrew writings human beings are made not in the image of the world but in the image of God. Accordingly, their role in the world is not to be well functioning, ordered units but in one sense to be "above" the world, to perform an ongoing creational and "cultural" role—husbanding and cultivating the garden. Genesis continues:

Then God said, "Let us make humankind in our image, according to our likeness; and let them have dominion over the fish of the sea, and over the birds of the air, and over the cattle, and over all the wild animals of the earth, and over every creeping thing that creeps upon the earth."

So God created humankind in his image,
  in the image of God he created them;
  male and female he created them.

God blessed them, and God said to them, "Be fruitful and multiply, and fill the earth and subdue it; and have dominion over the fish of the sea and over the birds of the air and over every living thing that moves upon the earth." God said, "See, I have given you every plant yielding seed that is upon the face of all the earth, and every tree with seed in its fruit; you shall have them for food. And to every beast of the earth, and to every bird of the air, and to everything that creeps on the earth, everything that has the breath of life, I have given every green plant for food." And it was so. And God saw everything that he had made, and indeed, it was very good. And there was evening and there was morning, the sixth day.

Chapter 2

Thus the heavens and the earth were finished, and all their multitude. And on the seventh day God finished his work that he had done, and he rested on the seventh day from all his work that he had done. So God blessed the seventh day and hallowed it, because on it God rested from all the work that he had done in creation.

Here following chapter 2, verse 3, there is an important narrowing of focus. The Author of creation, who so far has been called simply God ("Elohim") is now referred to as the LORD God ("Yahweh"), a more intimate name, less abstract, and more expressive of the relationship between the Maker and the human creatures with whom he speaks. Typical of the style of Genesis, just before the narrative "zooms in" on a detail of the larger picture already painted, the narrative backs up a little and recapitulates what has already been presented, here especially the intimate nature of the LORD God's creation of human beings.

These are the generations of the heavens and the earth when they were created.

In the day that the LORD God made the earth and the heavens, when no plant of the field was yet in the earth and no herb of the field had yet sprung up—for the LORD God had not caused it to rain upon the earth, and there was no one to till the ground; but a stream would rise from the earth, and water the whole face of the ground—then the LORD God formed man from the dust of the ground, and breathed into his nostrils the breath of life; and the man became a living being.

Later in Genesis (and elsewhere in the Old Testament) the stars are mentioned within a simile bespeaking incalculability. Modern commentators have found this locution interesting given that from ancient Babylonian astronomy (which held that the stars numbered about 3000) until the invention of the telescope, the number of the stars was considered in principle to be countable. In Genesis 15:5 it is said that the LORD "brought [Abraham] outside and said, 'Look toward heaven, and count the stars, if you are able to count them.' Then he said to him, 'So shall your descendants be.'" Elsewhere the innumerable stars are paired with the countless sands of the sea, as in Genesis 22:17 ("I will make your offspring as numerous as the stars of heaven and as the sand that is on the seashore") and Jeremiah 33:22 ("Just as the host of heaven cannot be numbered and the sands of the sea cannot be measured, so I will increase the offspring of my servant David").

THE HEAVENS DECLARE

In the poetry of the Old Testament, we find exalted meditations on the majesty of the sky, though the purpose of the poetry is clearly religious rather than scientific. It emphasizes "who," not "how"; it reminds the speaker or the audience that God is "above" the sky and is their maker; and it leads to a humble, awed response on the human side of the relationship between creator and creature.

In the book of Job (38:4–12), the LORD interrogates its longsuffering main character:

> Where were you when I laid the foundation of the earth?
>   Tell me, if you have understanding.
> Who determined its measurements—surely you know!
>   Or who stretched the line upon it?
> On what were its bases sunk,
>   or who laid its cornerstone,
> when the morning stars sang together
>   and all the heavenly beings shouted for joy?
>
> Or who shut in the sea with doors
>   when it burst forth from the womb?—
> when I made clouds its garment,
>   and thick darkness its swaddling band,
> and prescribed bounds for it,
>   and set bars and doors,
> and said, "Thus far shall you come, and no farther,
>   and here shall your proud waves be stopped"?

Have you commanded the morning since your days began,
    and caused the dawn to know its place?

Although Job is said to "repent in dust and ashes," he is nevertheless author-
itatively declared to be the LORD's *servant who has "spoken . . . what is*
right" (42:6–7).
    Similarly in the Psalms, the poet gives voice to an almost paradoxical sense
of being at once humbled and exalted when he considers the sky, which God
has created:

O LORD, our Sovereign,
    how majestic is your name in all the earth!

You have set your glory above the heavens.
    Out of the mouths of babes and infants
you have founded a bulwark because of your foes,
    to silence the enemy and the avenger.

When I look at your heavens, the work of your fingers,
    the moon and the stars that you have established;
what are human beings that you are mindful of them,
    mortals that you care for them?

Yet you have made them a little lower than God,
    and crowned them with glory and honor.
You have given them dominion over the works of your hands;
    you have put all things under their feet,
all sheep and oxen,
    and also the beasts of the field,
the birds of the air, and the fish of the sea,
    whatever passes along the paths of the seas.
O LORD, our Sovereign,
    how majestic is your name in all the earth! (Psalm 8)

Again, in Psalm 19, a poetic meditation on the sky sets the stage for recogni-
tion of the magnificence of God as creator:

The heavens are telling the glory of God;
    and the firmament proclaims his handiwork.
Day to day pours forth speech,
    and night to night declares knowledge.

There is no speech, nor are there words;
   their voice is not heard;
yet their voice goes out through all the earth,
   and their words to the end of the world.

In the heavens he has set a tent for the sun,
which comes out like a bridegroom from his wedding canopy,
   and like a strong man runs its course with joy.
Its rising is from the end of the heavens,
   and its circuit to the end of them;
   and there is nothing hid from its heat.

This reflection on the communicative, created nature of the heavens and the earth flows into a meditation on the perfection of "the law of the LORD" as engaged at a personal, human level, and the poet's response is thus confessional, as signalled by his "performative" vocative address to the Creator/ Lawgiver: "Let the words of my mouth and the meditation of my heart be acceptable to you, O LORD, my rock and my redeemer" (19:14).

Two further, lesser-known but beautiful cosmological "hymns" appear in the biblical Apocrypha, in the book of Sirach (or Ecclesiasticus). These harmonize with the psalmic presentation of the creation as a sign of the majesty of God. In the first of them, from chapter 1, the divine law is personified as Wisdom:

All wisdom is from the Lord,
   and with him it remains forever.
The sand of the sea, the drops of rain,
   and the days of eternity—who can count them?
The height of heaven, the breadth of the earth,
   the abyss, and wisdom—who can search them out?
Wisdom was created before all other things,
   and prudent understanding from eternity.
The root of wisdom—to whom has it been revealed?
   Her subtleties—who knows them?
There is but one who is wise, greatly to be feared,
   seated upon his throne—the Lord.
It is he who created her;
   he saw her and took her measure;
   he poured her out upon all his works,
upon all the living according to his gift;
   he lavished her upon those who love him.

Sirach, chapter 43, returns to something like the "significance/insignificance"
paradox of Psalm 8: Human beings may revel in the spectacle of the sky and
may praise their creator even while recognizing with awe the inadequacy of
both human praise and comprehension.

The pride of the higher realms is the clear vault of the sky,
  as glorious to behold as the sight of the heavens.
The sun, when it appears, proclaims as it rises
  what a marvelous instrument it is, the work of the Most High.
At noon it parches the land;
  and who can withstand its burning heat?
A man tending a furnace works in burning heat,
  but three times as hot is the sun scorching the mountains;
it breathes out fiery vapors,
  and its bright rays blind the eyes.
Great is the Lord who made it;
  at his orders it hurries on its course.

It is the moon that marks the changing seasons,
  governing the times, their everlasting sign.
From the moon comes the sign for festal days,
  a light that wanes when it completes its course.
The new moon, as its name suggests, renews itself;
  how marvelous it is in this change,
a beacon to the hosts on high
  shining in the vault of the heavens!

The glory of the stars is the beauty of heaven,
  a glittering array in the heights of the Lord.
On the orders of the Holy One they stand in their appointed places;
  they never relax in their watches.
Look at the rainbow, and praise him who made it;
  it is exceedingly beautiful in its brightness.
It encircles the sky with its glorious arc;
  the hands of the Most High have stretched it out.
                    · · ·
We could say more but could never say enough;
  let the final word be: "He is the all."
Where can we find the strength to praise him?
  For he is greater than all his works.
Awesome is the Lord and very great,

and marvelous is his power.
Glorify the Lord and exalt him as much as you can;
    for he surpasses even that.
When you exalt him, summon all your strength,
    and do not grow weary, for you cannot praise him enough.
Who has seen him and can describe him?
    Or who can extol him as he is?
Many things greater than these lie hidden,
    for we have seen but few of his works. (1–12, 27–32)

In Him all things hold together

At the center of the New Testament of the Bible is Jesus Christ, and one of his roles is seen as coinciding with the creative "speech-acts" of God in Genesis. Put theologically, the authority and efficacy of Christ as Redeemer are intimately linked to his "authorship" and agency as Creator. The classic statement of the doctrine that Jesus Christ is the very "speech" or the Word of God (Greek: logos) is found in the famous prologue to the Gospel of John, which begins with a deliberate echo of the first words of Genesis:

In the beginning was the Word, and the Word was with God, and the Word was God. He was in the beginning with God. All things came into being through him, and without him not one thing came into being. What has come into being in him was life, and the life was the light of all people. The light shines in the darkness, and the darkness did not overcome it.

There was a man sent from God, whose name was John. He came as a witness to testify to the light, so that all might believe through him. He himself was not the light, but came to testify to the light. The true light, which enlightens everyone, was coming into the world.

He was in the world, and the world came into being through him; yet the world did not know him. He came to what was his own, and his own people did not accept him. But to all who received him, who believed in his name, he gave power to become children of God, who were born, not of blood or of the will of the flesh or of the will of man, but of God.

And the Word became flesh and lived among us, and we have seen his glory, the glory as of a father's only son, full of grace and truth.

That Christ ("the Son") is the embodiment and manifestation of the creative power and glory of God is reasserted in a brief "cosmic" passage in St. Paul's letter to the Colossians (1:13–20). The Son, against the backdrop of Genesis 1, is here seen as replacing darkness with light and as consummating humankind's creation in God's image.

[God] has rescued us from the power of darkness and transferred us into the kingdom of his beloved Son, in whom we have redemption, the forgiveness of sins.

He is the image of the invisible God, the firstborn of all creation; for in him all things in heaven and on earth were created, things visible and invisible, whether thrones or dominions or rulers or powers—all things have been created through him and for him. He himself is before all things, and in him all things hold together. He is the head of the body, the church; he is the beginning, the firstborn from the dead, so that he might come to have first place in everything. For in him all the fullness of God was pleased to dwell, and through him God was pleased to reconcile to himself all things, whether on earth or in heaven, by making peace through the blood of his cross.

The agency of Christ in creation is declared once more in the book of Hebrews (chapter 1), where again the Christian teaching is tied carefully to that of the Old Testament, here by means of direct echo of the Psalms:

[God appointed his Son] the heir of all things, through whom he also created the worlds. He is the reflection of God's glory and the exact imprint of God's very being, and he sustains all things by his powerful word. When he had made purification for sins, he sat down at the right hand of the Majesty on high, having become as much superior to angels as the name he has inherited is more excellent than theirs.

For to which of the angels did God ever say,
> "You are my Son,
>   today I have begotten you?"

And,

> "In the beginning, Lord, you founded the earth,
>   and the heavens are the work of your hands;
> they will perish, but you remain;
>   they will all wear out like clothing;
> like a cloak you will roll them up,
>   and like clothing they will be changed.
> But you are the same,
>   and your years will never end."

SOURCE: *New Revised Standard Version Bible*, Catholic Edition, Nashville: Thomas Nelson, 1993.

Twice into the Same River?

Heraclitus and Parmenides

Most of us have heard A. N. Whitehead's remark that the history of (western) philosophy is "a series of footnotes to Plato." We may adapt this generalization to western cosmology: In its first two millennia, at least—leaving aside continued biblical influence—it is a series of footnotes to Plato's immediate Greek precursors, known collectively as the Presocratics. They were the first thinkers to engage cosmology as theory, as physics. The Babylonians and ancient Hebrews had a literature embodying stories or expressing awe about the heavens and the earth and about the religious significance of those stories and that awe. But the Presocratics made the first attempts to assemble the conceptual tools with which one might begin to answer the question how. How did the world, the order, the arrangement of things which we see and of which we are a part come into being?

One of the most fundamental problems they tackled—a problem still encountered by Big Bang cosmology when it tries to describe the very beginning—concerns the relationship between things that do not change and things that do. The first category usually includes truths of mathematics and laws of physics. The second category includes all observable phenomena including human beings, their bodies, and their institutions. This is the problem of being and becoming, of permanence and mutability, and, some would say, of eternity and time. Moreover, it raises the further question, Which comes first? Which is fundamental? Which, if any, is the governing reality? A further question still is, What is the nature of their relationship?

All of these questions were raised in one form or another by the Presocratics, two of whom—Heraclitus and Parmenides—provide opposite answers.

HERACLITUS

Heraclitus (fl. c. 5th c. B.C.) is best known for his aphorisms, which empha-size strife and change.

Everything flows and nothing abides; everything gives way and nothing stays fixed.

You cannot step twice into the same river, for other waters and yet others go ever flowing on.

It is in changing that things find repose.

Homer was wrong in saying, "Would that strife might perish from amongst gods and men." For if that were to occur, then all things would cease to exist.

There is exchange of all things for fire and of fire for all things, as there is of wares for gold and of gold for wares.

If there were no sun, the other stars would not suffice to prevent its being night.

Nature loves to hide.

Diogenes Laertius presents Heraclitus as a sort of deconstructionist observer of his day, now delivering profound insight, now confusing the reader by deliber-ately exemplifying the paradox and contradiction he sees as constituting reality.

Heraclitus held that fire was the element, and that all things were an ex-change for fire, produced by condensation and rarefaction. But he explains nothing clearly. All things were produced in opposition, and all things were in flux like a river.

The all is finite and the world is one. It arises from fire, and is consumed again by fire alternately through all eternity in certain cycles. This happens according to fate. Of the opposites, that which leads to the becoming of the world is called War and Strife; that which leads to the final conflagration is Concord and Peace.

He called change the upward and the downward path, and held that the world comes into being in virtue of this. . . .

He does not make it clear what is the nature of that which surrounds the world. He held, however, that there were bowls in it with the concave sides turned towards us, in which the bright exhalations were collected and produced flames. These were the heavenly bodies.

The flame of the sun was the brightest and warmest; for the other heavenly bodies were more distant from the earth, and for that reason gave less light and heat. The moon, on the other hand, was nearer the earth; but it moved through an impure region. The sun moved in a bright and unmixed region, and at the same time was at just the right distance from us. That is why it gives more heat and light. The eclipses of the sun and moon were due to the turning of the bowls upwards, while the monthly phases of the moon were produced by a gradual turning of the bowl.

Day and night, months and seasons and years, rains and winds, and things like these, were due to the different exhalations. The bright exhalations, when ignited in the circle of the sun, produced day, and the preponderance of the opposite exhalations produced night. The increase of warmth proceeding from the bright exhalations produced summer, and the preponderance of moisture from the dark exhalation produced winter. He assigns the causes of other things in conformity with this.

As to the earth, he makes no clear statement about its nature, any more than he does about that of the bowls.

PARMENIDES

Whereas Heraclitus emphasized strife and contingency, his contemporary Parmenides declared the primacy of Being and of Necessity.

One path only is left for us to speak of, namely that *It is*. In this path are very many tokens that what is is uncreated and indestructible; for it is complete, immovable, and without end. Nor was it ever, nor will it be; for now *it is*, all at once, a continuous one. . . .

Moreover, it is immovable in the bonds of mighty chains, without beginning and without end; since coming into being and passing away have been driven afar, and true belief has cast them away. It is the same, and it rests in the selfsame place, abiding in itself. And thus it remains constant in its place; for hard necessity keeps it in the bonds of the limit that holds it fast on every side. Wherefore it is not permitted to what is, to be infinite; for it is in need of nothing; while, if it were infinite, it would stand in need of everything. . . .

Since, then, it has a furthest limit, it is complete on every side, like the mass of a rounded sphere, equally poised from the center in every direction; for it cannot be greater or smaller in one place than another. . . .

Here shall I close my trustworthy speech and thought about the truth. Henceforward learn the beliefs of mortals, giving ear to the deceptive ordering of my words. . . .

Thou shalt know the substance of the sky, and all the signs in the sky, and the resplendent works of the glowing sun's pure torch, and whence they arose. And thou shalt learn likewise of the wandering deeds of the round-faced moon, and of her substance. Thou shalt know, too, the heavens that surround us, whence they arose, and how Necessity took them and bound them to keep the limits of the stars . . . how the earth, and the sun, and the moon, and the sky that is common to all, and the Milky Way, and the outer-most Olympus, and the burning might of the stars arose.

The narrower bands were filled with unmixed fire, those next them with night, and in the midst of these rushes their portion of fire. In the midst of these is the divinity that directs the course of all things.

The brilliant and controversial novelist-historian Arthur Koestler (1905–1983) sees Aristotle as brokering a deal between Heraclitus and Parmenides. Aristotle's cosmology—which was superseded only with the advent of Copernicus and Galileo—says Koestler in The Sleepwalkers *(1959), represents "a compromise between two opposite trends in philosophy."*

On the one side there was the "materialistic" trend, which had started with the Ionians, and was continued by men like Anaxagoras . . . ; by Heraclitus, who regarded the universe as a product of dynamic forces in eternal flux; and culminated in Leucippus and Democritus, the first atomists. The opposite tendency, which originated with the Eleatics, found its extreme expression in Parmenides, who taught that all apparent change, evolution and decline, were illusions of the senses, because whatever exists cannot arise from anything that does not, or is different from it; and that the Reality behind the illusion is indivisible, unchangeable, and in a state of static perfection. Thus for Heraclitus Reality is a continuous process of Change and Becoming, a world of dynamic stresses, of creative tensions between opposites; whereas for Parmenides Reality is a solid, uncreated, eternal, motionless, changeless, uniform sphere.

The preceding paragraph is, of course, a woeful oversimplification . . . but my purpose is merely to show how neatly the Aristotelian model of the universe solved the basic dilemma by handing over the sub-lunary region to the Materialists, and letting it be governed by Heraclitus's motto "all is change"; whereas the rest of the universe, eternal and immutable, stood in the sign of the Parmenidean "nothing ever changes."

Once again, it was not a reconciliation, merely a juxtaposition, of two world-views, or "world-feelings," both of which have a profound appeal to the minds of men. This appeal was increased in power when, at a later stage, mere juxtaposition yielded to *gradation* between the opposites; when the original Aristotelian two-storey universe—all basement and loft—was superseded by an elaborately graded, multi-storeyed structure; a cosmic hierarchy where every object and creature had its exact "place" assigned to it, because its position in the many-layered space between lowly earth and high heaven defined its rank on the Scale of Values, in the Chain of Being. ... This concept of a closed-in cosmos graded like the Civil Service (except that there was no advancement, only demotion) survived for nearly a millennium and a half.

In spite of the fact that Koestler's antithesis between the Heraclitean and the Parmenidean is an oversimplification (as is his account of a Civil Service universe with "no advancement"—witness Dante), the polarity of "world-feelings" provides a useful insight into the human motivations that underlie world-views. Hélène Tuzet (fl. mid-20th century), author of a large study of cosmos and the imagination (Le cosmos et l'imagination, 1965), further expounds the psychological dimension of the polarity.

The Parmenidean places himself outside of time and takes the side of the eternal. Underneath his choice, one can detect perhaps a fear, a recoil from whatever is transformed, crumbles, decays . . . ; in short, he recoils from the biological laws which include decomposition as an integral part. Because he fears death, the Parmenidean does not love life. But there is something more: an aesthetic taste, a choice of an idea, and at times, a religious motivation.

The forms of cosmic pathos to which the follower of Parmenides is susceptible are those which have come to terms with the Eternal, attracted by the purity and rigidity of an incorruptible substance. Everything enters into a clear and stable harmony: the Pythagorean aesthetic of numbers and configurations, the circle and sphere as types of perfection, and as the divine type of motion a steady eternal rotation, equivalent to the immoveable. With a greater degree of complexity, the Music and Dance of the planets appear in a harmony of numbers and combination of configurations in a similarly experienced duration and in strictly determined limits.

For this aesthetic of the Eternal is an aesthetic of the Finite: what is perfect or complete necessarily had limits. It is also the aesthetic of Discontinuity and of Hierarchy: the Scale of Being is fixed with distinct levels in the Parmenidean cosmos. Each thing has its place, and the thinking man enjoys the pleasure of feeling that he is in his right place. It is an aesthetic of immutable Unity, and not of a process of Fusion.

The Parmenidean thinker is more or less susceptible . . . to the pathos of Unity in explanation, of simplicity in basic assumptions, and of implacable rigor in formulated laws. There is also the pathos of ideal exactness in the appropriation and coherence of a well-knit network of logical correspondences and relations which take in the whole of creation and leave nothing out.

As for the Heraclitean, he is susceptible to the pathos of Becoming, and in order that it may unfold and reveal itself, he needs the Unlimited. If we seek any deep motivation, we discover a taste for life which accepts everything which life implies, including death as a condition for a new birth. There is a boldness in his outlook which rejects protection and authority, and assumes a willingness to take risks of all sorts. The appeal of the Heraclitean kind of pathos to instinctive forces and to the Unconscious is naturally greater than it is in the Parmenidean family of minds.

The Heraclitean type includes everything arising out of the fascination of change, and transfers to the cosmic plane whatever is integral to the cycle of life. There are dreams of life's genesis: the pathos of Birth and its original freshness, the pathos of continuous Creation and its inexhaustible onward surge. There are dreams of life's evolution: the pathos of continuity and of the flow of the forms of life. Opposing the Parmenidean pathos of Unity is that of Variety: the taste for profusion and even disorder; the taste for the irregular, the original, the unique which will feed the dream of the plurality of worlds. In opposition to the joy of feeling satisfied with being "in one's place," there is the intoxication of being lost in the swarming proliferation of universal Being. In order to accommodate all these wonderful things, the true Heraclitean requires Plenitude, a fullness within the Infinity of space, akin to the infinity of God and to the unlimited capacity of the soul of man.

Whereas the Parmenidean accepts hierarchy and its hemmed-in gradations, the Heraclitean, on the other hand, is alive to the pathos of absolute freedom; and in certain eras, he experiences the pathos of liberation, of transport, and of flight without thought of return. He is a traveler in the mind. Lastly, the science of motion for him is not mechanics but dynamics. Cosmic energies are absorbed in vital forces; he is receptive not to steady and completely smooth rotation but welcomes the conflict of opposites, tension, and effort, so that his Universe tends to be polarized.

SOURCES: Fragments from Heraclitus cited from *The Presocratics*, ed. Philip Wheelwright, New York: Odyssey Press, 1966; Diogenes Laertius and Parmenides cited from Thomas L. Heath, *Greek Astronomy*, London: J. M. Dent, 1932; Arthur Koestler, *The Sleepwalkers: A History of Man's Changing Vision of the Universe*, London: Hutchinson, 1959; Hélène Tuzet, "Cosmic Images," in *Dictionary of the History of Ideas*, vol. 1, ed. Philip P. Wiener, New York: Scribner's, 1973.

3

The Things of the Universe Are Not Sliced Off with a Hatchet

Empedocles and Anaxagoras

What is the relationship between things that do not change and things that do? The question is still with us. The fifth-century B.C. *Presocratic philosophers Empedocles and Anaxagoras, who are sometimes classified as "qualitative pluralists," begin with the changing qualities on the face of things and seek their unchanging roots.*

EMPEDOCLES (C. 484–C. 424 B.C.)

Among Empedocles' sayings are some directly astronomical statements, such as "it is the earth that makes the night by getting in the way of the sun's beams." Empedocles is probably best known, however, for identifying four "elements" or "roots" of physical reality: fire, water, earth, and air. His teaching is not just physics and not just poetry, but poetical physics.

Come now, hearken to my words; learning will enlarge your mind. . . . I shall tell of a twofold process. For at one time there grew to be a single One out of many, while at another time there came to be many by division out of One—fire, water, earth, and the lofty height of air. Apart from these and in balanced relation to them is dreadful Strife; while Love resides in their midst, throughout their length and breadth. Envision her with your mind, instead of sitting with glazed eyes. Mortals can know and recognize her, for she is implanted within their bodies. It is thanks to her that mortals enjoy thoughts of amity and do works of peace. They call her by the names Joy and Aphrodite. . . .

All of these are equal and of the same age, but each has its own kind of activity and its own character, and each gains ascendancy when its time comes round. Nothing is added to them nor taken away from them. For if they were continually perishing, they would at last no longer exist. And since there is nothing else, how could anything be added that would cause them to increase? And how could anything perish, since there is nothing empty? No, these are the only things that are; and by interpenetrating they become one thing in one place and another in another. . . .

Sunbeams and earth, sky and sea, are at one with the parts that compose them, even though thrown in different guises to mortals' apprehension. . . .

Come now, look at the things that bear witness to what I said formerly, in case there was anything defective in my earlier account. Behold the sun, sending warmth and brightness everywhere, and the countless things perpetually bathed by his radiance; there is also the rain-cloud, dark and cold on all sides; and there is the earth, from which solid bodies, the foundations of things, come forth. When Hostility is at work, all these things are distinct in form and separated; but they come together in love, and are desired by one another. Thence have sprung all the things that ever were, are, or shall be. . . . In reality there are only the basic elements, but interpenetrating one another they mix to such a degree that they assume different characteristics.

When painters wise and skilled in their craft are preparing sumptuous votive altars in a temple, they use pigments of many colors and blend them judiciously, now a little more of this and now of that; thereby they produce likenesses of all things—of trees, of men and women, of beasts and birds and water-dwelling fishes, and even of such honored beings as the long-lived gods. The way in which all the actual things of the world have come into existence, although they are incalculably more numerous, is essentially no different from this. . . .

These two forces, Strife and Love, existed in the past and will exist in the future; nor will boundless time, I believe, ever be empty of the pair.

Now one prevails, now the other, earth in its appointed turn, as change goes incessantly on its course. These alone truly are, but interpenetrating one another they become men and tribes of beasts. At one time they are brought together by Love to form a single order, at another they are carried off in different directions by the repellant force of Strife; then in course of time their enmity is subdued and they all come into harmony once more. Thus in the respect that by nature they grow out of many into one, then divide from one into many, they are changing things and their life is not lasting, but in respect of their perpetual cycle of change they are unalterable and eternal.

As things came together in harmony, Strife withdrew to the outermost region.

In that condition neither can the sun's swift limbs be distinguished, no, nor shaggy mighty earth, nor the sea; because all things are brought so close together in the perfect circularity of the Sphere.

Equal on all sides and utterly unlimited is the Sphere, which rejoices in its circular solitude. There is no discord and no unseemly strife in his limbs. There is no pair of wings branching forth from his back. He has no feet, no nimble knees, no genitals. He is spherical and equal on all sides.

When, in the fullness of time set by the primordial oath, Strife had grown to greatness in the limbs [of the Sphere] and was flaunting his demands for honors and privileges, . . . then all of God's limbs in turn began to quake.

ANAXAGORAS (fl. MID-5TH CENTURY B.C.)

Anaxagoras, like Empedocles, uttered the occasional intriguing astronomical aphorism. He states, for example, that "it is the sun that puts brightness into the moon." But also like Empedocles, he struggled at a philosophical level to give voice to the dynamic tension between change and continuity, and between reality and appearance.

Because of the weakness of our senses, we are not able to judge the truth. Appearances are a glimpse of the unseen. The Greeks do not rightly understand what they call coming-to-be and perishing. A real thing does not come-to-be or perish; occurrences that are so called are simply the mixing and separating of real entities. . . .

The things of the universe are not sliced off from one another with a hatchet, neither the hot from the cold nor the cold from the hot. In everything there is a portion of everything else, except of mind; and in some things there is mind also.

Some of Anaxagoras's reflections are fascinating for the expression they give to incipient ideas of the infinitesimal, a notion important not only for calculus but also ultimately for cosmology.

Since the great and the small share equally with respect to the number of parts they possess, here is a further reason why everything must possess a portion of everything else. Thus nothing exists apart; everything has a share of everything else. For since there is no smallest amount, it is impossible for a complete isolation to be brought about, and it is equally impossible for anything to come-to-be out of not-being. In everything there is always a multiplicity of different ingredients; and there are as many ingredients separable from the lesser as there are ingredients separable from the greater.

In the small there is no least, but always a lesser; for being cannot be defined by reference to non-being. Likewise there is always something bigger than what is big. The large and the small are thus equal in amount. And each thing taken from its own standpoint is large and small simultaneously.

Anaxagoras postulates a single physical source of all things. Yet, separate from this he appears to assume the existence of something transcendent, the creative or creating principle, which he calls Mind (nous).

All things were together, unlimited both in number and in smallness, for smallness too was unlimited. And when all things were together, none of them could be distinguished because of their smallness. . . . When all things were together, before any separating had taken place, not even any color was discernible. This was because of the utter mixture of all things—of moist with dry, hot with cold, bright with dark. And there was a great quantity of earth in the mixture, as well as seeds which were unlimited in number and of the utmost variety. . . . Neither in speculation nor in actuality can we ever know the number of things that are separated out. . . .

While other things have a share in the being of everything else, Mind is unlimited, autonomous, and unmixed with anything, standing entirely by itself. For if it were not by itself but were mixed with anything else whatever, it would thereby participate in all that exists; because, as I have said before, in everything that exists there is a share of everything else. If Mind were to share in the universal mixture, the things with which it was mixed would prevent it from having command over everything in the way that it now does, whereas the truth is that Mind, because of its exceptional fineness and purity, has knowledge of all that is, and therein it has the greatest power.

For Anaxagoras, the exercise of Mind is preeminently in the creation and ordering of things.

Mind took charge of the cosmic situation, so that the universe proceeded to rotate from the very beginning. At first the rotation was small, but by now it extends over a larger space, and it will extend over a yet larger one. Both the things that are mingled and those that are separated and individuated are all known by Mind.

And Mind set in order all that was to be, all that ever was but no longer is, and all that is now or ever will be. This includes the revolving movements of the stars, of the sun and moon, and of the air and aether as they are being separated off. It was the rotary movement that caused the separation—a separation of the dense from the rare, the hot form the cold, the bright from the dark, and the dry from the moist.

When Mind first set things in motion, there began a process of separation in the moving mass; and as things were thus moving and separating, the process of separation was greatly increased by the rotary movement.

The rotation and separation are characterized by force and swiftness. The swiftness makes the force. Such swiftness is not like the swiftness of anything known to us, but is incalculably greater.

SOURCE: *The Presocratics*, ed. Philip Wheelwright, New York: The Odyssey Press, 1966.

4

Atoms and Empty Space

Leucippus, Democritus,
Epicurus, Lucretius

From the point of view of modern cosmology, some of the most interesting an-
cient thinkers are those known as the atomists. The Greek Presocratic philoso-
phers Leucippus and Democritus provided the basis of Epicurus's physics,
which in turn was disseminated through the Roman Lucretius's great poem On
the Nature of Things (De Rerum Natura).

Much of what we know about Leucippus, Democritus, and Epicurus has
come down to us in mere fragments and in the writings of others. In the fol-
lowing account from Lives of Eminent Philosophers *by Diogenes Laertius (fl.*
3rd century A.D.*) we catch a glimpse of the thought of Leucippus (fl. 440*
B.C.*?), the founder of atomism, and of his effort to account for the universe*
in terms of the physical interaction of material things moving in space.

Leucippus was born at Elea, but some say at Abdera and others at Miletus.
He was a pupil of Zeno. His views were these. The sum of things is unlimited, and they all change into one another. The All includes the empty as well
as the full. The worlds are formed when atoms fall into the void and are entangled with one another; and from their motion as they increase in bulk
arises the substance of the stars. The sun revolves in a larger circle round the
moon. The earth rides steadily, being whirled about the center; its shape is
like that of a drum. Leucippus was the first to set up atoms as first principles. . . .

Although mathematical conceptions such as infinity cannot precisely be read
back into the vocabulary of ancient Greek thought, Leucippus is clearly in-

volved in forging ideas essential for a grasp of the physical universe—such as those of empty space, of the unlimited (or the boundless), and of the physically irreducible or indivisible (which is the root meaning of atom*).*

He declares the All to be unlimited, as already stated. But of the All, part is full and part empty, and these he calls elements. Out of them arise the worlds unlimited in number and into them they are dissolved. This is how the worlds are formed. In a given section many atoms of all manner of shapes are carried from the unlimited into the vast empty space. These collect together and from a single vortex, in which they jostle against each other and, circling round in every possible way, separate off, by like atoms joining like. And, the atoms being so numerous that they can no longer revolve in equilibrium, the light ones pass into the empty space outside, as if they were being winnowed. The remainder keep together and, becoming entangled, go on their circuit together, and form a primary spherical system. This parts off like a shell, enclosing within it atoms of all kinds. And, as these are whirled round by virtue of the resistance of the center, the enclosing shell becomes thinner, the adjacent atoms continually combining when they touch the vortex.

In this way the earth is formed by portions brought to the center coalescing. And again, even the outer shell grows larger by the influx of atoms from outside, and, as it is carried round in the vortex, adds to itself whatever atoms it touches. And of these some portions are locked together and form a mass, at first damp and miry, but, when they have dried and revolve with the universal vortex, they afterwards take fire and form the substance of the stars.

DEMOCRITUS (C. 460– C. 370 B.C.) AND ZENO (fl. MID-5ᵀᴴ CENTURY B.C.)

Democritus, Leucippus's student, became much better known than his teacher, even though it is unclear to which of the two his teachings actually belong. In any case, Democritus became a conduit for atomism's basic ideas. As we have already seen in Leucippus, part of the challenge was to forge conceptual tools that would account for continuity, and hence physical movement, in space.

The paradoxes of Zeno are the most famous expressions of the mind's difficulty in conceptualizing becoming, of which simple physical motion is one species. Zeno argued: "If anything is moving, it must be moving either in the place in which it is or in a place in which it is not. However, it cannot move in the place in which it is [for the place in which it is at any moment is of the same size as itself and hence allows it no room to move in], and it cannot move in the place in which it is not. Therefore movement is impossible." In the history of mathematics and physics, such conundrums are not at all trivial, as the struggle to develop the calculus attests.

Zeno's paradox indicates perhaps why a concept of space or void in the sense of "room to move" was so important to the atomists and others. It is worth noting that in Diogenes Laertius's summary of Democritus's teachings, the concept of atoms and the concept of space or void go together:

The first principles of the universe are atoms and empty space. Everything else is merely thought to exist. The worlds are unlimited [or boundless]. They come into being and perish. Nothing can come into being from that which is not nor pass away into that which is not. Further, the atoms are unlimited in size and number, and they are borne along in the whole universe in a vortex, and thereby generate all composite things—fire, water, air, earth. For even these are conglomerations of given atoms. And it is because of their solidity that these atoms are impassive and unalterable. The sun and the moon have been composed of such smooth and spherical masses [i.e., atoms], and so also the soul, which is identical with reason.

To glimpse Democritus's radical materialism is to understand why atomistic teachings, physical as well as moral, were subsequently seen as such a threat by proponents both of Platonism and of Christianity. Atomism is reductionist: It reduces things of the soul and the spirit fundamentally to matter in motion. The All, accordingly, is all there is; and there is nothing that is not physical. This is worth remembering when we read, in one of Democritus's fragments, that "Man is a small 'ordered world' [kosmos]." Although the idea of the macrocosm and the microcosm became a commonplace of European thought within a Christian context, in Democritus it is founded on an analogy between large and small physical "conglomerations," nothing more.

EPICURUS (341–270 B.C.)

It was under the name of Epicurus that atomistic physics and especially its accompanying "this-worldly" ethical teaching were bequeathed to subsequent history. Diogenes Laertius quotes at length from a letter by Epicurus himself that summarizes his physics:

Epicurus to Herodotus, greeting.

For those who are unable to study carefully all my physical writings or to go into the longer treatises at all, I have myself prepared an epitome of the whole system, Herodotus, to preserve in the memory enough of the principal doctrines, to the end that on every occasion they may be able to aid themselves on the most important points, so far as they take up the study of physics. . . .

To begin with, nothing comes into being out of what is non-existent. For in that case anything would have arisen out of anything, standing as it would in no need of its proper germs. And if that which disappears had been destroyed and become nonexistent, everything would have perished, that into which the things were dissolved being nonexistent. Moreover, the sum total of things was always such as it is now, and such it will ever remain. For there is nothing into which it can change. For outside the sum of things there is nothing which could enter into it and bring about the change.

Further, the whole of being consists of bodies and space. For the existence of bodies is everywhere attested by sense itself, and it is upon sensation that reason must rely when it attempts to infer the unknown from the known. And if there were no space (which we call also void and place and intangible nature), bodies would have nothing in which to be and through which to move, as they are plainly seen to move. Beyond bodies and space there is nothing which by mental apprehension or on its analogy we can conceive to exist. When we speak of bodies and space, both are regarded as wholes or separate things, not as the properties or accidents of separate things.

Again, of bodies some are composite, others the elements of which these composite bodies are made. These elements are indivisible ["*atoma*"] and unchangeable, and necessarily so, if things are not all to be destroyed and pass into non-existence, but are to be strong enough to endure when the composite bodies are broken up, because they possess a solid nature and are incapable of being anywhere or anyhow dissolved. It follows that the first beginnings must be indivisible, corporeal entities.

Again, the sum of things is infinite [or boundless]. For what is finite has an extremity, and the extremity of anything is discerned only by comparison with something else. Now the sum of things is not discerned by comparison with anything else. Hence, since it has no extremity, it has no limit. And since it has no limit, it must be unlimited or infinite.

Moreover, the sum of things is unlimited both by reason of the multitude of the atoms and the extent of the void. For if the void were infinite and bodies finite, the bodies would not have stayed anywhere but would have been dispersed in their course through the infinite void, not having any supports or counter-checks to send them back on their upward rebound. Again, if the void were finite, the infinity of bodies would not have anywhere to be.

Furthermore, the atoms, which have no void in them—out of which composite bodies arise and into which they are dissolved—vary indefinitely in their shapes. For so many varieties of things as we see could never have arisen out of a recurrence of a definite number of the same shapes. The like atoms of each shape are absolutely infinite. But the variety of shapes, though indefinitely large, is not absolutely infinite.

The atoms are in continual motion through all eternity.

LUCRETIUS (C. 99–C. 55 B.C.)

The greatest vehicle by which the atomism of Democritus and Epicurus was conveyed from Greek to Roman culture was a poem entitled On the Nature of Things, *written by Lucretius in about 50 B.C. Although in our age we do not use poems to teach physics, Lucretius's artistry played a role in introducing and popularizing atomistic thought not only in Latin and in the Roman Empire, but also in the English language at the dawn of modern science. In the last half of the seventeenth century prominent writers such as John Evelyn and John Dryden translated selections of Lucretius into English. And just as Lucretius had to forge a new vocabulary to translate atomistic thought into Latin, so poets and scientists in the late seventeenth century were engaged in the creation of language that would serve and embody new concepts. In general this effort led away from poetry to the plain style associated with the Royal Society. However, poetry's power of word-building and aphorism played a role that should not be overlooked.*

Thomas Creech's translation of all six books of De Rerum Natura, *first published in 1682, encapsulates majestically the mind's effort to conceptualize change and order in the world:*

> I treat of things abstruse, the Deity,
> The vast and steady motions of the sky;
> The rise of things, how curious Nature joins
> The various seed and in one mass combines
> The jarring principles; what new supplies
> Bring nourishment and strength; how she unties
> The Gordian knot, and the poor compound dies;
> Of what she makes, to what she breaks the frame,
> Called "seeds" or "principles," though either name
> We use promiscuously, the thing's the same.

Although in Paradise Lost *(1667) Milton uses the phrase "embryon atoms" to describe the elements of Chaos, Creech's use of "seed" and "principle" suggests that the term "atom" was still not yet firmly established in English. The Latin* individuum, *which translated the Greek* atomos, *entered English*

as "individual," a word that has subsequently had a political and psychological rather than physical or chemical history. Nevertheless, seventeenth century "atomic" teaching could still evoke politics or religion. In a verse
published with Evelyn's translation of book 1 of De Rerum Natura, Edmund
Waller suggests a parallel between democracy (with a pun on Democritus's
name) and the chaos of primordial atoms:

> Lucretius . . .
> Comes to proclaim in English verse
> No monarch rules the universe;
> But chance and atoms make this All
> In order Democratical,
> Where bodies freely run their course,
> Without design, or fate, or force.

Waller, like many before and since, perceives the antireligious and perhaps
politically disruptive tenor of atomistic materialism. (Democritus was the
subject of Karl Marx's doctoral thesis). And Lucretius's poetry expresses concisely a common theme—sometimes a prejudice—of historians concerning
what Andrew Dixon White, writing in the early twentieth century, calls the
warfare of science with theology.

> Long time men lay oppressed with slavish fear,
> Religion's tyranny did domineer,
> Which being placed in Heaven looked proudly down,
> And frighted abject spirits with her frown.
> At length a mighty one of Greece [namely, Epicurus] began
> To assert the natural liberty of Man,
> By senseless terrors and vain fancy led
> To slavery; straight the conquered phantoms fled.
> Not the famed stories of the Deity,
> Not all the thunder of the threatening sky
> Could stop his rising soul. Through all he passed
> The strongest bounds that powerful Nature cast.
> His vigorous and active mind was hurled
> Beyond the flaming limits of this world
> Into the mighty space, and there did see
> How things begin, what can, what cannot be;
> How all must die, all yield to fatal force,
> What steady limits bound their natural course.
> He saw all this, and brought it back to us.

> Wherefore by his success our right we gain,
> Religion is our subject, and we reign.

The fabric of the universe as envisaged by Lucretius, in keeping with Democritus's and Epicurus's teaching, is strikingly simple: again, atoms and empty space. "For if 'tis tangible, and hath a place, / 'Tis body; if intangible, 'tis space." As Lucretius charmingly argues, unless we presume the existence of space we have no alternative but to imagine universal gridlock.

> Yet bodies do not fill up every place:
> For besides those there is an empty space,
> A void. This known, this notion framed aright
> Will bring to my discourse new strength and light,
> And teach you plainest methods to descry
> The greatest secrets of philosophy.
> A void is space intangible, thus proved:
> For were there none, no body could be moved;
> Because where'er the brisker motion goes,
> It still must meet with stops, still meet with foes,
> 'Tis natural to bodies to oppose.
> So that to move would be in vain to try
> But all would fixed, stubborn, and moveless lie,
> Because no yielding body could be found
> Which first should move, and give the other ground.
> But every one now sees that things do move
> With various turns in earth and heaven above;
> Which, were no void, not only we'd not seen,
> But the bodies too themselves had never been:
> Ne'er generated, for matter all sides pressed
> With other matter would for ever rest.

Lucretius's teaching thus not only popularizes the categories of matter and space in the age of Newton but also reinforces the rejection of geocentrism: "For since the void is infinite, the space / Immense, how can there be a middle place?" Finally, from a modern perspective, one of atomism's most intriguing contributions is its attempt to conceptualize what we might call structure: the way in which abounding qualitative variety may arise from the recombination of simple elements. In this attempt to conceive complexity within simplicity, alphabetically structured language itself provides the analogy. Derrick de Kerckhove comments that, "without the slightest shred of ev-

idence, [Democritus] came to his conclusion that the elements of matter must be like the indivisible phonemes of the alphabet, thus inventing the very notion that one day in 1945 threatened to destroy the world" (The Skin of Culture, 1995, p. 35). *In Lucretius's words as translated into English:*

> And hence, as we discoursed before, we find
> It matters much with what first seeds are joined,
> Or how, or what position they maintain,
> What motion give, and what receive again:
> And that the seeds remaining still the same,
> Their order changed, of wood are turned to flame.
> Just as the letters little change affords
> *Ignis* and *Lignum* ["fire" and "wood"], two quite different words.

SOURCES: Diogenes Laertius, *Lives of Eminent Philosophers*, trans. R. D. Hicks, 2 vols, London: Heinemann, 1925; fragments from Zeno and Democritus quoted from *The Presocratics*, ed. Philip Wheelwright, New York: The Odyssey Press, 1966; *T. Lucretius Carus . . . De Natura Rerum Done into English Verse*, trans. Thomas Creech, 2nd edition, Oxford, 1683.

5

The Moving
Image of Eternity

Plato

Although the Timaeus *is more conspicuously about cosmology than is any of the other dialogues of Plato (427–347 B.C.), it has been debated whether we can attach Plato's authority or unambiguous assent to the cosmology which the work presents. The main speaker in the dialogue is not Socrates but Timaeus, a Pythagorean philosopher, and physical cosmology is not a subject that Plato invests in heavily elsewhere in the dialogues. Nevertheless, no one questions the* Timaeus's *influence, especially since it alone among the dialogues survived in a Latin translation through late antiquity and the Middle Ages.*

Recognizing that cosmology and cosmogony are directly related to theology, and that firm knowledge in these areas is not to be had, Timaeus sets out cautiously. Although he reasons in the analogical manner familiar to readers of the Republic, *his method explicitly pursues not certainty but probability.*

Was the heaven or the world, whether called by this or any other more acceptable name . . . always in existence and without beginning, or created and having a beginning? Created, I reply, being visible and tangible and having a body, and therefore perceptible; and all perceptible things apprehended by opinion and sense are in a process of creation and created.

Now that which is created must of necessity be created by a cause. But how can we find out the father and maker of all this universe? And when we have found him, to speak of his nature to all men would be impossible. Yet one more question has to be asked about him: Which pattern had the artificer in view when he made the world—the pattern which is unchangeable, or that which is created? If the world is indeed fair and the artificer good, then plainly he must have looked to that which is eternal. But if what cannot be said without blasphemy were true, then he looked to the created pattern. Every one will see that he must have looked to the eternal, for the world is the fairest of creations and he is the best of causes.

Having been created in this way, the world has been framed with a view to that which is apprehended by reason and mind and is unchangeable, and must (if this is admitted) of necessity be the copy of something. Now that the beginning of everything should be according to nature is a great matter. And in speaking of the copy and original we may assume that words are akin to the matter which they describe: when they relate to the lasting and permanent and intelligible, they ought to be lasting and unfailing, and, as far as their nature permits, irrefutable and immovable—nothing less. But when they express only the copy or image and not the eternal things themselves, they need only be probable and analogous to the real words. As being is to becoming, so is truth to belief. If then, Socrates, amid the many opinions about the gods and the generation of the universe, we are not able to convey notions which are in every way exact and consistent with one another, do not be surprised. It is enough if we adduce probabilities as likely as any others, for we must remember that I who am the speaker, and you who are the judges, are only mortal men, and we ought to accept the tale which is probable and not enquire further.

Although the four elements of earth, air, fire, and water are popularly associated mainly with the teaching of Aristotle, they originated with the teaching Empedocles, and the cosmology of the Timaeus *employs them too, though with a distinctively Pythagorean, geometrical flavor.*

Now that which is created is of necessity corporeal, and also visible and tangible. Now nothing is visible where there is no fire, and nothing is tangible which is not solid, and there is no solidity without earth. Therefore, God in the beginning of creation made the body of the universe to consist of fire and earth. But two things cannot be held together without a third; they must have some bond of union. The fairest bond is that which most completely fuses and is fused into the things which are bound; and geometrical proportion is best adapted to effect such a fusion. . . . If the universal frame had been created a surface only and without depth, one mean would have sufficed to bind together itself and the other terms; but now, as the world must be solid, and solid bodies are always compacted not by one mean but by two, God placed water and air in the mean between fire and earth, and made them to have the same proportion so far as was possible (as fire is to air, so is air to water; and as air is to water, so is water to earth). Thus he bound and put together a visible and palpable heaven. For these reasons, and out of such elements which are four in number, the body of the world was created in harmony and proportion, and so possessing the spirit of friendship. And being at unity with itself, it was indissoluble by the hand of any other than that of the framer.

The creation took up the whole of each of the four elements. For the creator compounded the world out of all the fire and all the water and all the air and all the earth, leaving no part of any of them nor any power of them outside. He intended first that the living being should be as far as possible a perfect whole and of perfect parts, and should be one, leaving no remnants over from which another such world might be created, and also that it should be free from old age and unaffected by disease. . . . And he gave to the world the figure which was suitable and also natural. What was suitable to the living being which was to contain all living beings was that figure which contains within itself all other figures. Therefore he made the world in the form of a globe, round as from a lathe, in every direction equally distant from the center to the extremes, the most perfect and the most like itself of all figures.

Already in the Timaeus *questions of cosmology are closely linked to those concerning the nature of time. The profoundly self-reflexive nature of Plato's reasoning is glimpsed in part in the recognition that all human reasoning, which functions in the time-bound medium of language, is severely limited in its powers to define time or, even more so, that which exists beyond time.*

When the father and creator saw the creature which he had made moving and living, the created glory of the eternal gods, he was delighted, and in his joy determined to conform the work to the original still more. As this was eternal, he sought to make the universe eternal, as far as might be. Now the nature of the intelligible being is eternal, but to attach eternity to the creature is impossible. Therefore he resolved to make a moving image of eternity, which he made when he set in order the heaven moving according to number, while eternity rested in unity.

This moving image we call time. For there were no days and nights and months and years before the heaven was created, but when he created the heaven he created them also. They are all parts of time, and the past and future are created species of time, which we unconsciously but wrongly ascribe to the eternal being. For we say indeed "he was," "he is," and "he will be"; but the truth is that only "he is" is truly spoken of him. "Was" and "will be" only apply to becoming in time, for they are motions. But that which is immovably the same cannot become older or younger by time. Nor ever did he, nor has he become, nor hereafter will he grow older; nor is he subject at all to any of those

states of generation which affect the movement of perceptible things. These are
the forms that time exhibits as it imitates eternity, moving in a circle measured
by number. Moreover, when we say that what has become has become, and
what is becoming is becoming, and what will become will become, and that
that which is not is not—all these are inaccurate modes of expression. But perhaps this is not the place for us to discuss minutely such matters.

*Although generations of "synchretizing" Christian commentators interpreted
the creation as depicted by the* Timaeus *in conformity with the creation account of Genesis, chapter 1 (and vice versa), the differences between Plato
and the Bible are as notable as are the similarities. By contrast with biblical
monism and the doctrine of* creatio ex nihilo *which grew from it, the teaching of the* Timaeus *is tied to the analogy of a human architect whose relative
success in making something is inevitably limited by the nature of available
raw materials that are* not *of the architect's own making. The presence of
that "necessity" as a potentially intractable raw material renders Plato's account—for all its Christianizable features—fundamentally dualistic.*

Thus far in what we have been saying, with small exceptions, the works of
intelligence have been set forth. Now we must place beside them the things
done from necessity. For the creation is mixed, being made up of necessity
and mind. Mind, the ruling power, persuaded necessity to bring the greater
part of created things to perfection, and thus in the beginning, when the influence of reason got the better of necessity, the universe was created. But if
one will truly tell of the way in which the work was accomplished, one must
include the other influence of the variable cause as well. Therefore, we must
return again and . . . consider the nature of fire, and water, and air, and earth,
which were prior to the creation of the heavens, and what happened before
they were elements. For no one has explained them, yet we speak of fire and
the rest, whatever they really mean, as if people knew their natures, and we
treat them as the letters or elements of the whole, when they cannot reasonably be compared to the syllables or first compounds by any sensible person.
Moreover, let me say this: I will not speak of the first principle or principles
of all things, or by whatever name they should be called, because it is difficult to express my opinion according to the mode of discussion which we are
at present employing. . . . Therefore, when I speak of the beginning of each
and all, I shall observe the rule of probability with which I began. . . . Once
more, then, I call upon God at the beginning of my discourse and beg him to
see us safely through a strange and unusual enquiry, and to bring us to probability. So now let us begin again.

This new beginning of our discussion of the universe requires a fuller division than the former. For then we posited two classes; now a third must be

added. The two sufficed for the former discussion. One (we assumed) was a pattern intelligible and always the same. The second was only the imitation of the pattern, generated and visible. We did not then distinguish a third, considering that two would be enough. But now the argument seems to require that we account for another kind, which is hard to explain and obscure. What powers and what nature shall we attribute to this new kind of being? We reply that it is the receptacle, and in a manner the nurse, of all generation.

Timaeus is of course right that the third something he wishes to account for is a difficult entity to conceptualize, although in general we can see it as the material (if abstract) substrate of creation, not the formal pattern to which the divine architect looks as he creates, and not the product of creation. It is apparently akin to the "necessity" mentioned earlier, that which is imprinted with the forms of creation.

Let me make one more attempt to explain my meaning more clearly. Suppose a person makes all kinds of figures of gold and never stops transforming them out of one form into all the others. Somebody points to one of them and asks, "What is that?" We answer most truly and safely, "That is gold"— not "That is a triangle" or any other figure formed in the gold. . . .

In the same way, that which is to receive perpetually and through its whole extent the resemblances of eternal beings ought to be destitute of any particular form. Therefore, the mother and receptacle of all created and visible and perceptible things is not to be termed earth, or air, or fire, or water, or any of their compounds, or any of the elements out of which they are composed, but is an invisible and formless being which receives all things and attains in a mysterious way a portion of the intelligible.

Timaeus summarizes the three classes, which we can call (1) Form, (2) Copy or Imitation, and (3) the "Nurse" or Receptacle, which is now called space:

The third nature is space, and is eternal, and admits not of destruction, and provides a home for all created things, and is perceived without the help of sense, by a kind of spurious reason, and is hard to believe in. We behold it as if in a dream and say that all existence must of necessity be in some place and occupy a space—and that what is neither in heaven nor in earth has no existence. These and other such things which are related to the true and waking reality of nature we apprehend only in such a dreamlike manner that we are unable to arouse ourselves to describe them or to determine them truly. But an image, not possessing the essence of that of which it is an image, and existing as a constantly changing shadow of something else, must inhere in that third nature, space, if it is to participate in reality to any degree at all—if it is not to be nothing.

The poetic pinnacle of Timaeus's account comes with the description—which includes an almost Homeric epic simile—of the elements in their chaotic pre-creational state. It is not surprising that interpreters of Genesis 1:2, with its parsimonious description of the earth "without form and void," should turn to this account for an imaginative amplification of the narrative of creation. One may emphasize again, however, that Plato's dualism entails a quite different theodicy—a contrasting account of the origin of evil in the world—from the monistic and largely moral picture presented by Genesis. The God of the Timaeus *is simply not almighty in any biblical sense: In the face of necessity, he made things good only "as far as possible."*

Thus I have given concisely the results of my thinking. My opinion is that being and space and generation, these three realities, existed before the heaven. The nurse of generation, moistened by water and inflamed by fire, and receiving the various forms of earth and air, and experiencing all the other accidents that attach to them, took a variety of shapes. But there was in it no equilibrium or homogeneity of powers, nowhere any state of equipoise. Thus it swayed unevenly to and fro, shaken by them, and by its motion it shook them in turn. And the elements when moved were divided like the grain shaken and winnowed by fans and other instruments used in the threshing of corn, when the close and heavy particles are borne away and settle in one direction while the loose and light particles are blown away in another. In this manner the four kinds or elements were shaken by the recipient vessel, which, moving like a winnowing machine, scattered far away from one another the elements most unlike, and forced the most similar elements into the closest contact. The elements too, therefore, had different places before the universe that was formed out of them came into being.

At first all things were without reason and measure. But when the world began to become ordered, first fire and water and earth and air, having only certain faint traces of themselves, and being altogether such as everything may be expected to be in the absence of God—this being their nature—then God fashioned them by form and number. Let us always, in all that we say, affirm that God made things as far as possible most fair and good out of things which were not fair and good.

SOURCE: Adapted from Plato, *Timaeus*, in *The Dialogues of Plato*, trans. B. Jowett, 2nd ed., vol. 3, Oxford, 1875.

The Potency of Place

Aristotle

*No thinker or writer exerted a greater influence on pre-Copernican notions
of cosmology than did Aristotle (384–322 B.C.). Aristotelian assumptions
about place, space, matter, motion, and time served as the foundation for the
Ptolemaic system, which held sway in the west for more than a thousand
years. The plainness and unstrained authoritativeness of Aristotle's style may
give us a glimpse into the sources, both rhetorical and philosophical, of his
authority.*

In the following widely cited passage from book 4 of the Physics *we see
how, for Aristotle, definitions of place precede those of space; and we encounter the notion—a startling one for those whose education is Newtonian
or post-Newtonian—that place itself has "potency," and that places in and of
themselves are qualitatively different from each other.*

The physicist must have a knowledge of Place . . . namely, whether there is
such a thing or not, and the manner of its existence and what it is, both because all suppose that things which exist are *somewhere* (the non-existent is
nowhere—where is the goat-stag or the sphinx?), and because "motion" in
its most general and primary sense is change of place, which we call "locomotion." . . .

Further, the typical locomotions of the elementary natural bodies, namely
fire, earth and the like, show not only that place is something, but also that it
exerts a certain influence. Each is carried to its own place, if it is not hindered, the one up, the other down. Now these are regions or kinds of place—
up and down and the rest of the six directions. Nor do such distinctions (up
and down and right and left, etc.) hold only in relation to us. To *us* they are

not always the same but change with the direction in which we are turned: that is why the same thing may be both right *and* left, up *and* down, before *and* behind. But in *nature* each is distinct, taken apart from itself. It is not every chance direction which is "up," but where fire and what is light are carried; similarly, too, "down" is not any chance direction but where what has weight and what is made of earth are carried—the implication being that these places do not differ merely in relative position, but also as possessing distinct potencies. . . .

These considerations then would lead us to suppose that place is something distinct from bodies, and that every sensible body is in place. Hesiod too might be held to have given a correct account of it when he made chaos first. At least he says, "First of all things came chaos to being, then broad-breasted earth," implying that things need to have space first, because he thought, with most people, that everything is somewhere and in place. If this is its nature, the potency of place must be a marvelous thing, and take precedence of all other things. For that without which nothing else can exist, while it can exist without the others, must needs be first; for place does not pass out of existence when the things in it are annihilated.

Accordingly, for Aristotle place is not isotropic (identical in all directions) as it is for Newton; however, it is absolute and independent of matter in the same way that it is for Newton (but not for Einstein). Time, by contrast, is relative to things, to matter, in a way that place is not:

Not only do we measure the movement by the time, but also the time by the movement, because they define each other. The time marks the movement, since it is its number, and the movement the time. We describe the time as much or little, measuring it by the movement, just as we know the number by what is numbered, e.g. the number of the horses by one horse as the unit. For we know how many horses there are by the use of the number; and again by using the one horse as unit we know the number of the horses itself. So it is with the time and the movement; for we measure the movement by the time and vice versa. It is natural that this should happen; for the movement goes with the distance and the time with the movement, because they are quanta and continuous and divisible. The movement has these attributes because the distance is of this nature, and the time has them because of the movement. And we measure both the distance by the movement and the movement by the distance; for we say that the road is long, if the journey is long, and that this is long, if the road is long—the time, too, if the movement, and the movement, if the time.

Time is a measure of motion and of being moved.

The principles laid out in the Physics, *especially concerning the importance of place and the explanation it provides for what we call gravity—why certain substances naturally seek certain places—are applied to the universe itself in* On the Heavens.

First, however, we must explain what we mean by "heaven" and in how many senses we use the word, in order to make clearer the object of our inquiry. (a) In one sense, then, we call "heaven" the substance of the extreme circumference of the whole, or that natural body whose place is at the extreme circumference. We recognize habitually a special right to the name "heaven" in the extremity or upper regions, which we take to be the seat of all that is divine. (b) In another sense, we use this name for the body continuous with the extreme circumference, which contains the moon, the sun, and some of the stars; these we say are "in the heaven." (c) In yet another sense we give the name to all body included within the extreme circumference, since we habitually call the whole or totality "the heaven." The word, then, is used in three senses.

Now the whole included within the extreme circumference must be composed of *all* physical and sensible body, because there neither is nor can come into being any body outside the heaven. . . . The world as a whole, therefore, includes all its appropriate matter, which is, as we saw, natural perceptible body. So that neither are there now, nor have there ever been, nor can there ever be formed more heavens than one, but this heaven of ours is one and unique and complete.

It is therefore evident that there is also no place or void or time outside the heaven. For in every place body can be present; and void is said to be that in which the presence of body, though not actual, is possible; and time is the number of movement. But in the absence of natural body there is no movement.

Aristotle turns his attention subsequently to the earth's position, shape, and rest or motion within the universe. His discussion by no means operates in a philosophical vacuum but accounts briefly for other views in competition with his own. The line of argument is instructive for any who have imbibed the old cliché according to which geocentric cosmology is said to locate the earth in the place of greatest importance in the universe.

As to earth's *position* there is some difference of opinion. Most people—all, in fact, who regard the whole heaven as finite—say it lies at the center. But the Italian philosophers known as Pythagoreans take the contrary view. At the center, they say, is fire, and the earth is one of the stars, creating night and day by its circular motion about the center. . . . There are many others who would agree that it is wrong to give the earth the central position, looking for confirmation rather to theory than to the facts of observation. Their view is that the most precious place befits the most precious thing. But fire, they say, is more precious than earth, and the limit than the intermediate, and the circumference and the center are limits. Reasoning on this basis they take the view that it is not earth that lies at the center of the sphere, but rather fire. The Pythagoreans have a further reason. They hold that the most important part of the world, which is the center, should be most strictly guarded, and name it, or rather the fire which occupies that place, the "Guard-house of Zeus," as if the word "center" were quite unequivocal, and the center of the mathematical figure were always the same with that of the thing or the natural center.

But it is better to conceive of the case of the whole heaven as analogous to that of animals, in which the center of the animal and that of the body are different. For this reason they have no need to be so disturbed about the world, or to call in a guard for its center. Rather, let them look for the center in the other sense and tell us what it is like and where nature has set it. That center will be something primary and precious; but to the mere position we should give the last place rather than the first. For the middle is what is defined, and what defines it is the limit, and that which contains or limits is more precious than that which is limited, seeing that the latter is the matter and the former the essence of the system. . . .

There are similar disputes about the *shape* of the earth. Some think it is spherical, others that it is flat and drum-shaped. For evidence they bring the fact that, as the sun rises and sets, the part concealed by the earth shows a straight and not a curved edge, whereas if the earth were spherical the line of section would have to be circular. In this they leave out of account the great distance of the sun from the earth and the great size of the circumference, which, seen from a distance on these apparently small circles, appears straight. Such an appearance ought not to make them doubt the circular shape of the earth. . . .

Some have been led to assert that the earth below us is infinite, saying, with Xenophanes of Colophon, that it has "pushed its roots to infinity"—in order to save the trouble of seeking for the cause. Hence the sharp rebuke of Empedocles, in the words "If the deeps of the earth are endless and endless the ample ether—such is the vain tale told by many a tongue, poured from the mouths of those who have seen but little of the whole." Others say the earth rests upon water. This indeed is the oldest theory that has been preserved, and is attributed to Thales of Miletus. It was supposed to stay still because it floated like wood and other similar substances, which are so constituted as to rest upon water but not upon air. As if the same account had not to be given of the water which carries the earth as of the earth itself! It is not the nature of water, any more than of earth, to stay in mid-air: it must have something to rest upon. Again, as air is lighter than water, so is water than earth. How then can they think that the naturally lighter substance lies below the heavier?

Aristotle's continuous assumption concerning the potency of place, whereby substances seek their proper or natural location, underlies his conclusions concerning both the shape and the movement or rest of the earth. The still-current popular notion that geocentrism identifies earth as the center of the universe (as distinct from placing it, on account of its mere heaviness, at the center) dissolves before the sophisticated clarity of Aristotle's account.

Further, the natural movement of the earth, part and whole alike, is to the center of the whole—whence the fact that it is now actually situated at the center. But it might be questioned, since both centers are the same, which center it is that portions of earth and other heavy things move to. Is this their goal because it is the center of the earth or because it is the center of the whole? The goal, surely, must be the center of the whole. For fire and other light things move to the extremity of the area which contains the center. It happens, however, that the center of the earth and of the whole is the same. Thus they do move to the center of the earth, but accidentally, in virtue of the fact that the earth's center lies at the center of the whole. . . .

From what we have said, the explanation of the earth's immobility is also apparent. If it is the nature of earth, as observation shows, to move from any point to the center . . . then it is impossible that any portion of earth should move away from the center except by constraint. For a single thing has a single movement, and a simple thing a simple. . . . If then no portion of earth can move away from the center, obviously still less can the earth as a whole so move. For it is the nature of the whole to move to the point to which the part naturally moves. Since then it would require a force greater than itself to move it, it must needs stay at the center. . . .

Earth's shape must necessarily be spherical. For every portion of earth has weight until it reaches the center, and the jostling of parts greater and smaller would bring about not a waved surface, but rather compression and convergence of part and part until the center is reached. The process should be conceived by supposing the earth to come into being in the way that some of the natural philosophers describe. Only they attribute the downward movement to constraint, and it is better to keep to the truth and say that the reason of this motion is that a thing which possesses weight is naturally endowed with a centripetal movement. When the mixture, then, was merely potential, the things that were separated off moved similarly from every side towards the center. Whether the parts which came together at the center were distributed at the extremities evenly, or in some other way, makes no difference. If there were a similar movement from each quarter of the extremity to the single center, it is obvious that the resulting mass would be similar on every side. For if an equal amount is added on every side the extremity of the mass will be everywhere equidistant from its center, i.e., the figure will be spherical.

Given the commonplace picture of Aristotle as anti-empirical, one is perhaps surprised at how often he appeals to experience, as he does in supporting his contention that the earth is spherical.

Again, our observations of the stars make it evident not only that the earth is circular, but also that it is a circle of no great size. For quite a small change of position to south or north causes a manifest alteration of the horizon. . . . Indeed there are some stars seen in Egypt and in the neighborhood of Cyprus which are not seen in the northerly regions; and stars, which in the north are never beyond the range of observation, in those regions rise and set. All of which goes to show not only that the earth is circular in shape, but also that it is a sphere of no great size: for otherwise the effect of so slight a change of place would not be so quickly apparent. Hence one should not be too sure of the incredibility of the view of those who conceive that there is continuity between the parts about the pillars of Hercules and the parts about India, and that in this way the ocean is one.

As further evidence in favor of this they quote the case of elephants, a species occurring in each of these extreme regions, suggesting that the common characteristic of these extremes is explained by their continuity. Also, those mathematicians who try to calculate the size of the earth's circumference arrive at the figure of 400,000 stades [perhaps 40,000 miles]. This indicates not only that the earth's mass is spherical in shape, but also that as compared with the stars it is not of great size.

SOURCE: Aristotle, *Physics* and *On the Heavens*, in *The Works of Aristotle*, ed. vol. 2, W. D. Ross, Oxford: Clarendon Press, 1930.

7

He Supposes the Earth to Revolve

Aristarchus and Archimedes

Not everyone before Copernicus believed that the planets and the sun re-volved around the earth. The first astronomer known to have proposed a he-liocentric rather than a geocentric model was Aristarchus of Samos (c. 310–c. 230 B.C.). Heliocentrism appears nowhere in his extant works, but we have authoritative attributions of the idea to him in other ancient sources. For ex-ample, Plutarch mentions Aristarchus's "attempt to save the phenomena by supposing the heaven to remain at rest, and the earth to revolve in an oblique circle, while it rotates, at the same time, about its own axis."

The most detailed reference, however, appears in the writings of Aristarchus's younger contemporary Archimedes (287–212 B.C.). The work in which Archimedes mentions Aristarchus's ideas is interesting for two fur-ther reasons. In it Archimedes is applying his development of an exponential system to express very large numbers—numbers which we quite naturally re-fer to as "astronomical" and which, without an exponential system, we would be at a loss to express. Archimedes asks, in what modern scientists would call a "thought experiment," how many grains of sand the universe it-self might hold. The discussion also provides an occasion for worrying about the size and very definition of the term universe.

There are some . . . who think that the number of the sand is infinite in mul-titude; and I mean by the sand not only that which exists about Syracuse and the rest of Sicily, but also that which is found in every region, whether inhab-ited or uninhabited. Again, there are some who, without regarding it as infinite, yet think that no number has been named which is great enough to exceed its multitude. And it is clear that they who hold this view, if they imagined a mass made up of sand as large in size as the mass of the earth, including in it all the seas and the hollows of the earth filled up to a height equal to that of the highest mountain, would be many times further still from recognizing that any number could be expressed which exceeded the multitude of the sand so taken.

But I will try to show you, by means of geometrical proofs, which you will be able to follow, that of the numbers named by me . . . some exceed not only the number of the mass of sand equal in size to the earth filled up in the way described, but also that of a mass equal in size to the universe.

Now you are aware that "universe" is the name given by most astronomers to the sphere the center of which is equal to the straight line between the center of the sun and the center of the earth. This you have seen in the treatises written by astronomers.

But Aristarchus of Samos brought out a book consisting of certain hypotheses, in which the premises lead to the conclusion that the universe is many times greater than that now so called. His hypotheses are that the fixed stars and the sun remain motionless, that the earth revolves about the sun in the circumference of a circle, the sun lying in the middle of the orbit, and that the sphere of the fixed stars, situated about the same center as the sun, is so great that the circle in which he supposes the earth to revolve bears such a proportion to the distance of the fixed stars as the center of the sphere bears to its surface.

Archimedes rightly objects that the ratio of a point to a circumference is no ratio at all, for a point has no dimension whatsoever. Therefore he adjusts the parallel thus: Aristarchus must mean that the ratio of the size of the earth to the size of the universe as Archimedes has defined it must be equal to the ratio of the size of the sphere of earth's orbit to the size of the universe as Aristarchus defines it. In short, the Aristarchan theory entails an increase in the size of the universe by many orders of magnitude.

Now it is easy to see that this is impossible; for since the center of the sphere has no magnitude, we cannot conceive it to bear any ratio whatever to the surface of the sphere. We must, however, take Aristarchus to mean this: since we conceive the earth to be, as it were, the center of the universe, the ratio which the earth bears to what we describe as the "universe" is the same as the ratio which the sphere containing the circle in which he supposes the earth to revolve bears to the sphere of the fixed stars. For he adapts the proofs of the phenomena to a hypothesis of this kind, and in particular he appears to suppose the size of the sphere in which he represents the earth as moving to be equal to what we call the "universe."

I say then, that, even if a sphere were made up of sand to a size as great as Aristarchus supposes the sphere of the fixed stars to be, I shall still be able to prove that . . . some [numbers] exceed in multitude the number of the sand which is equal in size to the sphere referred to.

As T. L. Heath comments in Greek Astonomy *(p. 108), after much "sheer calculation" Archimedes finds "that the number of grains of sand that would be contained in a sphere of the size attributed to the universe is less than the number which we should express as* 10^{63}.*"*

SOURCE: Thomas L. Heath, *Greek Astronomy,* London: J. M. Dent, 1932.

8

A Geometrical Argument

Eratosthenes

A strangely persistent modern myth is that before about 1492, people thought the earth was flat. On the contrary, the ancient Greeks and others knew the earth to be spherical. Visual evidence could be found in lunar eclipses and in the fact that, as one travels southward, the pole star and others appear lower in the sky while new stars on the southern horizon come into view. But given that the earth is a sphere, and that the ancients had no opportunity to circumnavigate the globe, how large did they imagine this sphere to be?

Eratosthenes (c. 275 – c. 195 B.C.) became famous for an achievement that is geo-metrical *in the most literal sense of the term: He measured the earth. The second-century* B.C. *astronomer Cleomedes wrote the principal extant account of what he calls Eratosthenes' "geometrical argument," but the version of the story as recounted by Robert Osserman (b. 1926) is unsurpassed for clarity.*

How could one measure the whole earth, when the immense expanses of the oceans formed impenetrable barriers to travel? A most ingenious answer was provided by Eratosthenes of Alexandria.

Alexandria was founded at the delta of the Nile in northern Egypt, where the river empties into the Mediterranean Sea, by Alexander the Great, who wanted a city to match the grandeur of his own ambitions. He succeeded to an astonishing degree. Ancient Alexandria attracted the most outstanding literary, scholarly, and scientific talent of the day, in part because of its library—the most comprehensive in the world. The head of the library in the latter half of the third century B.C. was Eratosthenes, one of the greatest scientific talents in Alexandria as well as the author of books of poetry and literary criticism.

Eratosthenes' method for determining the size of the earth rested on three elements. The first was a bit of elementary geometry which will be explained in a moment. The second involved a serendipitous geographic fact regarding a city on the Nile River in southern Egypt called Syene in those days, now known as Aswan. The third was an absurdly simple apparatus called a *gnomon*.

The gnomon had been in use for a very long time. It consisted of a vertical stick placed on a level piece of ground. The gnomon was a device that allowed one to follow the sun's shadow as the sun moves across the sky. Although the gnomon cannot be used to tell time in the manner of its more advanced cousin, the sundial, it does provide a surprising amount of useful information.

First, the gnomon gives the exact time once a day, at the moment that the sun is highest in the sky and the shadow of the gnomon is the shortest—at noon. In addition, it acts as a compass, since [in the northern hemisphere] the shadow at noon points due north. . . .

The gnomon also serves as a primitive calendar, determining two key days of each year: the summer and winter solstices. If one places a mark where the shadow ends at noon on each day of the year, one finds that in winter, when the sun is low in the sky, the shadows are longer, while in the summer, with the sun high in the sky, the shadows are shorter. The shadow at noon goes through a yearlong cycle, from the shortest noon shadow in summer, gradually reaching its greatest length six months later, and then shortening again over the succeeding six months. The day on which the noon shadow is shortest, and the sun is highest, is called the *summer solstice*. The day six months later when the sun is lowest and the noon shadow is longest is known as the *winter solstice*. Counting the number of days from solstice to solstice also provided one of the earliest accurate measurements of the length of the year.

Finally, the gnomon could be used to determine the altitude of the sun— that is, the angular distance of the sun above the horizon at any given moment (at least on sunny days). All one had to do was measure the length of the shadow and the length of the stick. By drawing a right triangle to scale with those measurements, one can measure the angle opposite, and that angle will indicate how far off the sun's direction is from an overhead, vertical direction. (See page 48.)

These uses of the gnomon were well known to Eratosthenes and his contemporaries. But it was the fortuitous geographical properties of Aswan that gave Eratosthenes his inspiration for determining the size of the earth. Aswan is almost due south of Alexandria. It also enjoys the special privilege of having

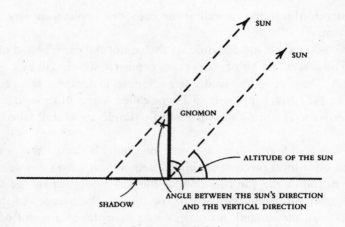

Gnomon and shadow.

the sun pass directly overhead at one moment of each year: at noon on the
summer solstice. At that one moment each year, a gnomon in Aswan casts no
shadow at all. (Aswan lies almost exactly on the Tropic of Cancer. . . .)

By combining these facts with some simple but clever geometric reasoning,
Eratosthenes was able to produce his remarkable pièce de résistance: the circumference of the earth. At noon on the summer solstice, he simply used his
gnomon to determine the angle between the sun and the vertical direction at
Alexandria. Since the sun at that moment is directly overhead in Aswan, he
thereby knew the angle between the vertical direction at Alexandria and at
Aswan. He found the angle to be 1/50 of the circumference of a circle. That
meant that the entire circumference of the earth is 50 times the distance between Alexandria and Aswan. Since the distance from Aswan to Alexandria
is roughly 500 miles, by today's measurements, the earth must be approximately 25,000 miles around.

The brilliant simplicity of Eratosthenes' method is not diminished by the
fact that his estimate involves several inaccuracies and uncertainties: first,
measuring the angle between the direction of the sun and the vertical direction could be done only approximately; second, Aswan is not exactly due
south of Alexandria, but only roughly so; third, it would have been difficult
or impossible to obtain an accurate measure of the distance between the two
cities; and finally, there is considerable uncertainty about how to interpret
ancient units of measurement in modern terms. Large distances were given in
terms of *stades*—the length of a stadium. According to Eratosthenes, the circumference of the earth was 250,000 stades. The length of a "stade" was
standardized at 600 "feet," but the length of a foot was not standard, and
varied by 10 percent or more. The figure of 25,000 miles for the earth's cir-

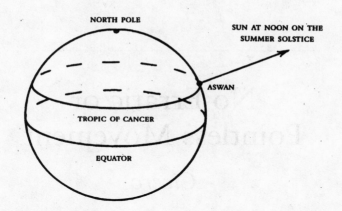

Tropic of Cancer is the name of a circle of latitude about 23.5 degrees above the equator.

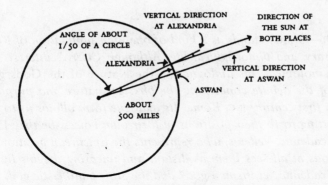

Eratosthenes' method for measuring the earth: when the sun is directly overhead at Aswan, measure the angle between the sun and the vertical direction at Alexandria using the shadow of a vertical pole.

cumference results from choosing a value at the low end of the scale for the length of a stade. The net effect was that Eratosthenes' calculation might on several counts be termed a "ballpark estimate" rather than a scientifically precise measurement. Nevertheless, it provides dramatic testimony to the ability of simple but ingenious geometric reasoning to succeed where a direct approach—involving traversals of two polar regions and an ocean—was well beyond the realm of possibility.

SOURCE: Robert Osserman, *Poetry of the Universe: A Mathematical Exploration of the Cosmos*, New York: Anchor Books, 1995.

No Erratic or Pointless Movement

Cicero

Cicero (106–43 B.C.) stands as a kind of one-man summation of Roman political, literary, and philosophical life. Although Cicero's interests were not mainly cosmological, his dialogue On the Nature of the Gods *gives us a snapshot of the debate concerning the physical nature and purpose of the universe in first century B.C. Rome. Its dialogue form allows us to hear proponents setting forth their positions in their own voices, be they Epicurean, Stoic, or Academic. Velleius, who represents the Epicurean position, which if it posits gods at all sees them as distant and uncaring, begins by mocking both the Academic "craftsman-god" and the Stoic pantheistic god.*

"What you are going to hear are no airy-fairy, fanciful opinions, like the craftsman-god in Plato's *Timaeus* who constructs the world, or the prophetic old lady whom the Stoics call Pronoia, and whom in Latin we can term *Providentia*. I am not going to speak of the universe itself as a round, blazing, revolving deity endowed with mind and feelings. These are the prodigies and wonders of philosophers who prefer dreaming to reasoning. I ask you, what sort of mental vision enabled your teacher Plato to envisage the construction of so massive a work, the assembling and building of the universe by the god in the way which he describes? What was his technique of building? What were his tools and levers and scaffolding? Who were his helpers in so vast an enterprise? How could the elements of air and fire, water and earth knuckle under and obey the will of the architect? . . .

"The question . . . is this: why did these world-builders [the Platonic Demiurge and the Stoic Providence] suddenly emerge after lying asleep for countless generations? For the non-existence of the universe does not necessarily imply absence of periods of time. . . . So what I am asking, Balbus [the Stoic], is this: why did your Pronoia remain idle throughout that boundless length of time? Was she avoiding hard work? But hard work does not impinge upon a god, and in any case there was no such labor, for all the elements of sky, stars, lands, and seas obeyed the divine will. . . .

"As for those who have maintained that the world itself possesses life and wisdom, they have totally failed to see into what shape the nature of intelligent mind could be installed. . . . I shall merely express surprise at the slow-wittedness of those who would have it that a living creature endowed with both immortality and blessedness is spherical in shape, merely because Plato maintains that no shape is more beautiful than the sphere. In my view, the cylinder, the cube, the cone, the pyramid are more beautiful. And what sort of life is assigned to this rotund god? Why, to be spun at speed the like of which cannot even be imagined; I cannot envisage mental stability or a life of happiness resident in that!"

Having set up straw deities, Velleius appeals—as materialists still do—to the "natural" production of the world that has no recourse to divine explanation.

"But we Epicureans define the life of blessedness as residing in the possession of untroubled minds and relaxation from all duties. Our mentor who has schooled us in all else has also taught us that the world was created naturally, without the need for a craftsman's role, and the process which in your view cannot be put in train without the skillful touches of a god is so straightforward that nature has created, is now creating, and will continue to create innumerable worlds. Because you Stoics do not see how nature can achieve this without being endowed with mind, you behave like poets of tragedy, unable to draw the plot to its close, and having recourse to a *deus ex machina*.

"You would surely have no need of the activity of such a figure if you would only observe how unlimited, unbounded tracts of space extend in all directions. When the mind strains and stretches itself to observe these distances, it journeys abroad so far that it can observe no ultimate limit at which to halt. It is in this boundless extent of breadth, length, and height, then, that innumerable atoms in infinite quantity flit around. There is space between them, yet they latch on to each other. In gripping each other they form a chain, as a result of which are fashioned the shapes and forms of things which you Stoics believe cannot be created without bellows and anvils. So you have implanted in our heads the notion of an external lord whom we are to fear day and night; for who would not stand in awe of a god who is a prying busybody, who foresees and reflects upon and observes all things, believing that everything is his business?"

In The Nature of the Gods, *however, Cicero gives pride of place to the Stoic position. Balbus, the Stoic speaker, sketches an "argument from design" to support belief in the existence of deity.*

"What can be so obvious and clear, as we gaze up at the sky and observe the heavenly bodies, as that there is some divine power of surpassing intelligence by which they are ordered? If this were not the case, how could Ennius have won general assent with the words

> Behold this dazzling vault on high, which all
> Invoke as Jupiter!

and not merely as Jupiter, but also as the lord of creation, governing all things by his nod, and (to exploit Ennius's words again) as "father of gods and men," an attentive and supremely powerful God? I completely fail to understand how anyone who doubts this can avoid also doubting whether the sun exists or not—for in what way is the sun's existence more obvious than God's? If this realization was not firmly implanted in our minds, such steadfast belief would not have endured nor been strengthened in the course of time, nor could it have become securely lodged in succeeding generations and ages of mankind."

Although some of Balbus's reasoning may sound like that of St. Paul in the epistle to the Romans, the pantheism of the Stoical position is unmistakable: the universe itself is divine. Balbus then goes on to make the observation, one reemphasized in our own day by supporters of the "anthropic principle," that if rational, self-aware beings are produced by and part of the universe, then the universe itself must in some sense be seen as rational and self-aware.

"It can be established that the universe is wise, and blessed, and eternal, for all embodiments of these attributes are superior to those without them, and nothing is superior to the universe. This will lead to the conclusion that the universe is God.

"Zeno [the Stoic] also produced this argument: 'Nothing which is devoid of sensation can contain anything which possesses sensation. Now some parts of the universe possess sensation; therefore the universe is not devoid of sensation.' He goes further, pressing the argument more closely: 'Nothing which lacks a vital spirit and reason can bring forth from itself a being endowed with both life and reason. Therefore the universe is endowed with life and reason.' He also pressed home his argument with his favorite technique of the simile, like this: 'If flutes playing tunefully were sprouting on an olive-tree, you would surely have no doubt that the olive-tree had some knowledge of flute-playing; again, if plane-trees bore lutes playing in tune, you would likewise, I suppose, judge that plane-trees were masters of the art of music. Why then is the universe not accounted animate and wise, when it brings forth from itself creatures which are animate and wise?'"

Many of the features of the universe as Bálbus describes them in Cicero's dialogue persisted as cosmological stock-in-trade for many centuries to come, including a preoccupation with circles and spheres. In chiding Velleius for his earlier flip comments about geometrical solids, Balbus betrays how deeply aesthetic one's choice of world view may be.

"Velleius, please do not parade the utter ignorance of learning of your school. You say that you regard the cone and cylinder and pyramid as shapes more beautiful than the sphere. In this you betray the same curious judgement in aesthetics which you show in all else. Let us suppose, however, that these shapes are more beautiful in appearance. This is not a view that I share; for what can be more beautiful than the shape which alone embraces and gathers in all other shapes, which can exhibit no rough surface, no jagged projection, no angular indentations or bends, no protuberances or yawning gaps? There are two shapes which excel all others: in solid bodies, the globe (*globus* is the word I use to render the Greek *sphaera*), and in planes the circle or orb, the Greek word for which is *kuklos*. These two shapes alone are closely similar in all their parts, with the circumference equidistant from the center at all points. Nothing can be better ordered than that. Still, if you Epicureans do not realize this because you have never traced diagrams in the dust of the schools, could you natural philosophers not have grasped even this, that the uniform movement and regular positions of the stars could not have been preserved in any other shape? So nothing could be more ignorant than the usual assertion of your school; for you claim that it is not certain that this universe of ours is round, since it has possibly another shape, and you maintain that there are countless other worlds of varying shapes. . . .

"Now there are two types of heavenly bodies. The first type travels from east to west over the same unchanging regions, never at any time making the slightest alteration to their course. The second type covers the same expanse and the same route in two revolutions without a break between them. The two types reveal to us both the rotatory movement of the universe, achievable only because it is spherical in shape, and the circling revolutions of the stars.

"Take the sun first, which has pride of place among the heavenly bodies. In its course it first fills the lands with abundant light, and then shrouds them successively in shade, for night results when the earth's shadow blocks the sunlight. Its journeys in darkness have the same regularity as those in daylight. The sun also regulates the limits of cold and heat by drawing slightly nearer and retiring slightly further. The round of the year is complete by some 365 1/4 daily circuits by the sun; and by adjusting its course now northward, now southward, the sun creates summer and winter, and the two seasons which follow the tail-ends of winter and summer. The transformation of the four seasons ensures the birth and the rationale of all things which are begotten on land and sea.

"Next, the moon in her monthly circuit traverses the course over which the sun takes a year. When she draws nearest to the sun, her light becomes dimmest, and her orb is fullest when she is most distant. Not merely do her appearance and shape change, as she waxes and then by gradual diminution returns to her original form, but she alters her position in the sky. Her position in north or south creates in her course the equivalent of the winter and summer solstices; she is the source of the many effluences which result in the nurture and growth of living creatures, and which cause the plants which sprout from the earth to swell and ripen."

Balbus's survey of the heavens makes explicit two important astronomical definitions. He is critical of the term used for the planets, for the Greek plan-etes means "wanderer," which to the Stoic ear in any case wrongly suggests "error" or "going astray." The dialogue also provides the first extant reference to the concept of the Great Year (the enormous period of time required for all heavenly bodies to return to a given initial position—like all three hands of a clock returning to a vertical position at the stroke of midnight).

"Most remarkable, too, are the movements of the five planets, mistakenly labelled 'those which stray'; mistakenly, because nothing can be said to 'go astray' which through all eternity maintains in a steady, predetermined pattern its various movements forward, backward, and in other directions. What is all the more remarkable in these bodies under discussion is that at one moment they disappear, and at another reappear; now they draw close, and now retire; at one time they draw ahead, and at another lag behind; they alternatively accelerate and decelerate; on occasion, they cease to move at all, and remain still for some time. Mathematicians have exploited the varying movements of the planets to calculate the length of the Great Year, which is accomplished once the sun, moon, and five planets have completed all their revolutionary courses, and have returned to the same relative positions.

"The actual length of the Great Year is a difficult question, but it is undoubtedly a fixed and delineated period. The planet bearing the name of Saturn, which the Greeks call *Phaenon* ('shining') is furthest from the earth. It takes about thirty years for it to complete its journey, in the course of which it does many remarkable things. It goes ahead, and then falls back; it disappears during the hours of evening and shows itself again in the matutinal hours; and yet in age after age throughout eternity it never varies, but behaves identically at the identical times. Below Saturn and closer to the earth the planet Jupiter speeds on its way; men call it *Phaethon* ('blazing'). Jupiter completes the same circuit through the twelve signs of the zodiac every twelve years, and in its course it indulges in the same variations as Saturn does. The nearest circuit below this is covered by *Pyroeis* ('fiery'), which is called the star of Mars; this planet completes in, I think, some six days short of twenty-four months the same round as the other two. Below Mars lies the star of Mercury, called *Stilbon* ('gleaming') by the Greeks, which takes about a year to circle through the zodiac; it never distances itself from the sun more than one sign's length, sometimes leading ahead and sometimes falling behind. Lowest of the five planets and nearest to the earth is the star of Venus, in Greek called *Phosphoros* ('light-bringing'), and in Latin when it precedes the sun, Lucifer, but *Hesperus* ('at evening') when it follows behind. Venus completes its course in a year; it traverses the breadth as well as the length of the zodiac as also do the planets above it, and it never departs more than two signs' distance from the sun, sometimes lying ahead, and sometimes behind.

"This is why I cannot envisage such regular behavior in the stars, and such remarkable coincidence of timing in their varied paths throughout eternity, as existing without intelligence, reason, and planning; and since we observe these qualities in heavenly bodies, it is impossible for us not to number them among the gods.

"It is the same with the so-called fixed stars: they too evince the same intelligence and foresight. Every day they revolve with due and dependable regularity. It is not that they merely revolve with the aether, or that they cling close to the firmament, as is assumed by many who are ignorant of the laws of physics; for the composition of the aether is not such that it grips the stars and twists them round by its force, since it is rarefied and diaphanous and endowed with uniform heat. Thus it seems unsuited to be a receptacle for the stars. Accordingly the fixed stars have their own spheres, separated from and free of attachment to the aether. Their perennial and unceasing journey, traversed with a wondrous regularity beyond belief, makes manifest the divine force and intelligence which resides within them. Hence anyone who fails to realize that they possess the power of gods seems incapable of any kind of observation.

In Balbus's Stoic picture of the universe we find also the notion, a common-place of literature and cosmology for the better part of two millennia, that change and irregularity are confined to the sub-lunary sphere (for the moon, with its changing phases, marks the boundary between the earthly realm, where things are conspicuously unstable, and the heavenly realm, where reg-ularlity and even "fixity" prevail).

"So the heavens contain no chance or random element, no erratic or point-less movement; on the contrary, all is due order and integrity, reason, and regularity. All that lacks these qualities, and misleads with falsehood and abounds in error, belongs to the vicinity of earth below the moon, the lowest of the heavenly bodies, and to earth itself. Therefore, any person who imag-ines that the heavens are mindless, when their remarkable order and regular-ity beyond belief ensure the total preservation and well-being of everything in the universe, must himself be regarded as out of his mind."

Although physically confined to this mutable realm beneath the moon, hu-man beings nevertheless have a special ability to contemplate the universe be-yond. And this ability in turn argues a special place for human beings in any analysis of cosmic purpose—of "teleology"—an issue that remains actively discussed among cosmologists to this day.

"Has not our human reason advanced to the skies? Alone of living creatures we know the risings, settings, and courses of the stars. The human race has laid down the limits of the day, the month, the year; they have come to rec-ognize eclipses of the sun and moon, and have foretold the extent and the date of each occurrence of them for all the days to come. Such observation of the heavens allows the mind to attain knowledge of the gods, and thus gives rise to religious devotion, with which justice and the other virtues are closely linked. These virtues are the basis of the blessed life which is equivalent and analogous to that enjoyed by the gods. . . .

"It remains finally for me to show . . . that all things in this universe of ours have been created and prepared for us humans to enjoy. So first, the uni-verse itself was made for the benefit of gods and men. All that is in it has been provided and devised for us to enjoy; for the universe is, so to say, the shared dwelling of gods and men, or a city which houses both, for they alone enjoy the use of reason, and live according to justice and law. So just as we must believe that Athens and Sparta were founded for the Athenians and Spartans, and that everything in those cities is rightly claimed to belong to those peoples, so all that exists in the entire universe must be regarded as the possession of gods and men.

"Again, the revolutions of sun, moon, and other heavenly bodies admittedly form part of the organic structure of the universe, but they also offer a spectacle to the human race. This is supremely the sight of which we never tire; it is more beautiful and reflects greater reason and intelligence than all others, for by measuring their courses, we become aware of the due arrival of the seasons, and of variations and changes in them. Since only humans have this awareness, we must infer that this is a dispensation made for their sake."

SOURCE: Marcus Tullius Cicero, *The Nature of the Gods*, trans. P. G. Walsh, Oxford: Clarendon Press, 1997.

Turning the Universe Upside Down

Plutarch

Since the beginning of human history and probably before, people have won-
dered about the moon. What is it made of? How does it shine? How does it
affect us?

Plutarch (A.D. 46?–c. 120), the great first-century biographer of Greeks
and Romans, gives us in his dialogue on The Face Which Appears on the Orb
of the Moon *an intriguing taste of some of the issues debated in the Roman*
Empire of his time. This dialogue also set the agenda for investigations by
Galileo and Kepler in the seventeenth century. In particular it raised the
question of what causes the mottled appearance of the moon's face. One
view discussed is that of the Stoics, "that the moon is a mere mixture of air
and mild fire, that the air grows dark on its surface, as a ripple courses over a
calm sea, and so the appearance of a face is produced." The response of
Lamprias, who also narrates the dialogue, combines a desire to explain the
matter scientifically with an effort to speak of the moon respectfully.

I said, ". . . It is a slap in the face to the moon when [the Stoics] fill her with
spots and black patches, addressing her in one breath as Artemis and
Athena, and in the very same describing a congealed compound of murky
air and charcoal fire, with no kindling or light of its own, a nondescript
body smoking and charred like those thunderbolts which poets address as
'lightless' and 'sooty.' . . . But if the moon is fire, where does all this air inside it come from? For this upper region, always in circular motion, is com-

posed not of air but of some nobler substance, which has the property of refining and kindling all things. If air has been generated, how can it not have been vaporized by the fire and so changed into some other form, instead of being preserved near the fire all this time, like a nail fitted into the same place and wedged there for ever? If it is rare and diffused, it should not remain stable, but be displaced. On the other hand, it cannot subsist in a solidified form, because it is mingled with fire, and has no moisture with it, and no earth, the only agents by which air can be compacted. . . . The Stoics are displeased with Empedocles when he describes the moon as a mass of air frozen like hail and enclosed within her globe of fire. Yet they themselves hold that the moon is a globe of fire which encloses air variously distributed, even though they do not allow that she has clefts in herself, or depths and hollows, for which those who make her an earth-like body find room. Instead they clearly suppose that the air lies upon her convex surface, an absurd idea given the problem of its stability, and impossible given what we see at full moon. For if they were right, we ought not to be able to distinguish black parts and shadow. Either everything there should be dull and shrouded, or else everything should radiate equally when the moon is caught by the sun. . . ."

Here Pharnaces [the Stoic] interrupted me: "There it is again, the old trick of the Academy brought out against us! They amuse themselves with arguing against others but never offer their own views to be examined. . . . You won't draw me on today to answer your charges against the Stoics unless we first hear an account of your behavior in turning the universe upside down."

Lucius smiled: "Yes, my friend," he said, "only do not threaten us with the writ of heresy that Cleanthes used to think the Greeks should have served upon Aristarchus of Samos for shifting the hearth of the universe. For that great man attempted 'to save the phenomena' with his hypothesis that the heavens are stationary, while our earth moves round in an oblique orbit, at the same time whirling about her own axis. It's true we Academics have no views of our own, but tell me: How do those who assume that the moon is an earth turn things upside down any more than you do when you fix the earth where she is, suspended in mid air, a body considerably larger than the moon? At least mathematicians tell us so, calculating the magnitude of the obscuring body from what takes place in eclipses, and from the passages of the moon through the shadow. . . . Yet you have fears for the moon lest she should tumble, while as for our earth, Aeschylus [525–456 B.C.] has perhaps satisfied you that Atlas

> 'Stands, and the pillar which parts Heaven and Earth
> His shoulders prop, no load for arms t'embrace!'

Then you think that under the moon there circulates light air, quite inadequate to support a solid mass, while the earth, in Pindar's words, 'is compassed by pillars set on adamant.' And this is why for his part Pharnaces has no fear of the earth's falling. . . . Yet the moon does have something to prevent her from falling, the very speed and swing of her passage round, just as objects placed in slings are kept from falling by the whirl of their rotation. For everything is borne on in its own natural direction unless this is changed by some other force. Therefore, the moon is not drawn down by her weight, since that downward tendency is counteracted by her circular movement. There would perhaps be greater grounds for wonder if she were entirely at rest as the earth is. But as things are, the moon has a powerful cause that prevents her from being borne down upon us. The earth, by contrast, not having any movement, might naturally be moved by its own weight. It is heavier than the moon not merely in proportion to its greater bulk, but because the moon has been rarefied by heat and fire. It would actually seem that the moon, if she is a fire, needs earth all the more, as a solid substance to move about and to cling to, so feeding and sustaining the force of her flame. For it is impossible to conceive fire unless it is maintained with fuel. But you Stoics say that our earth stands firm without foundation or root."

"Of course," said Pharnaces, "it keeps its proper and natural place, as being the essential middle point, that place around which all weights press and bear, converging towards it from all sides. But all the upper region, even if it does receive an earth-like body thrown up with force, immediately thrusts it out hitherward, or rather lets it go, to be borne down by its own momentum."

In spite of both sides' accusations that their opponents are trying to turn the world upside down, all the participants in this dialogue hold a geocentric view. However, the Academics such as Lucius see the moon as essentially earth-like, and therefore solid and heavy. Their difficulty is thus how to account for the fact that the moon remains up there. (Lucius invokes something that Huygens and Newton would later call centrifugal force.) By contrast, the Stoics see the moon as akin to the other planets and the stars, which are up there because they are composed of light material, air and especially fire. Their problem is accordingly to explain the evidence (such as eclipses and the new moon) that would indicate the moon has no light of its own.

The other crucial physical disagreement to emerge from this discussion is about gravity. There is confusion as well as insight on both sides of the argument. The Stoics are Aristotelian in their explanation of gravity: Heavy things fall towards the center of the universe, and the earth lies at the center because it is the heaviest thing. The Academics display greater affinity with Newton or even Einstein: Bodies fall not towards a particular place but towards other bodies—and the greater the mass, the greater the gravity. Yet it is the Stoics' view that produces the most coherent description of how gravity actually behaves—for example how it would affect a body at, or passing through, the center of the earth. It is a description the Academics like Lamprias consider ridiculous:

I said, ". . . But philosophers must not be listened to, if they choose to meet paradoxes with paradoxes, and if, when contending against strange views, they invent views which are even more strange and wonderful. Here are these Stoics with their 'tendency towards the middle!' Is there any paradox which is not implicit there? That our earth, with all its depths and heights and inequalities, is a sphere? That there are people at our antipodes who live like timber-worms or lizards, their lower limbs turned uppermost as they plant them on earth? That we ourselves do not keep perpendicular as we move, but remain on the slant, swerving like drunkards? That masses of a thousand talents weight, borne through the depth of the earth, stop when they reach the middle point, though nothing meets or resists them; or, if mere momentum carry them down beyond the middle point, they wheel round and turn back of themselves? . . . This is to make

'Up down, down up, where Topsy-Turvy reigns,'

all from us to the center down, and all below the center becoming up in its turn! So that if someone, out of 'sympathy' with earth, were to stand with the central point of his own body touching the center, he would have his head up and his feet up too! . . .

"Such are the monstrous paradoxes which they bear on their backs and trail along behind them . . . nothing but a conjuror's stock-in-trade and show-booth. And then they call others triflers for placing the moon, which is an earth, up above, and not where the middle point is. Yet if every weighty body converges to the same point with all its parts, then the earth will claim the heavy objects not so much because she is middle of the whole, as because they are parts of herself. And the inclination of falling bodies provides evidence not of any property of earth as middle of the universe, but rather of a community and fellowship between earth and her own parts, once ejected but now drawn back to her. For as the sun draws into himself the parts of which he has been composed, so earth receives the stone as belonging to her, and draws it towards herself. . . . It is not proved that earth is the middle of the universe. Moreover, the way in which bodies here are collected and

drawn together towards the earth suggests how bodies which have fallen together onto the moon may reasonably be supposed to keep their place with
respect to her."

*The rejection of place as explanation for gravity is part of an effort to replace
an Aristotelian finite universe with one that is infinite. Yet this somehow remains a universe in which "up" and "down" are fundamental categories.*

"Look at the question broadly. In what sense is the earth 'middle,' and middle of what? For The Whole is infinite. Now the infinite has neither beginning nor limit, so it ought not to have a middle, for a middle is in a sense
itself a limit, whereas infinity is a negation of limits. It is amusing to hear
someone labor to prove that the earth is the middle of the universe, not of
The Whole, forgetting that the universe itself is subject to the same difficulties; for The Whole, in its turn, left no middle for the universe. 'Heartless
and homeless' it is borne across an infinite void towards nothing which it can
call its own. Or, if it finds some other cause for remaining, it stands still, but
not because of the nature of the place. Much the same can be speculated concerning the earth and the moon: if the one stands here unshaken while the
other moves, it is because of a difference of soul and of nature rather than of
place. And in addition to all this, hasn't one other important point escaped
them? If anything, however great, which is outside the center of the earth is
'up,' then no part of the universe is 'down.' Earth is 'up,' and so are the
things on the earth. Absolutely all bodies lying or standing about the earth
become 'up' and one thing only is 'down,' namely, that incorporeal point
which of necessity must resist the pressure of the whole universe, if 'down' is
naturally opposed to 'up.' Nor is this the only absurdity. [At the center]
weights lose the cause of their downward tendency and motion, since there is
no body below towards which they move. That the incorporeal should have
so great a force as to direct all things towards itself, or hold them together
about itself, is not probable."

*The Academics, in keeping with their general Platonism, mistrust merely
physical explanation and do not accept nature alone as a sufficient reason
why things should be as they are. For them, order in the world is attributable
not to nature but to the Good. Nature alone, as Hobbes pointed out in the
seventeenth century in regard to political and social affairs, may produce
nothing more than a war of each against all. Lamprias continues:*

"Consider well, my friend, whether, when you shift all things about and remove each to its 'natural' place, you are not framing a system that will dissolve the universe, and introducing Empedoclean strife, or rather stirring up

the old Titans against Nature, in your eagerness to see once more the dreadful disorder and discord presented by the myth. All that is heavy in a place by itself, and all that is light in another,

> 'Where neither sun's bright face is separate seen,
> Nor Earth's rough brood, nor Ocean any more,'

as Empedocles says! Earth had nothing to do with heat, nor water with wind; nothing heavy was found above, nothing light below; without commixture, without affection were the elements of all things, mere units, each desiring no intercourse with each or partnership, performing their separate scornful motions in mutual flight and aversion, a state of things which must always be, as Plato teaches, where God is absent, the state of bodies deserted by intelligence and soul. So it was until the day when Providence brought Desire into Nature, and Friendship was engendered there, and Aphrodite and Eros, as Empedocles tells us and Parmenides too and Hesiod, so that things might change their places, and receive faculties from one another in turn, and, from being bound under stress, and forced, some to be in motion, some to rest, might all begin to give in to the Better, instead of the Natural, and shift their places and so produce harmony and communion of The Whole."

The picture of nature without government, especially without providential government, is akin to the pictures of chaos presented earlier not only by Plato but also by the Atomists, and centuries later by Milton:

> [a] wild Abyss,
> The womb of nature, and perhaps her grave,
> Of neither sea, nor shore, nor air, nor fire,
> But all these in their pregnant causes mixed
> Confusedly, and which thus must ever fight,
> Unless the Almighty Maker them ordain
> His dark materials to create more worlds.

Returning to the specific issue of the moon, including the question of how best to honor her, Lucius the Academic argues that to treat the moon as earth-like actually saves her from the embarrassment of being a rather shabby star. A millennium and a half before its confirmation by Galileo, Lucius thus presents a picture of an earth-like moon reflecting the sun's borrowed rays.

"Regarding the other stars, and the heavens in general, when you [Aristotelians] assert that they have a nature which is pure and transparent, and

removed from all changes caused by passion, and when you introduce a circle of eternal and never ending revolution, perhaps no one would want to
contradict you, at least for the present, although there are countless difficulties. But when the theory comes down and touches the moon, the moon no
longer retains the freedom from passion and the beauty of form the others
possess. Quite apart from her other irregularities and points of difference,
this very face which appears upon her must have been caused either by some
passion peculiar to the moon herself or by admixture of some other substance. Indeed, mixture implies passion, since a body loses its own transparency when it is forcibly filled with something inferior to itself. Consider
her own torpor and dullness of speed, and her faint ineffectual heat . . . : to
what are we to attribute this except weakness in herself and affection, if affection can ever reside in an eternal and Olympian body?

"It comes down to this . . . : Look on her as earth, and she appears a very
beautiful object, venerable and highly adorned. But as a star, or light, or any
divine or heavenly body, I'm afraid she may be judged lacking in form and
grace, and do no credit to her beautiful name, if out of all the multitude in
heaven she alone goes round begging light of others, as Parmenides says,

'For ever peering toward the Sun's bright rays.'

Now when our friend, in his exposition, had explained the claim of Anaxagoras, that 'the sun places the brightness in the moon,' he was highly applauded.
But . . . I will gladly pass on to the remaining points. It is probable, therefore,
that the moon is illuminated not as glass or crystal by the sunlight shining in
and through her, nor by way of accumulation of light and rays, as torches
multiply their light. For then we should have full moon at the beginning of the
month just as much as at the middle, if she does not conceal or block the sun
but lets him pass through because of her rarity, or if he intermingles his rays
with the light around her and helps to kindle it with his own. For we cannot
say she bends or swerves aside, as when she is at half moon or when she is
gibbous or crescent. Being then, as Democritus puts it, 'plumb opposite' to the
body illuminating her, she receives and admits the sun, so that we should expect to see her shining herself and also allowing him to shine through her. But
she is very far from doing this. At those times she is herself invisible, and she
often hides him out of our sight. As Empedocles says,

'So from above for men
She quenched his beams, shrouding a slice of earth
Wide as the compass of the glancing moon.'

These words would suggest that the sun's light had fallen not upon another star but upon night and darkness. Empedocles also implies that the illumination which we get from the moon arises in some way from the reflection of the sun falling upon her. Hence her light reaches us without heat or lustre, whereas we should expect both heat and lustre if there were a kindling of the moon by the sun or an intermingling of lights. But as voices return an echo weaker than the original sound, and missiles which glance off strike with weaker impact,

> 'E'en so the ray which smote the moon's white orb'

reaches us in a feeble and exhausted stream, because the force is dissipated by reflection."

SOURCE: Adapted from Plutarch, *The Face Which Appears on the Orb of the Moon*, trans. A. O. Prickard, Winchester: Warren and Son, 1911.

Part Two

PTOLEMY, MIDDLE EARTH, MIDDLE AGES

The Peculiar Nature of
the Universe

Claudius Ptolemy

Claudius Ptolemy (c. 100–c. 175) was the author of the single most influential astronomy textbook ever written. Known as Almagest—*which means "the greatest"—this book was originally (and forgettably) entitled* Mathematical Systematic Treatise. *It is truly an advanced technical and mathematical work encompassing enormous numbers of diagrams, charts, and equations. It appeared first in Greek, probably shortly after the year 150, in the world's greatest center of learning at that time, Alexandria. In the entire Mediterranean area and in Europe east and west,* Almagest *became the standard authority on astronomy for well over a thousand years.*

The adjective Ptolemaic *is still used as synonymously with "geocentric," and it is still subject to the misunderstandings that sometimes cluster around that word. Clichés notwithstanding, it does not imply, for example, anthropocentric. Moreover, some who have not read Ptolemy—or Aristotle, upon whose physics Ptolemy's system is based—may assume that Ptolemaic cosmology arises from mere authority or abstract philosophical thought devoid of observation. However, as we can see from Ptolemy's attempt to reconstruct the process by which civilization arrived at a geocentric concept, he takes physical evidence very seriously—particularly mutually corroborating evidence obtained from different locations.*

THE HEAVENS MOVE LIKE A SPHERE

It is plausible to suppose that the ancients got their first notions on these topics from the following kind of observations. They saw that the sun, moon, and other stars moved from east to west along circular paths which were always parallel to each other, that they started by rising up from below the earth itself as it were, gradually achieving their ascent, and then kept circling in the same way and getting lower, until, seeming to fall to earth, they vanished completely. Then, after remaining invisible for some time, they rose and set once more. And they saw that the intervals between these motions, and also the locations of the rising and setting, were on the whole determined and regular.

The main phenomenon that led them to the idea of a sphere was the revolution of the ever-visible stars. They observed that this revolution was circular as well as continuous about a single common center. Naturally they considered that point to be the pole of the heavenly sphere. For they saw that the closer were stars to that point, the smaller were their circles. And the farther were stars from it, the greater were their circles—right out to the limit where stars became invisible. But here too they saw that some heavenly bodies near the ever-visible stars remained visible for only a short time, while some farther away remained invisible for a long time, again depending on how far away they were from the pole. So they arrived at the idea of the heavenly sphere merely from this kind of inference. But from then on, in subsequent investigations, they found that everything else fit with this notion, and that absolutely all appearances contradicted any alternative notion that was proposed.

For suppose that the stars' motion takes place in a straight line towards infinity, as some have thought. How then could one explain their appearing to set out from the same starting-point every day? How could the stars return if their motion were towards infinity? Or, if they did return, would not the straight-line hypothesis be obviously wrong? For according to it, the stars would gradually have to diminish in size until they disappeared, whereas in fact they appear greater at the very moment of their disappearance, at which point they are obstructed and cut off, as it were, by the earth's surface.

It is also absurd to imagine the stars ignited as they rise out of the earth and extinguished again as they fall to earth. Just suppose that the strict order in their size and number, their intervals, positions, and periods could be restored by such a random and chance process, and that one whole region of earth has igniting properties, and another has extinguishing properties—or rather that the same region ignites stars for one set of observers and extinguishes them for another set, and that the same stars are already ignited or extinguished for some observers while they are not yet for others! Even on this ridiculous supposition, what could we say about the ever-visible stars, which neither rise nor set? The stars that are ignited and extinguished ought to rise and set for observers everywhere, while those that are not ignited and extinguished should always be visible to observers everywhere. How would we explain the fact that this is not so? We can hardly say that stars that are ignited and extinguished for some observers never undergo this process for other observers. Yet it is utterly obvious that the very same stars that rise and set in certain regions of the earth neither rise nor set in other regions.

Finally, to assume any motion at all other than spherical motion would entail that the distances of stars measured from the earth upwards must vary, regardless of where or how we assume the earth itself is situated. Hence the apparent sizes of the stars and the distances between them would necessarily vary for the same observers during the course of each revolution, for their distances from the objects of observation would be now greater, now lesser. Yet we see that no such variation occurs. And the apparent increase in their sizes at the horizon is caused not by a decrease in their distances but by the exhalations of moisture surrounding the earth. These intervene between the place from which we observe and the heavenly bodies. In the same way, objects placed in water appear bigger than they really are, and the lower they sink, the bigger they appear.

When he turns to the shape of the earth, Ptolemy likewise relies on physical evidence: He reasons from phenomena, from things which are "sensible," that is to say, apprehended by the senses. Then he supports his conclusion by engaging in thought experiments, imagining for example how the phenomena would be different if the earth, instead of being spherical, were concave, flat, or cylindrical.

THE EARTH TOO, TAKEN AS A WHOLE, IS SENSIBLY SPHERICAL

That the earth, too, taken as a whole, is sensibly spherical can best be grasped from the following considerations. To repeat, we see that the sun, moon, and other stars do not rise and set simultaneously for everyone on earth, but do so earlier for those towards the east and later for those towards the west. And eclipses, especially lunar eclipses, take place simultaneously for all observers yet are not recorded by all observers as occurring at the same *hour* (that is, at an equal distance from noon). Rather, the hour recorded by observers in the east is always later than that recorded by those in the west. And we find that the differences in the recorded hour are proportional to the

distances between the places of observation. Hence, one can reasonably conclude that the earth's surface is spherical, because its evenly curving surface
(for so it is when considered as a whole) cuts off the heavenly bodies for each
set of observers in a manner that is gradual and regular.

This would not happen if the earth's shape were other than spherical, as
one can see from the following arguments. If the shape were concave, the stars
would be seen rising first by those more towards the west; if it were a plane,
they would rise and set simultaneously for everyone on earth; if it were triangular or square or any other polygonal shape, similarly they would rise and
set simultaneously for all those living on the same planar surface. Yet clearly
nothing like this takes place. Nor could the earth be cylindrical, with the
curved surface in the east-west direction, and the flat sides towards the poles
of the universe, as some might suppose more plausible. For to those living on
the curved surface none of the stars would be ever-visible. Either all stars
would rise and set for all observers, or the same stars, for an equal celestial
distance from each of the poles, would always be invisible for all observers. In
fact, however, the further we travel toward the north, the more of the southern stars disappear and the more of the northern stars become visible. Clearly,
then, here too the curvature of the earth cuts off the heavenly bodies in a regular fashion in a north-south direction and demonstrates the sphericity of the
earth in all directions.

Moreover, if we sail towards mountains or elevated places from whatever
direction, north, south, east or west, we observe them to increase gradually
in size as if rising up from the sea itself in which they had previously been
submerged. This is due to the curvature of the surface of the water.

*Ptolemy goes on to argue, for reasons largely based on observation, that the
earth is in the center of the world (note: not "is the center" but "is in the center"). Having done so, he then states "that the earth has the ratio of a point
to the heavens." This claim can be confusing if we take it in a mathematical
sense, for in Euclidean geometry a point has no dimension whatsoever. Thus,
if the earth does have some dimension, then Ptolemy can appear to imply
that the heavens are infinitely large, which he does not. His claim does make
sense, however, when we consider the qualification which he adds: "to the
senses." That is to say, to all appearances the earth is a point in relation to
the heavens. Again, then, his appeal is to visible evidence. Normally when we
view a distant object from two different vantage points we see the object
from different angles and in altered relations to its surroundings (the phenomenon known as parallax). But when we view the stars from widely different points on earth, we perceive none of these usual variations, no
parallax—in other words, as if our different perspective points were in fact
one and the same point.*

THE EARTH HAS THE RATIO OF
A POINT TO THE HEAVENS

The earth has, to the senses, the ratio of a point to the distance of the sphere of the so-called fixed stars. This is strongly indicated by the fact that the sizes and distances of the stars at any given time appear equal and the same from any and every place on earth. Observations of the same celestial objects from different latitudes are found to have not the least discrepancy from each other. Moreover, gnomons set up in any part of the earth whatever, and likewise the centers of armillary spheres, operate like the real center of the earth. . . .

Another clear demonstration of the above proposition is that a plane drawn through the observer's line of sight at any point on earth—we call this plane one's "horizon"—always bisects the whole heavenly sphere. This would not happen if the earth were of perceptible size in relation to the distance of the heavenly bodies. In that case only the plane drawn through the center of the earth could exactly bisect the sphere, and a plane through any point on the surface of the earth would always make the section of the heavens below the plane greater than the section above it.

Even when Ptolemy argues for a proposition we know to be mistaken, namely that the earth is immobile, he begins with a pretty impressive account of how falling objects behave on earth and of how up *and* down *are merely relative terms. He then moves to arguments—again a kind of thought experiment—that are difficult for us to follow simply because they have been so decisively disposed of. They reappear, nevertheless, in various kinds of literature on into the seventeenth century; and they indicate how nearly impossible it was for even a brilliant critical mind such as Ptolemy's to take seriously, much less accept, the concept of the earth's mobility.*

NEITHER DOES THE EARTH
HAVE ANY MOTION FROM PLACE TO PLACE

One can show by arguments like the one above that the earth can have no motion in the directions mentioned, nor indeed can it ever move at all from its position at the center. For if it did move, the same phenomena would result as those that would follow from its having any position other than the central one. To me it seems pointless, therefore, to ask why objects move towards the center of the earth, once it has been so clearly established from actual phenomena that the earth occupies the middle place in the universe, and that all heavy objects are carried towards that place. The following fact alone amply supports this claim. Absolutely everywhere on the face of the earth—which has been shown to be spherical and in the middle of the universe—the direction and path of the motion (I mean proper, natural motion) of all heavy bodies is everywhere consistently at right angles to the plane that is tangent to the point of impact on the earth's surface. Clearly, therefore, if these falling objects were not stopped by the earth's surface, they would certainly reach the center of the earth itself, since any line drawn through the center of a sphere is always perpendicular to the tangent plane at the line's point of intersection with the sphere's surface.

As he proceeds, Ptolemy in effect acknowledges the way in which cosmology is still recognized to be unique among sciences: Its object, the universe, is one of a kind. One of the consequences of this recognition is that what we assume concerning something within the "local sphere" of our own experience often cannot validly be assumed of the whole.

Those who think it paradoxical that the earth, having such great weight, is not supported by anything and yet does not move, seem to me to be making the mistake of judging on the basis of their own experience instead of taking into account the peculiar nature of the universe. They would not, I think, consider this fact strange if they realized that the magnitude of the earth, when compared with the whole surrounding mass of the universe, has the ratio of a point to it. Given this way of thinking, it will seem quite consistent that (relatively speaking) the smallest of things should be overpowered and pressed in equally from all directions to a position of equilibrium by the greatest of things (which possess a uniform nature). For there is no up and down in the universe with respect to itself, any more than "up" and "down" make sense within a sphere. Rather, in the universe, the proper and natural motion of compound bodies is as follows: light and rarefied bodies drift outwards towards the circumference, but seem to move in the direction which is "up" for each observer, since the overhead direction for all of us, which we also call "up," points towards the surrounding surface. Heavy and dense bodies, on the contrary, are carried towards the middle and the center, but seem to fall downwards, again because the line of movement towards our feet, which we call "down," also points towards the center of the earth. These heavy bodies, as one would expect, settle about the center because of their mutual pressure and resistance, which is equal and uniform from all directions. For the same reason it is plausible that the earth, since its total mass is so great compared with the bodies which fall towards it, can remain motionless under the impact of these very small weights (for they strike it from all sides), and receive, as it were, the objects that fall upon it. . . .

Certain people, however, propose what they consider to be a more convincing model. They do not disagree with what I have said above, since they have no argument to bring against it. But they think no evidence prevents them from supposing, for example, that the heavens remain motionless and that the earth revolves from west to east about the same axis, making approximately one revolution each day. Or they suppose that both heaven and earth move by some amount, each about the same axis and in such a way as to preserve the overtaking of one by the other. However, they do not realize that, although there is perhaps nothing in the celestial phenomena to count against that simpler hypothesis, nevertheless what would occur here on earth and in the air would render such a notion quite ridiculous.

For the sake of argument, let us suppose that, contrary to nature, the most rare and light matter should either be motionless or else move in exactly the same way as matter with the opposite nature. . . . Suppose, too, that the densest and heaviest objects have a proper motion of the quick and uniform kind which they suppose (although, again, as everyone knows, earthly objects are sometimes not readily moved even by an external force). Even granted this supposition, they would have to admit that the revolving motion of the earth must be the most violent of all the motions they postulate, given that the earth makes one revolution in such a short time. Accordingly, all objects not actually standing on the earth would appear to have the same motion, opposite to that of the earth: neither clouds nor other flying or thrown objects would ever be seen moving towards the east, since the earth's motion towards the east would always outrun and overtake them, so that all other objects would seem to move backwards towards the west. Even if they claim that the air is carried around in the same direction and with the same speed as the earth, still the compound objects in the air would always seem to be left behind by the motion of both earth and air together. Or, if those objects too were carried around, fused as it were to the air, then they would never appear to have any motion either forwards or backwards. They would always appear still, neither wandering about nor changing position, whether they were things in flight or objects thrown. Yet we quite plainly see that they do undergo all these kinds of motion in such a way that they are not even slowed down or speeded up at all by any motion of the earth.

SOURCE: Adapted from Claudius Ptolemy, *Almagest*, trans. G. J. Toomer, New York: Springer-Verlag, 1984.

The Weaknesses of the Hypotheses

Proclus

Proclus (412–485), best known as a commentator on Plato, also wrote about the Ptolemaic system. His astronomical work gives us a glimpse, from the very twilight of Greek antiquity, of strains within this system that more than a thousand years later caused it to disintegrate. The following excerpt, written in epistolary style, also touches on an issue that mathematicians still discuss: Does theory give us access to a Platonic higher world whose truths may then be applied to physical reality, or must we work outwards and upwards from the physical?

My dear friend: The great Plato thinks that the real philosopher ought to study the sort of astronomy that deals with entities more abstract than the visible heaven, without reference to either sense perception or ever-changing matter. In that world of abstract entities he will come to know slowness itself and speed itself in their true numerical relationships. Now, I think, you wish to bring us down from that contemplation of abstract truth to consideration of the orbits on the visible heaven, to the observations of professional astronomers and to the hypotheses which they have devised from these observations, hypotheses which people like Aristarchus, Hipparchus, Ptolemy and others like them are always writing about. I suppose you want to become acquainted with their theories because you wish to examine carefully all the theories, as far as that is possible, with which the ancients, in their speculations about the universe, have abundantly supplied us.

Last year, when I was staying with you in central Lydia, I promised you that when I had time, I would work with you on these matters in my accustomed way. Now that I have arrived in Athens and heaven has freed me from those many unending troubles, I keep my promise to you and will . . . explain to you the real truth which those who are so eager to contemplate the heavenly bodies have come to believe by means of long and, indeed, endless chains of reasoning. In doing so I must, of course, pretend to myself to forget, for the moment at any rate, Plato's exhortations and the theoretical explanations which he taught us to maintain. Even so, I shall not be able to refrain from applying, as is my habit, a critical mind to their doctrines, though I shall do so sparingly, since I am convinced that the exposition of their doctrines will suggest to you quite clearly what the weaknesses of their hypotheses are, hypotheses of which they are so proud when developing their theories.

Despite the fact that Proclus applies "a critical mind" to the Ptolemaic doctrines, he apparently upholds the system as a whole because it embodies "the simplest hypotheses and the most fitting."

Before I end, I wish to add this: in their endeavor to demonstrate that the movements of the heavenly bodies are uniform, the astronomers have unwittingly shown the nature of these movements to be lacking in uniformity and to be the subject of outside influences. What shall we say of the eccentrics and the epicycles of which they speak so much? Are they only conceptual notions or do they have a substantial existence in the spheres with which they are connected? If they exist only as concepts, then the astronomers have passed, without noticing it, from bodies really existing in nature to mathematical notions and, again without noticing it, have derived the causes of natural movements from something that does not exist in nature. I will add further that there is absurdity also in the way in which they attribute particular kinds of movement to heavenly bodies. That we conceive of these movements, that is not proof that the stars which we conceive of moving in these circles really move anomalously.

On the other hand, if the astronomers say that the circles have a real, substantial existence, then they destroy the coherence of the spheres themselves on which the circles are situated. They attribute a separate movement to the circles and another to the spheres, and again, the movement they attribute to the circles is not the same for all of them; indeed, sometimes these movements take place in opposite directions. They vary the distances between them in a confused way; sometimes the circles come together in one plane, at other times they stand apart, and cut each other. There will, therefore, be all sorts of divisions, foldings and separations.

I want to make this further observation: the astronomers exhibit a very casual attitude in their exposition of these hypothetical devices. Why is it that, on any given hypothesis, the eccentric or, for that matter, the epicycle moves (or is stationary) in such and such a way while the star moves either in direct or retrograde motion? And what are the explanations (I mean the real explanations) of those planes and their separations? This they never explain in a way that would satisfy our yearning for complete understanding. They really go backwards: they do not derive their conclusions deductively from their hypotheses, as one does in the other sciences; instead, they attempt to formulate the hypotheses starting from the conclusions, which they ought to derive from the hypotheses. It is clear that they do not even solve such problems as could well be solved.

One must, however, admit that these are the simplest hypotheses and the most fitting for divine bodies, and that they have been constructed with a view to discovering the characteristic movements of the planets (which, in real truth, move in exactly the same way as they *seem* to move) and to formulating the quantitative measures applicable to them.

SOURCE: Proclus, *Hypotyposis astronomicarum positionum*, trans. A. Wasserstein, in *Physical Thought from the Presocratics to the Quantum Physicists*, ed. Shmuel Sambursky, London: Hutchinson, 1974.

13

Their Peculiar Behavior Confounds Mortals' Minds

Martianus Capella and Boethius

Martianus Capella (fl. 410–439) and Boethius (480?–524) are both notable for (among other things) their contributions to the development of "the Seven Liberal Arts." This classification of learning into Grammar, Rhetoric, and Logic (the Trivium) plus Arithmetic, Geometry, Music, and Astronomy (the Quadrivium) was the taxonomy of the realms of higher education throughout the Middle Ages. Martianus and Boethius likewise both exemplify the prominent medieval tendency to use personification and allegory not only in "literary" but also in philosophical writing.

Although Martianus may have "garbled, distorted, and misunderstood" his sources, as Edward Grant claims (A Source Book in Medieval Astronomy [Harvard UP, 1974], p. 822), his account of astronomy is intriguing for the ambiguity it injects into what we sometimes blandly assume to be the simplicity of medieval astronomy. First, despite the popular stereotype that the Middle Ages considered the earth to be the center of the universe, and hence as occupying the place of privilege, Martianus repeatedly refers to earth as "clinging to" or "standing at" the middle and bottom position. (It is contrary to such a view that Galileo and Kepler later consciously sought to exalt the position of earth in the scheme of the universe.) Moreover, Martianus adumbrates a version of the system of Heraclides of Pontus, according to which Mercury and Venus revolve not around the earth but around the sun.

Astronomy, personified, appears before an assembly of the gods to impart her wisdom.

Before their eyes a vision appeared, a hollow ball of heavenly light, filled with transparent fire, gently rotating, and enclosing a maiden within. Several planetary deities, especially those which determine men's destinies, were bathed in its glare, the mystery of their behavior and orbits revealed. . . . As she came into their midst many of the gods smiled at her; the others admired her radiant beauty. She began her discourse as follows:

". . . Inasmuch as I have at one time or other in my peregrinations come to be known by the Greeks, whatever has been written by Eratosthenes, Ptolemy, Hipparchus, and other Greeks ought to suffice here and relieve me of the burden of discoursing at greater length. However, . . . I shall not keep silent in the presence of you celestial ones, who will be surveying the courses of your own heavenly bodies.

"The universe is formed in the shape of a globe composed entirely of four elements. The heavens, swirling in a ceaseless and rotary motion, set the earth apart in a stationary position in the middle and at the bottom. . . .

"If each belt of the encompassing substances is found to be homogeneous, no circle can waver from its ethereal orbit. When we use the word 'circles' we do not intend to convey a notion of corporeal demarcations of a fluid substance; we are merely illustrating the risings and settings of planetary bodies as they appear to us. I myself do not consider an axis and poles, which mortals have fastened in a bronze armillary sphere to assist them in comprehending the heavens, as an authoritative guide to the workings of the universe. For there is nothing more substantial than the earth itself, which is able to sustain the heavens. Another reason is that the poles that protrude from the hollow cavity of the perforated outer sphere, and the apertures, the pivots, and the sockets have to be imagined—something that you may be assured could not happen in a rarefied and supramundane atmosphere.

"Accordingly, whenever I shall use the terms axis, poles, or celestial circles, for the purpose of gaining comprehension, my terminology is to be understood in a theoretical sense." . . .

"Now I shall take up the orbits of the planets. Not because of their errant motions—for their courses are defined in the same way as the sun's, and they do not admit of any error—rather, because their peculiar behavior confounds mortals' minds, I shall call them not 'errant bodies' (*planetae*) but 'confusing bodies' (*planontes*). . . .

"For in varying amounts of time the planets strive to make up the distance that they are carried backward by a single diurnal rotation: the moon in a month, the sun in a year, Saturn in thirty years, and the others in periods of time proportional to the amount of space that they traverse.

"Although all these bodies are seen to move toward the eastern horizon, they do not move counter to the universe in a straight and direct line; rather they plod along with sideways motions across the fixed stars of the zodiac. It is well that they do, for the universe could not endure a contrary motion of its parts. . . .

"There is one motion that is common to all seven planets—an easterly one. Another point to be noted is that they all differ in the times and circumstances of their periods. For five of the planets undergo stations and retrogradations, but the sun and the moon are propelled in a steady course. Moreover, these two luminous bodies eclipse each other in turn; but the other five are never eclipsed. Three of these, together with the sun and the moon, have their orbits about the earth, but Venus and Mercury do not go about the earth.

"This general observation must be made, that the earth is eccentric to the orbits of all the planets (that is, it is not located at the center of their circles); and a second observation must be made about all seven, that although the celestial sphere rotates with the same uniform motion, the planets make daily changes in their positions and orbits. . . .

"Now Venus and Mercury, although they have daily risings and settings, do not travel about the earth at all; rather they encircle the sun in wider revolutions. The center of their orbits is set in the sun. . . .

"Now Venus, which is sometimes called Phosphoros, was manifestly thoroughly investigated by Pythagoras of Samos and his pupils. It has been shown to complete its orbit in a period of about a year. . . . When it makes its risings in the early morning, ahead of the sun, it is called Lucifer; when it blazes forth after the setting of the sun, it is called Vesper or Vesperugo. Venus is the only one of the five planets, like the moon, to cast a shadow, and it is the only planet to be clearly discernible and not yielding for a long period of time to the splendor of the rising sun."

If Martianus exemplifies medieval awe and delight in the face of an astronomical order that is beautiful even while defying consistent description, then Boethius expresses the complementary human longing that the disorder of this lower world might be more thoroughly penetrated by the harmony of the cosmos at large. While awaiting torture and execution on trumped-up charges of treason, Boethius wrote his famous Consolation of Philosophy, *in which the goddess Philosophy identifies the principle that creates union and harmony in the universe at large, as well as among human beings. In Philosophy's poem, echoed eight hundred years later by Dante at the very end of his* Divine Comedy, *Boethius praises that unifying cosmic principle, whose name is* Love.

The world, always changing,
Persists in harmony;
A covenant secure
Unites the warring atoms.
The trail-blazing sun
Leads forth the rosy dawn;
The evening star makes way
For night, the moon's dominion.
The eager ocean currents
Do not transgress their bounds;
Safe fenced remains the earth
Against invading waters.
It is Love who joins all these,
Reigning over land and sea;
The universe itself is ruled by Love.

If Love let slip the reins,
Whatever now keeps peace
Would fall to constant warring:
Beauty, trust, harmony
Dissolving into discord.
Love consecrates the bond
Uniting diverse peoples;
In marriage too Love spins
The cords of holy union;
And Love again decrees,
Let faithful friendship be.
O human race, how happy—
If equally your minds were ruled
By Love, who rules the universe.

SOURCES: *Martianus Capella and the Seven Liberal Arts*, vol. 2, *The Marriage of Philology and Mercury*, trans. William Harris Stahl and Richard Johnson with E. L Burge, New York: Columbia UP, 1977; Boethius translated from the *Consolatio Philosophiae*, Book 2, poem 8.

14

We Consider Time a Thing Created

Moses Maimonides

Moses Maimonides, or "son of Maimon" (1135–1204), is widely acknowl-edged to be the most influential Jewish thinker of the Middle Ages, and his most famous work is The Guide of the Perplexed, *a wide-ranging work on science, philosophy, and scriptural interpretation that was originally written in Arabic. The Guide is a highly engaging and sometimes surprising work. While it embodies many commonplaces that have come to define our notions of medieval thought, common assumptions are just as often treated critically. For example, although Aristotelianism is everywhere evident in* The Guide, *Moses' use of Aristotle is judicious and nuanced. Above all his thought is Jewish, rejecting any temptation to place God within the limits of human reason or of simple analogy. Having drawn parallels between "the Universe" and "Man," Moses warns: "Bear in mind, however, that in all that we have noticed about the similarity between the universe and the human being, nothing would warrant us to assert that man is a microcosm." Furthermore:*

The faculty of thinking is a force inherent in the body, and is not separated from it, but God is not a force inherent in the body of the universe, but is separate from all its parts. How God rules the universe and provides for it is a complete mystery; man is unable to solve it. For, on the one hand, it can be proved that God is separate from the universe, and in no contact whatsoever with it; but, on the other hand, His rule and providence can be proved to exist in all parts of the universe, even the smallest. Praised be He whose perfection is above our comprehension.

Typically, in volume II of The Guide, *proceeding to a consideration of things cosmological—which of course he does within a broadly Ptolemaic framework—Maimonides raises criticisms against the completeness or consistency of the Ptolemaic system. As one might expect in a medieval discussion, Moses' view is deeply hierarchical, and it is saturated with the notion of "influences" and the agency of higher created beings. However, as one might not expect, given the now-prevalent confusion between geocentrism and anthropocentrism, his views are opposed to placing human interests or "transient earthly beings" upon a pinnacle.*

When a simple mathematician reads and studies . . . astronomical discussions, he believes that the form and the number of the spheres are facts established by proof. But this is not the case; for the science of astronomy does not aim at demonstrating them, although it includes subjects that can be proved; e.g., it has been proved that the path of the sun is inclined against the equator; this cannot be doubted. But it has not yet been decided whether the sphere of the sun is eccentric or contains a revolving epicycle, and the astronomer does not take notice of this uncertainty, for his object is simply to find a hypothesis that would lead to a uniform and circular motion of the stars without acceleration, retardation, or change, and which is in its effects accordant with observation. He will, besides, endeavor to find such a hypothesis which would require the least complicated motion and the least number of spheres. He will therefore prefer a hypothesis which would explain all the phenomena of the stars by means of three spheres to a hypothesis which would require four spheres. From this reason we adopt, in reference to the circuit of the sun, the theory of eccentricity, and reject the epicyclic revolution assumed by Ptolemy.

Maimonides also addresses the issue of whether (as Aristotle believed) the universe is eternal:

Those who follow the Law of Moses, our Teacher, hold that the whole universe, i.e., everything except God, has been brought by Him into existence out of non-existence. In the beginning God alone existed, and nothing else; neither angels, nor spheres, nor the things that are contained within the spheres existed. He then produced from nothing all existing things such as they are by His will and desire. Even time itself is among the things created; for time depends on motion, i.e., on an accident in things which move, and the things upon whose motion time depends are themselves created beings, which have passed from non-existence into existence. We say that God *existed* before the creation of the universe, although the verb *existed* appears to imply the notion of time; we also believe that He existed an infinite space of time before the universe was created; but in these cases we do not mean time in its true sense. We only use the term to signify something analogous or similar to time. For time is undoubtedly an accident [in the Aristotelian sense] and, according to our opinion, one of the created accidents, like blackness and whiteness. It is not a quality, but an accident connected with a motion. . . .

We consider time a thing created. It comes into existence in the same manner as other accidents and the substances which form the substratum for the accidents. For this reason, namely, because time belongs to the things created, it cannot be said that God produced the universe *in the beginning*. Consider this well, for he who does not understand it is unable to refute forcible objections raised against the theory of *creatio ex nihilo* [creation out of nothing]. If you admit the existence of time before the creation, you will be compelled to accept the theory of the eternity of the universe. For time is an accident and requires a substratum. You will therefore have to assume that something [beside God] existed before this universe was created, an assumption which it is our duty to oppose.

Maimonides goes on to highlight anomalies in the Aristotelian/Ptolemaic system—even though this is the system he largely accepts. The purpose of the critique is to establish the Aristotelian/Ptolemaic system as a tool, not as an absolute explanation idolatrously relied upon.

You know of Astronomy as much as you have studied with me, and learnt from the book *Almagest*; we had not sufficient time to go beyond this. The theory that [the spheres] move regularly, and that the courses of the stars are in harmony with observation, depends, as you are aware, on two hypotheses: we must assume either epicycles, or eccentric spheres, or a combination of both. Now I will show that each of these hypotheses is irregular, and totally contrary to the results of Natural Science. Let us first consider an epicycle, such as has been assumed in the spheres of the moon and the five planets, rotating on a sphere, but not round the center of the sphere that carries it. This arrangement would necessarily produce a revolving motion; the epicycle would then revolve, and entirely change its place. But that anything in the spheres should change its place is exactly what Aristotle considers impossible. . . . (1) It is absurd to assume that the revolution of a cycle has not the center of the universe for its center; for it is a fundamental principle in the order of the universe that there are only three kinds of motion—from the center, towards the center, and round the center. But an epicycle does not move away from the center, nor towards it, nor round it. (2) Again, according to what Aristotle explains in Natural Science, there must be something fixed, round which the motion takes place. This is the reason why the earth remains stationary. . . .

Consider, therefore, how many difficulties arise if we accept the theory which Aristotle expounds in *Physics*. For, according to that theory, there are no epicycles, and no eccentric spheres, but all spheres rotate round the center of the earth! How then can the different courses of the stars be explained? How is it possible to assume a uniform perfect rotation with the phenomena which we perceive, except by admitting one of the two hypotheses or both of them? The difficulty is still more apparent when we find—admitting what Ptolemy said as regards the epicycle of the moon, and its inclination towards a point different both from the center of the universe and from its own center—that the calculations according to these hypotheses are perfectly correct, within one minute; and that their correctness is confirmed by the most accurate calculation of the time, duration, and extent of the eclipses, which is always based on these hypotheses. Furthermore, how can we reconcile, without assuming the existence of epicycles, the apparent retrogression of a star with its other motions? How can rotation or motion take place round a point which is not fixed? These are real difficulties.

Maimonides contrasts the monotheistic doctrine of creation to the worship of the heavens by members of the Sabean religion. Abraham himself is seen as the bearer of the high view of creation in the very midst of those who worship the stars.

It is well known that the Patriarch Abraham was brought up in the religion and the opinion of the Sabeans that there is no divine being except the stars. . . . They consider the stars as deities, and the sun as the chief deity. . . . They say distinctly that the sun governs the world, both that which is above and that which is below. These are exactly their expressions. . . .

All the Sabeans thus believed in the eternity of the universe, the heavens being in their opinion God. . . . [But] when [Abraham] the "pillar of the World" appeared, he became convinced that there is a spiritual Divine Being, which is not a body, nor a force residing in a body, but is the author of the spheres and the stars; and he saw the absurdity of the tales in which he had been brought up. He therefore began to attack the belief of the Sabeans, to expose the falsehood of their opinions, and to proclaim publicly in opposition to them, "the name of the Lord, the God of the universe" (Gen. 21:33), which proclamation included at the same time the existence of God, and the creation of the universe by God.

Maimonides also engages—again with rather surprising results—the peren-
nial cosmological issue of teleology: what, if any, is the purpose of the uni-
verse?

Intelligent persons are much perplexed when they inquire into the purpose of
the Creation. I will now show how absurd this question is, according to each
one of the different theories. An agent that acts with intention must have a
certain ulterior object in that which he performs. This is evident, and no
philosophical proof is required. It is likewise evident that that which is produced with intention has passed over from non-existence to existence. It is
further evident, and generally agreed upon, that the being which has never
been and will never be without existence is not in need of an agent. . . . The
question, "What is the purpose thereof?" cannot be asked about anything
which is not the product of an agent. Therefore, we cannot ask what is the
purpose of the existence of God. He has not been created. According to these
propositions it is clear that the purpose is sought for everything produced in-
tentionally by an intelligent cause; that is to say, a final cause must exist for
everything that owes its existence to an intelligent being. But for that which
is without a beginning, a final cause need not be sought. . . . After this explanation you will understand that there is no occasion to seek the final cause of
the whole universe, neither according to our theory of the creation, nor according to the theory of Aristotle, who assumes the eternity of the
universe. . . .
   The existence of an ultimate purpose in every species, which is considered
as absolutely necessary by everyone who investigates into the nature of
things, is very difficult to discover. Still more difficult is it to find the purpose of the whole universe. . . . It is clear that man is the most perfect being
formed by matter; he is the last and most perfect of earthly beings, and in
this respect it can truly be said that all earthly things exist for man, i.e., that
the changes which things undergo serve to produce the most perfect being
that can be produced. Aristotle . . . need therefore not ask to what purpose
does man exist, for the immediate purpose of each individual being is, according to his opinion, the perfection of its specific form. . . . It seems therefore clear that, according to Aristotle . . . there is no occasion for the
question what is the object of the existence of the universe. But of those who
accept our theory that the whole universe has been created from nothing,
some hold that the inquiry after the purpose of the creation is necessary, and
assume that the universe was only created for the sake of man's existence,
that he might serve God. Everything that is done, they believe, is done for
man's sake; even the spheres move only for his benefit, in order that his
wants might be supplied. . . .

On examining this opinion as intelligent persons ought to examine all different opinions, we shall discover the errors it includes. Those who hold this view, namely, that the existence of man is the object of the whole creation, may be asked whether God could have created man without those previous creations, or whether man could only have existence after the creation of all other things. If they answer in the affirmative, that man could have been created even if, e.g., the heavens did not exist, they will be asked what is the object of all these things, since they do not exist for their own sake but for the sake of something that could exist without them? Even if the universe existed for man's sake and man existed for the purpose of serving God, as has been mentioned, the question remains, What is the end of serving God? He does not become more perfect if all His creatures serve Him and comprehend Him as far as possible; nor would he lose anything if nothing existed beside Him. It might perhaps be replied that the service of God is not intended for God's perfection; it is intended for our perfection—it is good for us, it makes us perfect. But then the question might be repeated, What is the object of our being perfect? We must in continuing the inquiry as to the purpose of the creation at last arrive at the answer, It was the Will of God, or His Wisdom decreed it. And this is the correct answer. . . .

You must not be misled by what is stated of the stars [that God put them in the firmament of the heavens] to give light upon the earth, and to rule by day and by night. You might perhaps think that here the purpose of their creation is described. This is not the case. We are only informed of the nature of the stars, which God desired to create with such properties that they should be able to give light and to rule. In a similar manner we must understand the passage, "And have dominion over the fish of the sea" (Gen. 1:28). Here it is not meant . . . that man was created for this purpose, but only that this was the nature which God gave man. But as to the statement in Scripture that God gave the plants to man and other living beings, it agrees with the opinion of Aristotle and other philosophers. It is also reasonable to assume that the plants exist only for the benefit of the animals, since the latter cannot live without food. It is different with the stars. They do not exist only for our sake, that we should enjoy their good influence. For the expressions "to give light" and "to rule" merely describe . . . the benefit which the creatures on earth derive from them.

I have already explained to you the character of that influence that continually causes the good to descend from one being to another. To those who receive the good flowing down upon them, it may appear as if the being existed for them alone that sends forth its goodness and kindness unto them. Thus some citizen may imagine that it was for the purpose of protecting his house by night from thieves that the king was chosen. To some extent this is correct; for when his house is protected, and he has derived this benefit through the king whom the country has chosen, it appears as if it were the object of the king to protect the house of that man. In this manner we must explain every verse, the literal meaning of which would imply that something superior was created for the sake of something inferior, namely, that it is part of the nature of the superior thing [to influence the inferior in a certain manner]. We remain firm in our belief that the whole universe was created in accordance with the will of God, and we do not inquire for any other cause or object. . . .

You must not be mistaken and think that the spheres and the angels were created for our sake.

SOURCE: Moses Maimonides, *The Guide of the Perplexed*, trans. M. Friedländer, 3 vols., London, 1881–1885.

15

From This Point
Hang the Heavens

Dante Alighieri

Probably no twentieth-century interpreter of the Middle Ages produced a more sympathetic account of the period's cosmology than did C. S. Lewis (1898–1963). In his primer on that topic for students of literature, The Discarded Image, *Lewis expounds Chalcidius, a fourth-century commentator on Plato's* Timaeus, *as a prototype of the medieval worldview.*

For Chalcidius, the geocentric universe is not in the least anthropocentric. If we ask why, nevertheless, the earth is central, he has a very unexpected answer. It is so placed in order that the celestial dance may have a center to revolve about—in fact, as an aesthetic convenience for the celestial beings. It is perhaps because his universe is already so well and radiantly inhabited that Chalcidius, though he mentions the Pythagorean doctrine (which peopled the moon and other planets with mortals), is not interested in it. . . .

Centuries later . . . Alanus ab Insulis [d. 1203] compares the sum of things to a city. In the central castle, in the Empyrean, the Emperor sits enthroned. In the lower heavens live the angelic knighthood. We, on earth, are "outside the city wall." How, we ask, can the Empyrean be the center when it is not only on, but outside, the circumference of the whole universe? Because, as Dante was to say more clearly than anyone else, the spatial order is the opposite of the spiritual, and the material cosmos mirrors, hence reverses, the reality, so that what is truly the rim seems to us the hub.

The exquisite touch which denies our species even the tragic dignity of being outcasts by making us merely suburban, was added by Alanus. In other respects he reproduces Chalcidius' outlook. We watch "the spectacle of the celestial dance" from its outskirts. Our highest privilege is to imitate it in such measure as we can. The Medieval Model is, if we may use the word, anthropoperipheral.

As Lewis indicates, one of the most imaginative works written within the world view of the middle ages is Dante Alighieri's The Divine Comedy (c.1310–1314). The first volume of The Comedy, The Inferno, provides a moral analogy to medieval cosmology's assumption that the center of the universe, which is occupied by the center of the earth, is a kind of cosmic sump where that which is grossest and heaviest accumulates if nothing grosser and heavier stands in its way. This center Dante (sounding quite Aristotelian) refers to as "the middle, / Where everything of weight unites together." In Dante's moral universe, the "circles" of hell are thus worse according to their proximity to the center. Moreover, it is cold, not heat, that characterizes the symbolic terrain encountered there. Dante, in his narration of the end of his journey towards the center, recalls:

> Then I beheld a thousand faces, made
> Purple with cold; whence o'er me comes a shudder,
> And evermore will come, at frozen ponds.
> (Inferno 32.70–72)

And in the very center of hell, which coincides with the center of the earth, one finds the perpetrator of the worst evil, Satan himself—yet (again) not in flames but in ice, which depicts the utter lack of vitality and dynamism that is the nature of evil.

After leaving hell, Dante is led up from the center of the earth, up Mount Purgatory, and up through the heavens to the ninth sphere, the Primum Mobile. However, once here, he finds himself in some profounder sense no longer looking out onto the Empyrean but looking in. As Robert Osserman puts it in his discussion of the "poetry of the universe," in Dante "we are to think of the Empyrean as somehow both surrounding the visible universe and adjacent to it" (see also Osserman on Dante and the "curved space" of Riemann, chapter 57).

> A point beheld I, that was raying out
> Light so acute, the sight which it enkindles
> Must close perforce before such great acuteness.

And whatsoever star seems smallest here
Would seem to be a moon, if placed beside it.
As one star with another star is placed.

Perhaps at such a distance as appears
A halo cincturing the light that paints it,
When densest is the vapor that sustains it,

Thus distant round the point a circle of fire
So swiftly whirled, that it would have surpassed
Whatever motion soonest girds the world;

And this was by another circumcinct,
That by a third, the third then by a fourth,
By a fifth the fourth, and then by a sixth the fifth;

The seventh followed thereupon in width
So ample now, that Juno's messenger
Entire would be too narrow to contain it.

Even so the eighth and ninth; and every one
More slowly moved, according as it was
In number distant farther from the first.

And that one had its flame most crystalline
From which less distant was the stainless spark,
I think because more with its truth imbued.

My Lady, who in my anxiety
Beheld me much perplexed, said: "From that point
Dependent is the heaven and nature all."

(*Paradiso* 28.16–42)

SOURCES: C. S. Lewis, *The Discarded Image*, Cambridge: Cambridge UP, 1964; Dante, *The Divine Comedy of Dante Alighieri*, trans. Henry Wadsworth Longfellow, New York [c. 1895].

16

If a Man Were in the Sky and Could See the Earth Clearly

Nicole Oresme

Nicole Oresme (c. 1325–1382) was a French bishop and Aristotelian scholar who nevertheless presented one of the most cogent pre-Copernican statements of the hypothesis that the earth, not the universe, rotates once every twenty-four hours. He also speculated about the possibility of other worlds. However, Oresme's brilliant logical mind earned him a place in the history of economics as well as astronomy, and the principle he seems to have applied in both fields is in fact a kind of relativity. In economics, he emphasized the need for a stable coinage so that goods could be valued relative to a fixed currency. Perhaps from this concept it was not a big step to the recognition that, in the physical universe, fixity could be attributed as easily to the heavens as to the earth. In a word, Oresme was attracted by this explanation's economy.

In a thought experiment, Oresme imagines a man trying to judge movement and direction from on board a moving ship—and in the process provides us with a thematic link backward to Ptolemy and forward to Einstein.

It seems to me that we might well affirm, subject to correction, . . . that it is the earth that makes a daily rotation, and not the heavens. And I would like to assert the impossibility of establishing the contrary claim first by means of any observation or, secondly, by means of any rational process. And thirdly I will give my own reasons why the earth's movement might indeed be supported.

As for observation, we see with our own eyes that the sun, the moon, and a number of stars do rise and set, day after day, while some stars revolve about the north pole. And this could not happen unless the heavens turned. . . . So it is the heavens that make a daily rotation.

A further observation is this: if it is the earth that turns, then it makes a complete revolution in one natural day. And accordingly trees, houses, and we ourselves are moving very quickly eastwards—so that it would seem the air and the wind should always blow very strongly from the east and make a rushing sound just as it does against a shaft shot from a crossbow, only much louder. But this is not at all what we do see.

The third observation is one cited by Ptolemy: if someone were on board a ship moving very rapidly eastwards and he shot an arrow straight up in the air, then it would not fall onto the ship but far off to the west. Likewise, if the earth were rotating very rapidly from west to east, then supposing someone threw a stone straight up into the air, it would not fall there where it started off but far off to the west. But that is not what we actually see.

What I shall say about these arguments can also, I think, be directed against all the rest that will be put forward on the same topic.

Accordingly, I assume first that the whole physical system—the whole mass of all physical bodies in the universe—is divided into two parts. One is the heavens with the sphere of fire and the upper region of the air, all of which, according to Aristotle's first book of *Meteorology*, makes a daily rotation. And the other part, everything else—namely, the middle and lower regions of the air, the water, the earth, and the composite bodies—all of this, according to Aristotle, is immobile, unaffected by any daily rotation.

Furthermore, I assume that movement from one place to another can be apprehended by the senses only insofar as we apprehend that one body changes its situation relative to another body. Thus, if a man is in a ship A moving smoothly, be it fast or slow, and sees nothing but another ship B moving in exactly the same manner as A in which he is located, then it will appear to this man that neither of the ships is moving. If A is at rest and B is moving, then it appears to him as if B is moving. And if A is moving and B is at rest, it still appears to him as if A is at rest and B is moving. Thus, if A remained at rest for an hour while B was moving, and if, conversely, during the very next hour A were moving and B at rest, then this man would be unable to apprehend the change, the variation. Rather, it would seem to him the whole time that B was moving. This is what experience tells us. The reason is that these two bodies A and B stand in the same relative position the one to the other whether A moves and B is at rest, or B moves and A is at rest. As is affirmed in Witelo's *Perspective* [ca. 1270], book 4, we apprehend movement only insofar as we apprehend the change of a body's position relative to that of another.

I assert accordingly that if, of the two parts of the universe mentioned earlier, the upper part today made a daily rotation—as it does—and the lower part did not, but tomorrow the situation were reversed so that the part down here made a daily rotation while the other, the heavens, did not, we would be able to apprehend nothing of this change. Rather, everything would still seem just the same tomorrow and today as far as this matter is concerned. It would continue to appear to us that our part stayed put while the other part kept on moving, just as to a man in a moving boat it appears that the trees outside are moving. Likewise, if a man were in the sky and moving along with it in its presumed daily rotation, and if he could see the earth clearly and make out mountains, valleys, rivers, cities, and castles distinctly, then it would seem to him that *the earth* made a daily rotation, just as it seems to us here on earth that the heavens do. Similarly, if the earth made a daily rotation and the heavens did not, then it would seem to us that the earth was at rest and that the heavens moved. This can easily be imagined by anybody with good sense.

Thus we have a clear rejoinder to the first argument. For we would assert that the sun and the stars seem accordingly to rise and to set, and the heavens to revolve, on account of the movement of the earth and of the elements which we inhabit.

And it would seem that the second argument is answered, according to this interpretation, by the claim that the earth moves not merely by itself but together, as already mentioned, with the air and the water, albeit the water and the air down here can also be given additional motion by the winds and other causes. The case is similar to that of a ship, in which the enclosed air seems to those who are in it not to be moving.

The third argument—concerning the arrow or stone launched straight up, etc.—looks like the hardest to answer. But we could say that the arrow shot upwards is carried swiftly eastwards together with the air through which it is passing and with the whole mass of earth's lower regions, as described earlier, which makes a daily rotation. And this is why the arrow falls back to the same place on earth from which it departed.

To see how this is possible, compare it with the case of a man on board a ship moving swiftly eastwards without his being aware of the movement. Now if he brings his hand straight down along the line of the ship's mast, it will seem to him that his hand has not moved other than straight down. And likewise, according to the view we are considering, it seems to us that the arrow moves straight up or straight down.

Similarly, on a ship moving in the manner described, there can be movements lengthways, sideways, up, down, and any other way, and they appear to take place exactly as if the ship were at rest. This is why, if a man on board this ship were to walk westwards less quickly than the ship was heading eastwards, it would seem to him that he was moving farther west when actually he was moving farther east. And likewise in the case we have been considering, all movements down here would appear just as if the earth were stationary.

To sharpen the answer to the third objection, let me add to this artificial example a more natural one, whose validity is endorsed by Aristotle. Suppose that, in the upper region of the air there is a bundle of pure fire called A which is very light, so that it rises as high as it can, up to a place called B near the concave surface of the heavens. Now the case here will be the same as that of the arrow mentioned earlier. In this case the movement of A will consist of a movement that is rectilinear *and*, in part, circular, for the region of the air and the sphere of fire that A passed through both move, according to Aristotle, in a circular manner. And if they did not move thus, A would rise straight up along the vertical line AB. But because, in keeping with the earth's daily rotation, B has in the meantime moved to place C, obviously A in its ascent describes the line AC, and the movement of A consists in a movement at once rectilinear and circular. And likewise the flight of the arrow.. . .

Therefore I conclude we could make no observation that would establish that the heavens make a daily rotation and that the earth does not.

SOURCE: Translated (with kind advice from Richard Holdaway) from Nicole Oresme, *Le Livre du ciel et du monde* (1377); in *Mediaeval Studies*, vol. 4, Toronto: Pontifical Institute of Mediaeval Studies, 1942.

17

A Single Universe
in Which Each Star
Influences Every Other

Nicholas Cusanus

*Nicholas Cusanus (1401–1464) was born in Kues (or Cusa) on the Moselle
River a year after the death of Geoffrey Chaucer. He led an active life as international ecclesiastical diplomat and was made Cardinal in 1448. The con-
templative side of his career culminates in the treatise* On Learned Ignorance
*(1440), whose title hints at the paradoxical and at times mystical nature of
its contents.*

*What is perhaps most startling about this work is the manner in which
Nicholas arrives at apparently prescient conclusions about the universe using
a methodology that is entirely abstract and speculative. Contrary to still-pop-
ular beliefs concerning the empirical nature of what was to become Copernican cosmology, the reevaluation of the Ptolemaic system in fact was
grounded on a critical refusal to accept the evidence of the senses. It was really Aristotle who was empirical. For philosophers like Nicholas, on the
other hand, a truly critical critique of physical reality is possible precisely because there is a higher Reality that the physical may imitate but does not
comprise. Not unlike Kepler almost two centuries later, Nicholas employs a
form of Platonic or Neoplatonist deduction to undermine Aristotelian/Ptole-
maic tenets concerning the shape and structure of the world. In his discussion of movement there is even a hint, as there is in Oresme, of what in the
twentieth century would come to be known as relativity.*

Clearly, it is actually this earth that moves, though to us it does not appear to do so; for we apprehend motion only relative to something motionless. Anyone on board ship but not knowing that the water is flowing, nor able to see the riverbanks—how would he, from midstream, apprehend that the ship was moving? This is why to anyone at all, whether he be on earth, or on the sun or another planet, it always seems as if he is in the center, immobile as it were, while everything else is in motion. Certainly one always establishes one set of fixed points relative to oneself, whether one inhabits the sun or the earth, the moon or any of the other planets. Thus it is as if the world system had its center everywhere and its circumference nowhere, for God is its circumference and center, and he is everywhere and nowhere.

Just as for Plato no physical table or man or just act perfectly participates in the Forms (respectively) of the Table, or Man, or Justice; so for Nicholas no physical object has a shape that perfectly conforms to the mathematical ideal toward which our language of shapes (circles, spheres, cubes, etc.) gestures. Nicholas's discussion also adumbrates a geometry that is more Riemannian than Euclidean (that is, based on surfaces of spheres rather than flat planes).

Even this earth is not spherical, as people have said it is, though it tends towards the spherical. The shape of the universe is limited in its parts, just as its motion is. But when an infinite line is considered as limited so that, as limited, it can be neither more perfect nor more capacious, then it is circular, for it is in a circle that beginning meets end. Thus the more perfect motion is also circular, and from this it follows that the more perfect solid shape is spherical. . . .

The earth, then, has a noble, spherical shape and a circular motion, but it could be more perfect. For as regards the perfection, motion, and shape of the world there is no maximum or minimum. So clearly it is wrong to call this earth most vile or base. For although it seems more central than other things in the universe, it is therefore also farthest from the center, as explained earlier.

One of the consequences of this paradoxical deconstruction of location is the qualitative "neutralizing" of place. In contrast to Aristotle or Dante, for Nicholas Cusanus there is no "dead center," no location that marks a body's grossness or baseness. The importance of this contrast with the standard medieval understanding of the place and nature of the earth is profound. Among other implications, it entails a revision of the medieval doctrine of "influences," whereby the power and quality of stars and planets are communicated downward to the earth. For if downward *becomes a relative term,*

*then influence may travel a two-way street, and the earth itself may be recon-
ceived as a star shedding its own influence.*

Therefore the earth is a magnificent star possessing light, heat, and influence
different and distinct from all other stars, just as each of these is unique as regards light, nature, and influence. Each star communicates light and influence to the next, though this is not its purpose. For all stars move and shine
in order to be most fully what they are, from which their sharing of influence
arises as a consequence. Likewise, light gives light because that is its nature
and not so that I may see, yet the sharing of light arises as a consequence
when I use it for purposes of seeing. And in this way holy God has created all
things: as each thing desires to preserve its own being as a gift from God, it
does so within a fabric of sharing with other things.

*Furthermore, just as physical location is no marker of excellence or baseness,
so physical size for Nicholas becomes a neutral matter. It is rather "intellectual nature" that constitutes excellence. Therefore, even though Nicholas's
speculations lead him briefly to contemplate the existence of extraterrestrial
life, he returns to that intellectual nature here on earth—and to a suggestion
that is actually more anthropocentric than any that raw Aristotelian geocentrism could ever have generated.*

The fact that the earth is smaller than the sun and receives influence from it is
no reason for calling it more contemptible, for the whole region of the earth,
reaching all the way to the outer sphere of fire, is huge. It is true, as we see
from its shadow in eclipses, that the earth is smaller than the sun; yet it is not
known by how much the region of the sun is greater or smaller than that of
the earth. In any case they cannot be precisely equal, for no star can be equal
to another. Nor is earth the smallest star, for, as we know from eclipses, it is
greater than the moon and even, some would say, than Mercury and perhaps
also the other planets. From its size, therefore, no argument can be constructed for the earth's inferiority. . . .

Nor can *place* support such a claim: namely, that this place in the universe
is the home of humans, animals, and plants which are of a less noble rank
than those dwelling in the sun and other planets. God is the center and circumference of all the starry realms, and from him proceed natures of manifold excellence that inhabit those realms. It is not fitting that such celestial
and stellar locations be empty while this perhaps little earth is inhabited. And
yet, in accordance with the intellectual nature inhabiting earth and its envi-
rons, it seems impossible to postulate one that is more perfect or more noble,
even if other planets be inhabited by beings of a different kind. Human be-

ings, indeed, desire not to take on a different nature but merely to achieve the perfection of their own nature. . . .

Since we know nothing of that whole realm, we likewise know nothing at all of its inhabitants. A similar pattern is observable even here on earth: animals of one species join together as it were to make their own domain, and they share this domain and its features among themselves, neither caring much nor indeed knowing much about outsiders. Animals of one species have no concept of outsiders other than that communicated in the form of vocal expression, and this in a rather minimal way that produces nothing better than mere opinion, even after long experience. How very much less, then, can we know of the inhabitants of other worlds.

If the earth is a star, however, shedding influence upon and receiving influences from other stars, then the medieval notion of earth and the "sublunary sphere" as a unique realm of mutability "quarantined" from the rest of the universe cannot stand. Although empirical proof of change in the heavens came only later, with the observation of comets and novas in the last half of the sixteenth century and with Galileo's subsequent account of sunspots, Nicholas in his own time radically undermined the division of the cosmos into two "zones" of mutability and immutability. To put it positively, he retheorized the unity *of the physical universe.*

Not even the corruptibility of things which we here experience is compelling evidence of earth's baseness. For given a single universe in which each star influences every other, we have no grounds for declaring anything to be utterly corruptible. It is better instead to conceive of corruption as one or another mode of being: where influences so to speak were once knit together, they now unravel, so that a thing's mode of being either this or that passes away. Thus death has no place, as Virgil says. Rather, death appears to be merely the dissolving of a composite thing into its components. And who can know whether such dissolution occurs only among things of this earth?

For Nicholas Cusanus, the whole physical universe radically falls short of perfection, if only because it is a created thing. Yet its glory and unity are not therefore diminished, for these reflect the glory of the Maker, whose creative skills Nicholas conceives of as combining those of an arithmetician, a geometrician, and a musician.

It is the unanimous judgment among the wise that the vastness, beauty, and order of the visible creation cause us to be astounded at God's artistry and excellence. Now having touched on some of the products of his marvelous skill, let us add a further brief word of wonder regarding the creation of the universe as far as the setting and composition of its components are concerned.

In creating the universe God employed arithmetic, geometry, music, and astronomy, arts that we too use when we investigate the structure of things, including their substance and motion. By means of arithmetic God joined things together. By means of geometry he shaped things according to the rank of each so as to produce solidity, stability, and mobility. By means of music he gave things proportion in such a way that there should be no more earth in earth than there is water in water, air in air, or fire in fire; accordingly, no element may be wholly resolved into another. And from this it follows that the world system cannot pass away. . . .

Thus God composed the elements in a wonderful and orderly manner, creating all things in number, weight, and measure—number in keeping with arithmetic, weight with music, measure with geometry. Heaviness is sustained and constrained by lightness. For example, fire suspends the heavy earth as it were in its midst, and lightness is supported by heaviness, as fire is by earth. Moreover, in ordaining these things, Eternal Wisdom employed an inexpressible symmetry. He foreknew by what degree each element should precede another, and he measured them in such a way that water should be as much lighter than earth, as air is than water, and as fire is than air; and at the same time that weight should be proportionate to size, and that a container should occupy a larger space than that which it contains. He linked things together in such interdependence that they are necessary to each other's existence. And thus the earth, as Plato says, is like an animal whose bones are stones, whose arteries are rivers, whose hairs are trees; and animals feed among those hairs just as mites do among the hairs of the animals. . . .

Who will not stand in awe of the Craftsman who employed this same skill in the spheres and stars and realms of stars? For thus he has made all things to blend in unceasing diversity and harmony. In one single universe he weighed out the multitudinous stars in their places, ordaining their motions and their distribution in such a way that, unless each region were precisely as it is, neither could it exist nor persist in its place and arrangement, nor could the universe itself continue to be. He gave to each star its unique brightness, influence, shape, and color, as well as heat which is transmitted along with brightness. And he so proportionately and harmoniously composes the proportions of parts, that in all things the motion of the part is relative to the whole—heavy things moving downward towards the center and light things rising upwards from the center, or else about the center, as we observe with the orbital movement of stars.

These matters are indeed full of wonder, variety, and contrast. Yet in them, learned ignorance teaches us what we have already heard: that we have not the capacity to fathom the reason for all of God's works; but we may stand in awe before them. For great is the Lord, and his greatness is without bounds. As he is absolute greatness, the author and comprehender of all his works, so too is he their end. In him all things *be*, and without him is nothing. He is the beginning, the middle, the end of all things, the center and circumference of all that is. Accordingly, it is he who is to be sought in all things, for apart from him all things are nothing. To possess God alone is to possess all things, for he is all. And to know him is to know all things, for he is the truth of all. It is his will that the system of this universe should cause us to stand in wonder; yet the more we wonder at it, the more he hides it from us, since it is *he* whom he would have us seek with all diligence and with all our heart.

SOURCE: Translated from *Nicolai Cusani De docta ignorantia libri tres*, Bari: Laterza, 1913.

Part Three

COPERNICUS TO NEWTON

18

Almost Contrary to Common Sense

Nicholas Copernicus

There is great pathos in the story of Nicholas Copernicus (1473–1543) being brought a freshly published copy of De Revolutionibus Orbium Caelestium *on his deathbed. We can only speculate what his thoughts then might have been about his book's future reception and impact. Yet perhaps his dedicatory letter to Pope Paul III gives us a clue. In this letter, Copernicus is most obviously defensive; he worries his opinions will be merely "shouted down." Less obviously, however, we sense that, while surely believing in the scientific veracity of his theory, Copernicus as a person is still reeling with amazement at what he hears himself propounding. Repeatedly he uses the term* absurd *to characterize his own teachings, and that is how they initially must have tasted even to the teacher himself, who nevertheless presents his views with undoubted if reluctant courage.*

To His Holiness Pope Paul III

Holy Father, I can guess already that some people, as soon as they find out about this book I have written on the revolutions of the universal spheres, in which I ascribe a kind of motion to the earthly globe, will clamor to have me and my opinions shouted down. Nor am I so pleased with my own work that I disregard others' judgments concerning it. I know that a philosopher's thoughts are beyond the reach of common opinion, because his aim is to search out the truth in all things—so far as human reason, by

God's permission, can do that. But I do think that completely false opinions are to be avoided.

My thoughts, then, were these: Those who know that the judgment of many centuries supports the view that the earth stands firm in the midst of the heavens—their center, as it were—will think it an absurd bit of theater if I on the contrary declare that the earth moves. So for a long time I wavered. Should I publish my argument showing that the earth moves? Or would I be better to follow the example of the Pythagoreans, and some others, who handed down the secrets of their philosophy only to relatives and friends—orally, not in writing—as the letter of Lysis to Hipparchus indicates. They did so, it seems to me, not (as some think) out of mere unwillingness to share their teachings, but out of a desire to protect beauties and profundities discovered by great men from the contempt of those who refuse to give any effort to literary accomplishment unless it turns a profit—or who, even if by the advice and example of others they do apply themselves freely to the study of philosophy, are, as a result of their native stupidity, among philosophers like drones among bees. Accordingly, when I contemplated the contempt I would face on account of the novelty and absurdity of my opinion, I almost gave up completely the work I had started.

And yet, although for a long time I hesitated and even resisted, my friends drew me along. Foremost among them was Nicholas Schönberg, Cardinal of Capua, famous in all fields of learning. Next to him was my dear friend Tiedemann Giese, Bishop of Kulm, a great student of the sacred writings and all good literature. For repeatedly he encouraged me, commanded me, sometimes sharply, to publish this book and let it see the light of day after lying buried and hidden, not for nine years but going on four times nine. More than a few other eminent and learned men advised me to do the same. They urged me to set aside my anxieties, abandon my reluctance, and share my work for the common good of astronomical learning. According to them, the more absurd my doctrine of the earth's motion appeared to most people, the greater would be their amazement and gratitude once my book was published and the clouds of absurdity had been dispersed by radiant proofs. These persuasions and this hope, therefore, finally convinced me to allow my friends, as they had long requested, to prepare an edition of this work.

However, Your Holiness, perhaps your amazement that I would publish these findings—after having so exerted myself in the work of thinking them through, even to the point of deciding to commit to writing my conclusions concerning the earth's motion—perhaps your amazement is not so great as your desire to hear why it would occur to me, contrary to the received opinions of astronomers and almost contrary to common sense, to dare to imagine any motion of the earth.

Accordingly, Your Holiness, I would have you know that what moved me to conceive a different model for explaining the motions of the universal spheres was merely my realization that the astronomers are not consistent among themselves regarding this subject. In the first place, they are so uncertain concerning the motions of the sun and the moon that they can neither observe nor predict even the constant length of a tropical year. Secondly, in calculating the motions of these as well as the other five planets, they do not use the same principles and assumptions, nor the same explanations for their apparent revolutions and motions. For while some use only concentric circles, others employ eccentrics and epicycles, from which however the desired results do not quite follow. Those relying on concentrics, though they may use these for modelling diverse motions, nevertheless have not been successful in using them to obtain firm results in perfect accordance with the phenomena. Yet those who have invented eccentric circles, while they seem for the most part to have solved apparent motion in a manner that is arithmetically consistent, at the same time also seem to have introduced several ideas that contradict the first principles of uniform motion. Nor have they been able to discover or deduce by means of their eccentrics the main point, which is to describe the form of the universe and the sure symmetry of its parts. Instead they have been like someone attempting a portrait by assembling hands, feet, head, and other parts from different sources. These several bits may be well painted, but they do not fit together to make up a single body. Bearing no genuine relationship to each other, such components, joined together, would compose a monster, not a man.

Thus in their process of demonstration—"method," as they call it—those employing eccentrics have either omitted something essential or else admitted something extraneous and irrelevant. This would not have happened if they had observed sound principles. For unless the hypothesis they adopted were fallacious, all the predictions following from them would be verifiable beyond dispute. (Even if what I am saying here is obscure, it will become clearer in its proper place.)

Thus I pondered for a long time this lack of resolution within the astronomical tradition as far as the derivation of the motions of the universal spheres is concerned. It began to irritate me that the philosophers, who otherwise scrutinized so precisely the minutiae of this world, could not agree on a more reliable theory concerning the motions of the system of the universe, which the best and most orderly Artist of all framed for our sake. So I set myself the task of rereading all the philosophers whose books I could lay my hands on, to find out whether anyone had ever held another opinion concerning the motion of the universal spheres than those asserted by the teachers of astronomy in the schools. Indeed, I found, first in Cicero, that Nicetus supposed the earth to move. And later I discovered in Plutarch that some others held the same opinion. . . .

Following their example, therefore, I too began to contemplate the possibility that the earth moves. To be sure, it seemed an absurd idea. Yet I knew that others before me had been accorded the liberty to imagine whatever circles they chose in order to explain the astronomical phenomena. Thus I presumed that I likewise would surely be permitted to test, given some motion of the earth, whether a more solid explanation of the revolutions of the heavenly spheres were possible than had so far been provided.

Accordingly, I posited the motion which later in this volume I assign to the earth. And by deep and extensive investigation I finally found that if the motion of the other planets is viewed in relation to the circular motion of the earth, and if this calculation is made for the revolution of each planet, then not only do the phenomena follow consistently, but also the orders and magnitudes of all the orbs and spheres and heaven itself are so interconnected that not one of its parts could be removed without throwing the other parts and the whole universe into confusion.

In the arrangement of this work, therefore, I have observed the following order. In the first book I set out all the positions of the spheres along with the motions I ascribe to the earth, so that this book comprises as it were the overall structure of the universe. And in the remaining books I relate the motions of the rest of the planets and all the spheres to the movement of the earth in order to show to what extent their appearances, if we do relate them to the earth's motion, can be saved.

I have no doubt that astute and learned astronomers will agree with me if, in keeping with the chief requirement of this discipline, they will study and examine—not superficially but in depth—the evidence for these matters which I set forth in this work. However, so that learned and unlearned alike may see I am a person who flees the judgment of no one at all, I have chosen to dedicate these my late-night studies to you, Your Holiness, rather than to anyone else. For even here in this remote corner of the earth which I inhabit, you are held to be the highest authority by virtue of your exalted office and your love for all literature, even astronomy. Thus by your authority and discernment you may easily repress the malice of slanderers, even if (as the proverb says) there is no remedy against the teeth of a backbiter.

Perhaps some idle talkers, thinking they can judge astronomy though completely ignorant of it, and distorting some passage of Scripture twisted to their purposes, will dare to criticize and censure my teaching. I shall not waste time on them; I have only contempt for their audacity. As is well known, Lactantius, otherwise a distinguished writer but no astronomer, speaks quite immaturely about the shape of the earth when he mocks those who assert that the earth is spherical. No scholar need be surprised, therefore, if such persons ridicule me likewise. Astronomy is written for astronomers—and they, if I am not mistaken, will see the value that these efforts of mine have for the ecclesiastical community over which Your Holiness now holds dominion. For not long ago, under Leo X, the Lateran Council raised the issue of emending the church calendar. No decision was then arrived at merely because the Council concluded that the lengths of the year and the month and the motions of the sun and the moon were not yet measured accurately enough. Since then, at the urging of that most eminent man Dr. Paulus, Bishop of Sempronia, who was in charge of the proceedings, I have concentrated on studying these matters with greater accuracy. What I have accomplished in this regard, however, I hand over to be judged by Your Holiness in particular and by all other learned astronomers. And lest Your Holiness should think I promise more regarding the usefulness of this volume than I can fulfill, I now proceed to the work itself.

CHAPTER 1. THE UNIVERSE IS SPHERICAL

The first thing for us to realize is that the universe is spherical. This is so either because, of all forms, the sphere is the most perfect, requiring no joins, and being an integrated whole; or because it is the most capacious of all forms, and so best fitted to enclose and preserve all things—or also because the most perfected parts of the universe such as the sun, the moon, and the stars display this shape; or because all things strive to be bounded thus, as we observe in drops of water and other liquids when they seek to be bounded within themselves. There can be no doubt, then, about the rightness of ascribing this shape to the heavenly bodies.

CHAPTER 2. THE EARTH TOO IS SPHERICAL

The earth also has the shape of a globe, because all of its parts tend towards its center. We do not immediately perceive it as a perfect sphere because the mountains are so high and the valleys so deep, and yet these hardly affect the overall sphericity of the earth. This is clear from the fact that if one travels northward, the pole of the diurnal rotation gradually rises, while the opposite pole sinks accordingly, and more stars in the northern sky seem never to set, while some in the south seem never to rise. Thus Italy does not see Canopus, which is visible from Egypt; and Italy does see the last star of the River, which up here in our frozen territory is unknown. Conversely, as one travels southward, such stars rise higher, while those which appear high to us sink lower. Also, the angle of elevation of the poles is everywhere constantly proportionate to the distance one thus travels across the earth, something that happens with no other shape than a sphere. Hence we see that the earth too is delimited by poles and thus shaped like a globe. Moreover, evening eclipses of sun or moon are not seen by inhabitants of the east, nor morning eclipses by inhabitants of the west; but those that occur in between are observed later by the former and earlier by the latter.

Seafarers know that the waters too conform to this shape, for land that is not visible from the ship is observed from the top of the mast. And if a bright light is placed at the top of the mast, then to those remaining on the shore it appears gradually to sink as the ship moves farther off from land. Finally, the light as it were sets and disappears. Also, like earth, water, in keeping with its nature as a fluid, always obviously seeks a lower level and so does not push farther inland than the curvature of the shore permits. Hence, one accepts that whatever land emerges from the ocean is higher than it is.

In chapter 3, "How the earth and the water together make up a globe," Copernicus almost belabors the point that the earth as a whole is spherical. His rejection of other alternatives—that the earth is flat, or bowl-shaped, or cylindrical—makes his own suggestion seem more novel than it in fact is. Fourteen centuries earlier Ptolemy had also declared authoritatively that the earth is a globe, and for similar reasons. Nevertheless, the extension of seafaring, and in particular the discoveries of Columbus only half a century earlier, lent freshness to the conception of earth as a sphere and extended Europeans' ideas about how opposite sides of it might be inhabited. Copernicus postulates (pretty accurately) that "geometrical reasons cause us to believe that America is located diametrically opposite to India of the Ganges."

Chapter 4 then moves from earth to the heavens. In it Copernicus makes his stunning (and stunningly simple) proposal that the irregularities we seemingly have to attribute to the heavenly bodies are a function of our viewpoint here on a body that is not at the center of things. But we also notice a certain arbitrary geometrical idealism that underlies Copernicus's argument: To his mind, only a circle brings a body back to where it started. It took the long labors of Kepler to dispel this wrong assumption.

CHAPTER 4. THE MOTION OF THE HEAVENLY BODIES IS UNIFORM, PERPETUAL, AND CIRCULAR, OR MADE UP OF CIRCULAR MOTIONS

Let us now recall that the motion of heavenly bodies is circular. For that movement which a sphere possesses is movement in a circle, by which action it expresses its own form as the simplest shape, in which no beginning or end is to be discerned, nor can these be distinguished from each other, while the sphere keeps on moving within its own bounds. Yet a multitude of motions applies to the various spheres. The most obvious of all is the daily rotation, which the Greeks call *nuchthemeron*, which measures the passage of a day and a night. By this motion, it is assumed the whole universe—except for the earth—glides round from east to west. This is recognized as the common measure of all motion, since we reckon even time itself mainly by counting days.

Next we see other, as it were contrary revolutions, moving from west to east, namely those of the sun, the moon, and the five planets. Thus the sun metes out our year and the moon our month, these being the other most common measures of time. In this way too each of the five planets completes its circuit. Yet there are differences among their various motions. First, these do not turn about the same axis as the primary motion but take a slantwise course through the zodiac. Secondly, they are not observed moving uniformly in their orbits. For we see that the sun and moon in their courses sometimes move slowly and sometimes more quickly. As for the other five planets, as we observe, sometimes they even come to a stop and retrace their steps. And while the sun always keeps strictly to its own pathway, these others wander in various ways, sometimes towards the south, sometimes towards the north—which is why they are called planets [from Greek *planetes*, "wanderer"]. Moreover, sometimes they are nearer the earth and said to be "in perigee"; at other times they are farther off and said to be "in apogee."

We must admit, nonetheless, that their motions are circular, or made up of several circles, because these nonuniformities conform to a consistent law and to the fact that the planets return to where they began, which could not be the case unless the motions were circular, for only a circle can replicate what has already taken place. For example, by a motion made up of circles the sun causes for us a repetition of unequal days and nights and of the four seasons. In this cycle we discern several motions, since no simple heavenly body can move irregularly in a single sphere. For such irregularity would have to result either from an inconstancy in the force of movement, whether arising internally or externally, or from some irregularity in the revolving body. But either alternative is abhorrent to reason. We must not ascribe any such indignity to things framed and governed optimally.

We must conclude, then, that their uniform motions appear to us as irregular either because they take place around different axes, or else because the earth is not at the center of their circles of revolution. For us on earth as we observe the movements of these planets, this is what happens: because of their nonuniform distances they appear larger when they are near us than when they are farther away (optics proves this principle). And similarly, across equal portions of their circumferences their motions over a given time will appear unequal because viewed from different distances.

Above all, then, I think we must examine carefully the relationship of the earth to the heavens. Otherwise, in our desire to investigate things of the highest order we may remain ignorant of what is nearest to us, likewise mistakenly attributing things that are earthly to things that are heavenly.

CHAPTER 5. IS CIRCULAR MOTION APPROPRIATE TO THE EARTH? AND WHAT IS EARTH'S LOCATION?

Chapter 5 begins by reformulating the questions contained in its title. Although "the authorities generally agree that the earth rests in the middle of the universe," Copernicus calls for closer consideration of the matter concerning earth's location.

Every apparent change of place is caused by the movement either of the observer or of the thing observed, or indeed by some unequal alteration in the position of both. (When observer and observed move uniformly relative to each other, no motion is perceived.) Yet it is from earth that we behold the circuit of the heavens; it is here that its spectacle is represented to us. Therefore if any motion is predicated of the earth, the same motion will appear in all that is beyond the earth, but in a contrary direction, as if everything were moving about it. The prime example of this is the daily rotation, whereby apparently the whole universe except for the earth itself is driven round. However, if you grant that the heavens have no part in this rotation, but that the earth itself turns from west to east, then considering the matter seriously you will find this is actually the case as far as the rising and setting of sun, moon, and stars are concerned. And since everything is contained within the heavens, which serve as the location and setting of all things, it is not immediately apparent why motion should be attributed to the container rather than the contents, to the location rather than the thing located. Indeed, this was the opinion of Heraclides and Ecphantus the Pythagoreans and, according to Cicero, of Nicetus of Syracuse, who held that earth rotates in the middle of the universe. For their judgment was that the stars set when the earth comes in the way and rise when it ceases to be in the way.

Having introduced the notion of merely the rotation of the earth, Copernicus now broaches the issue of the other main motion of the earth, namely its revolution not in the center but about the center.

Given this rotation, a further, equally important question follows concerning the earth's location. Admittedly, virtually everyone has been taught, and believes, that the earth is the center of the universe. However, anyone who denies that the earth occupies the center or midpoint may still assert that its distance from the center is negligible by comparison with that of the sphere of the fixed stars, yet noticeable and noteworthy relative to the spheres of the sun and other planets. He may consider that this is why their motions appear nonuniform, and that they are regular relative to some center other than that of the earth. In this way, perhaps, he can offer a not-so-inept explanation for the appearance of irregular motion. For the fact that we observe the planets sometimes nearer the earth and sometimes farther away is logical proof that the center of the earth is not the center of their orbits. . . .

CHAPTER 6. THE IMMENSITY OF THE
HEAVENS COMPARED TO THE SIZE OF THE EARTH

As evidenced by his respectful mention of Heraclides and the Pythagoreans, Copernicus is careful not to appear to be promoting mere novelties. Moreover, some of his arguments, though they lead in a non-Ptolemaic direction, begin with Ptolemaic materials. Chapter 6 is a good example, beginning with a virtual recapitulation of Ptolemy's argument regarding the immense—literally immeasurable—*disparity between the size of the universe and that of the earth. Like Ptolemy, Copernicus makes this case by pointing out that a horizontal plane tangent to the earth's surface appears to bisect the universe. Mathematically, a bisecting plane necessarily contains the center point. Even in a geocentric universe, however, to claim that the horizontal plane itself does so implies a contradiction, for a plane tangent to the earth's surface is at that point distant from the center by the length of the earth's radius. Ptolemy's point, and Copernicus's, is that earthly distance is as nothing compared to the size of the universe. Although neither believed that the universe is actually infinite, both considered that the immense extent of the heavens rendered earth, so far as our ability to measure is concerned, a mere point by comparison. In Copernicus's words: "So far as our senses can tell, the earth is related to the heavens as a point is to a body and as something finite is to something infinite."*

From this agreement with Ptolemy, Copernicus moves to the next critical step of his argument, which comprises two discernible parts. First (and this relates to the question of earth's rotation on its axis), given that the earth is a point within an immensity, why should we expect the immensity to turn while the point stands still? "How astonishing, if within the space of twenty-four hours the vast universe should rotate rather than its least point!" And second (this being the argument that de-centers the earth), if the surface of the earth merely appears *to be in the very center of the universe, as Ptolemy declares, then the same explanation can be applied regarding the location of earth if it is not perfectly central but moves in a sphere (or orbit) "near the center." For, the earth being near the center, its movement relative to the immensity of the universe will be very small indeed.*

Copernicus ends the chapter by bolstering his argument with a parenthetical analogy between the macroscopic and the microscopic (this, of course, more than half a century before the invention of telescopes and microscopes).

Things enclosed within a smaller orbit revolve more quickly than those turning in a larger circle. Thus Saturn, the highest of the planets, revolves in thirty years, while the moon, undoubtedly the nearest the earth, has a circuit of one month. Finally, one will presume, earth rotates within the space of a day and a night. Hence the question of the daily rotation arises once more.

And yet the question of earth's location remains uncertain, all the more so because of what was said above. For nothing has been proven except the indescribable size of the heavens compared to that of the earth. But the degree of that immensity remains unclear. (Consider that at the opposite extreme there are minuscule indivisible bodies called "atoms." Because they are imperceptible, when they are taken two or several at a time they do not make up any visible body. Yet they can be sufficiently multiplied to the point where there are enough of them to form a visible mass. The same thing applies to the position of the earth. Even though it is not in the center of the universe, its distance from the center is nevertheless inconsiderable when compared to the distance of the sphere of the fixed stars.)

Chapters 7 and 8 form a pair, as their titles indicate:

CHAPTER 7. WHY THE ANCIENTS BELIEVED THAT THE EARTH RESTS IN THE MIDDLE OF THE UNIVERSE AS IF IT WERE ITS CENTER

and

CHAPTER 8. EXPLANATION OF THESE REASONS AND OF THEIR INSUFFICIENCY

In short, chapter 7 summarizes the Aristotelian/Ptolemaic physics according to which earth, the heaviest of the elements, must seek the universal center or low-point, and according to which a daily rotation of the earth would cause

the earth to lose its coherence and "fly apart," an idea Copernicus dismisses
as "quite ridiculous."

He continues his refutation in chapter 8 by demonstrating in effect that
Ptolemy and his like employ circular reasoning. The notion of a stationary,
central earth is so firm a component of what Aristotelian physics conceives
of as natural that it rejects any movement of the earth precisely on the
grounds that it is unnatural. However, Copernicus simply undercuts this line
of argument by declaring that if the earth does indeed move, then it does so
in accordance with nature, not contrary to it.

In the course of this discussion, Copernicus momentarily looks beyond a
medieval, enclosed universe, though he quickly averts his gaze.

Ptolemy therefore has no reason to fear that earth and all things terrestrial
will fly apart on account of a rotation brought about by means of nature's
own operation, which is very different from anything artificial or devised by
human ingenuity.

Yet why is he not just as worried about the universe as a whole, whose
swiftness of motion must be that much greater in proportion as the heavens
are greater than the earth? Or are the heavens so immense precisely because
the ineffable force of their motion impels them away from the center? Would
they otherwise collapse if they did stand still? If this reasoning were sound,
then surely the magnitude of the heavens must expand to infinity. For the
higher they are impelled by the force of their motion, the faster their motion
will be on account of the continuously expanding circumference which has to
make its revolution every twenty-four hours. In turn, as the motion increased, so would the immensity of the heavens—speed thus increasing size,
and size increasing speed, *ad infinitum*. Yet according to that axiom of
physics, nothing that is infinite can be traversed nor moved by any means,
and so the heavens are necessarily at rest.

But beyond the heavens, it is said, there is no body, no place, no vacuity,
absolutely nothing, and so there is nowhere for the heavens to go. It is truly
miraculous, then, if something can be contained within nothing. However, if
the heavens are infinite, and finite only in their hollow interior, then perhaps
there will be greater reason to believe that outside the heavens there is nothing, for in this case every single thing, no matter how much space it takes
up, will be inside them. But the heavens will remain motionless. For the
strongest piece of evidence produced in support of the earth's finitude is its
motion. Whether the universe is finite or infinite, however—let us leave that
question for the natural philosophers to dispute while we hold firmly to the
belief that the earth is delimited by its poles and enclosed by a spherical surface.

Why, then, do we still hesitate to accept the earth's movement in keeping with the nature of its form instead of attributing motion to the whole universe, whose bounds are unknown and unknowable. As regards the daily rotation, why not grant that in the heavens is the appearance but in the earth is the reality? It is like the case spoken of by Vergil's Aeneas: "We sail forth from the harbor, and lands and cities draw backwards" [*Aeneid*, III.72]. For when a ship glides along smoothly, its passengers see its motion reflected by everything outside of the ship and, by contrast, suppose themselves and everything else on board to be motionless. No wonder, then, that the movement of the earth makes us think the whole universe is turning round.

CHAPTER 9. CAN THE EARTH BE SAID TO MOVE IN MORE WAYS THAN ONE? WHERE IS THE CENTER OF THE UNIVERSE?

Since therefore nothing precludes the earth's movement, I propose we now consider whether it may be thought to move in more than one way: can it be regarded as one of the planets?

For earth is not the center of all the revolutions. This claim is demonstrated by the apparently nonuniform motion of the planets and by their variable distances from the earth, which cannot be conceived as implying circles concentric to the earth. Therefore, there being numerous centers, it is worth asking whether the center of the universe, or some other, is the center of earthly gravity. In my view gravity is nothing but a certain natural desire which by divine providence the Creator of all has infused into the parts, whereby they draw themselves into a unity and an integrity in the form of a globe. The same desire may be credibly predicated also of the sun, the moon, and the other luminous planets; by its efficacy they persist in the rounded shape in which we behold them, although they pursue their own various orbits.

Therefore, if the earth too moves in other ways—about a center, for example—then this must similarly be reflected in many external things. Among them, it would seem, is the annual revolution. For if, granting immobility to the sun, we exchange earthly movement for solar movement, then the risings and settings of the constellations and the fixed stars which accompany morning and evening will appear just as they do. Furthermore, the stations as well as both the backward and forward motions of the planets will be seen not as their own motions but as earthly motion transmuted into apparent planetary motions. Finally, it will be accepted that the sun occupies the center of the universe.

We learn all these things by discerning the order whereby the planets follow one another and by the harmony of the entire universe—if only we examine these matters (as they say) with both eyes open.

CHAPTER 10. THE ORDER OF
THE HEAVENLY SPHERES

In the last half of chapter 10, Copernicus summarizes the model of the solar system which we most readily associate with his name. He begins with the moon, so long thought to circle the earth along with the other planets, but now together with the earth seen as forming a system which itself orbits the sun.

We should not be ashamed to admit that this whole domain encircled by the moon, with the center of the earth, traverses this great orbit amidst the other planets in an annual revolution around the sun, and that near the sun is the center of the universe; and moreover that, since the sun stands still, whatever motion the sun appears to have is instead actually attributable to the motion of the earth. Furthermore, although the distance between the earth and the sun is quite noticeable relative to the size of the other planetary orbits, it is imperceptible as compared with the sphere of the fixed stars—so great indeed is the size of the universe. I think it is a lot easier to accept this than to drive our minds to distraction multiplying spheres almost ad infinitum, as has been the compulsion of those who would detain earth in the center of the universe. Instead, it is better to follow the wisdom of nature, which just as it strongly avoids producing anything superfluous or useless, so it often prefers to endow a single thing with multiple effects.

This whole matter is difficult, almost paradoxical, and certainly contrary to many people's way of thinking. In what follows, however, God helping me, I shall make these things clearer than sunlight, at least to those not ignorant of the art of astronomy. And so, with the first principle firmly established (for nobody can propose one more fitting than that the magnitude of a planet's orbit is proportionate to its period of revolution), the order of the spheres is as follows, beginning with the highest:

First and highest of all is the sphere of the fixed stars, containing itself and all things, and therefore immovable, the very location of the universe, that to which the motion and position of all the other heavenly bodies is referred. . . . This is followed by the first of the planets, Saturn, which completes his circuit in thirty years. Then comes Jupiter, moving in a revolution with a twelve-year period. Next, the circuit of Mars is two years. Fourth comes the annual revolution in which, as mentioned earlier, the earth is carried along, with the moon as it were in an epicycle. Venus, in fifth place, circles round in nine months. And then in sixth place Mercury completes his course in the space of eighty days.

And behold, in the midst of all resides the sun. For who, in this most beautiful temple, would set this lamp in another or a better place, whence to illuminate all things at once? For aptly indeed do some call him the lantern—and others the mind or the ruler—of the universe. Hermes Trismegistus calls him the visible god, and Sophocles' Electra "the beholder" of all things. Truly indeed does the sun, as if seated upon a royal throne, govern his family of planets as they circle about him. Nor is the earth thus deprived of the moon's services; rather, as Aristotle asserts in his book on animals, the moon shares closest kinship with the earth. Meanwhile, the earth is impregnated by the sun, by whom is begotten her annual offspring.

Thus we discover in this orderly arrangement the marvelous symmetry of the universe and a firm harmonious connection between the motion and the size of the spheres such as can be discerned by no other means. For this model permits anyone who is diligent to comprehend why the progressions and regressions of Jupiter appear greater than those of Saturn and smaller than those of Mars, and again greater for Venus than for Mercury. And the reversals appear more frequently in Saturn than in Jupiter, and even more rarely in Mars and in Venus than in Mercury. . . . All these phenomena appear for the same reason: that the earth moves.

However, that none of these phenomena appears in the fixed stars proves that these are immensely distant, for which reason even the motion of the annual revolution, or the appearance thereof, vanishes from sight. For each visible thing has a certain limit of distance beyond which it becomes invisible, as demonstrated in optics. The sparkling of the stars shows what an enormous distance remains between their sphere and that of the highest planet, Saturn. This is principally what distinguishes them from the planets, for there had to be an enormous difference between that which moves and that which does not. So great, certainly, is the divine handiwork of Him who is himself the greatest and the best.

SOURCE: Translated from Nicholas Copernicus, *De Revolutionibus Orbium Caelestium*, Nuremberg, 1543, in consultation with *Nikolaus Kopernikus Gesamtausgabe*, vol. 2, Munich: Oldenbourg, 1949.

19

The Poetic Structure of the World

Fernand Hallyn and Thomas Kuhn

Historians—including historians of science—sometimes perform their best work not in telling us what we don't know but in causing us to reexamine what we think we do know. In fact, part of the task of this anthology is to let cosmologists such as Copernicus emerge from the fog of history's clichés and be heard afresh. When we do hear Copernicus, a number of those clichés begin to dissipate, among them the century-old construction of what Andrew Dixon White (1832–1918) called the "warfare of science with theology." While it is true that Copernicus had some theological qualms about his own teachings, it is also the case that his deepest motivations go far beyond any narrow sense of the "scientific" to include the religious, the literary, and the aesthetic.

The two historians excerpted here both make this point in relation to the famous passage in Book I, chapter 10, of De Revolutionibus *where Copernicus virtually sings a hymn praising the sun at the center of the universe and the harmony of the system as a whole:*

And behold, in the midst of all resides the sun. For who, in this most beautiful temple, would set this lamp in another or a better place, whence to illuminate all things at once? For aptly indeed do some call him the lantern—and others the mind or the ruler—of the universe. Hermes Trismegistus calls him the visible god, and Sophocles' Electra "the beholder" of all things. Truly indeed does the sun, as if seated upon a royal throne, govern his family of planets as they circle about him. . . .

Thus we discover in this orderly arrangement the marvelous symmetry of the universe and a firm harmonious connection between the motion and the size of the spheres such as can be discerned by no other means.

Fernand Hallyn (b. 1945), a European literary historian, comments as follows in his aptly titled The Poetic Structure of the World:

This is an often cited passage. [Alexandre] Koyré goes so far as to say that it identifies for us "the deepest motivation of Copernican thought." Frances Yates calls particular attention to the reference to Hermes Trismegistus and notes that "it is, in short, in the atmosphere of the religion of the world that the Copernican revolution is introduced." For Yates, the passage becomes an important argument affirming the role of the hermetic tradition in the birth of modern science.

Without a doubt, Edward Rosen is right to judge that one should not isolate the allusion to Hermes Trismegistus from its context. It appears, after all, in the midst of references to other figures—to Plato, Cicero, Sophocles, and Pliny—not all of whom can be classified as writing in the hermetic tradition. It is true moreover . . . that the geocentric cosmos did not in any way obstruct the development of a significant solar symbolism. The passage to Copernicanism was not absolutely necessary for the development of that theme.

Should we go farther and agree with Jean Bernhardt that the entire passage is "suspect of being purely literary"? Given that the Copernican sun plays no dynamic role in the motion of the planets, Bernhardt concludes that the metaphors likening the sun to a "ruler" or "governor" are exaggerated and unsatisfactory. Throughout the passage, Copernicus seeks solely "to provide external support for his arguments and to make the position that he assigns the sun appear less of an innovation than it really is."

An expression like "suspect of being purely literary" is of course itself suspect. It would be more neutral to speak of a stylistic marking, without introducing by choice of terms the idea of an inverse proportionality between the relevance of the information conveyed and the visibility that a style guarantees for the passage. On the other hand, although the sun does not effectively participate in a dynamic relation with the motion of the planets, it is nonetheless true that it "governs," that it is the "ruler," the center of the "symmetry" according to which the relationship between distance and time is organized. A point of reference for the cosmicality of the cosmos, it introduces into homogeneous space . . . a *perspectiva superior.*

The decision to treat the passage as an "exterior" element obviously arises from a modern prejudice concerning the separations among "science," "literature," "philosophy," and so on. Elsewhere, Copernicus wrote that astronomy leads to the "contemplation of the highest good." His meditation on the meaning of the sun's centrality invites precisely such an act of contemplation, and that explains the passage's lyricism and proliferation of devices such as the rhetorical question, enumeration, asyndeton, metaphor, and comparison. Rather than consider such passages as "literary," why not recognize in their coherence and insistence a specific constituent element in the comprehensiveness and unity of the Copernican enterprise?

Thomas Kuhn (1922–1996), perhaps the most famous late-twentieth-century historian of Copernicanism, likewise emphasized the importance (if not the sufficiency) of the aesthetic—of "evidence drawn from harmony"—both in the motivation and in the reception of Copernicus's model.

Throughout [the] crucially important tenth chapter, Copernicus's emphasis is upon the "admirable symmetry" and the "clear bond of harmony in the motion and magnitude of the Spheres" that a sun-centered geometry imparts to the appearances of the heavens. If the sun is the center, then an inferior planet cannot possibly appear far from the sun; if the sun is the center, then a superior planet must be in opposition to the sun when it is closest to the earth; and so on and on. It is through arguments like these that Copernicus seeks to persuade his contemporaries of the validity of his new approach. Each argument cites an aspect of the appearances that can be explained by either the Ptolemaic or the Copernican system, and each then proceeds to point out how much more harmonious, coherent, and natural the Copernican explanation is. There are a great many such arguments. The sum of the evidence drawn from harmony is nothing if not impressive.

But it may well be nothing. "Harmony" seems a strange basis on which to argue for the earth's motions, particularly since the harmony is so obscured by the complex multitude of circles that make up the full Copernican system. Copernicus' arguments are not pragmatic. They appeal, if at all, not to the utilitarian sense of the practicing astronomer but to his aesthetic sense and to that alone. They had no appeal to laymen, who, even when they understood the arguments, were unwilling to substitute minor celestial harmonies for major terrestrial discord. They did not necessarily appeal to astronomers, for the harmonies to which Copernicus' arguments pointed did not enable the astronomer to perform his job better. New harmonies did not increase accuracy or simplicity. Therefore, they could and did appeal primarily to that limited and perhaps irrational subgroup of mathematical astronomers whose

Neoplatonic ear for mathematical harmonies could not be obstructed by page after page of complex mathematics leading finally to numerical predictions scarcely better than those they had known before. Fortunately . . . there were a few such astronomers.

SOURCES: Fernand Hallyn, *The Poetic Structure of the World: Copernicus and Kepler,* trans. Donald M. Leslie, New York: Zone Books, 1990; Thomas S. Kuhn, *The Copernican Revolution: Planetary Astronomy in the Development of Western Thought,* Cambridge, Mass.: Harvard UP, 1957.

20

This Art Unfolds the Wisdom of God

John Calvin and Johannes Kepler

Especially since the late nineteenth century, the progress of Copernicanism has often been portrayed as the struggle of science against religion. Of course, the most famous case superficially supporting this view is that of Galileo and his persecution by the Inquisition.

A balanced telling of the story, however, should not neglect two important facts. First, and most simply, cosmologists such as Copernicus and Kepler were deeply religious Christians (compare also Campanella, chapter 27). And second, both the Renaissance and the Reformation promoted a scholarly and critical approach to the reading of texts, especially the Bible, that had its counterpart in how scientists were coming to read the book of nature. In this sense the work of Copernicus may be seen as akin to "systematic theology": as a re-reading of the text of the heavens, taking the literal surface (the appearances) very seriously indeed, but trying anew to understand their coherence at a level beyond the merely apparent.

This search for overall coherence is exemplified by the method that the reformer John Calvin (1509–1564) used to interpret the Bible. Sometimes called the "analogy of faith" (a term Calvin borrows from Romans 12:6), this method interprets any particular doubtful passage of Scripture in conformity with what is already firmly established by Scripture as a whole. However, if God is author of both Scripture and Nature, then of course the same principle may be applied to both "books" at once. Therefore, when a clear, well founded reading of Nature appears to conflict with a literalistic reading of the Bible, the solution is to pursue a scriptural interpretation which, like true science, penetrates the merely superficial and seeks consistency with what is already known to be true.

Calvin, though accepting a geocentric cosmology (and writing only a decade after the appearance in 1543 of De Revolutionibus), may thus be seen as a precursor, hermeneutically, of a Copernican such as Johannes Kepler (1571–1630). Both Calvin and Kepler explain the style of the Bible where it relates to the creation by referring to what ordinary people actually see, and both endorse the view of creation as a divine text/textile/fabric in which God is in some sense clothed for, and so revealed to, his creatures. In Calvin's words, "God . . . clothes himself, so to speak, with the image of the world in which he would present himself to our contemplation"; we may "behold him thus magnificently arrayed in the incomparable vesture of the heavens and the earth." Moreover, elsewhere in his commentary on Genesis (1554), Calvin draws a sharp distinction between the practical experience of the senses and the theoretical knowledge gained through astronomy, whose practice and value he defends.

1:15. "Let them be for lights." It is well again to repeat what I have said before, that it is not here philosophically discussed how great the sun is in the heaven, and how great, or how little, is the moon; but how much light comes to us from them. For Moses here addresses himself to our senses, that the knowledge of the gifts of God which we enjoy may not glide away. Therefore, in order to apprehend the meaning of Moses, it is to no purpose to soar above the heavens; let us only open our eyes to behold this light which God enkindles for us in the earth. . . . For as was appropriate for a theologian, he had respect to us rather than to the stars. Nor, in truth, was he ignorant of the fact that the moon had not sufficient brightness to enlighten the earth, unless it borrowed from the sun . . . .

1:16. "The greater light." I have said that Moses does not here subtly descant, as a philosopher, on the secrets of nature, as may be seen in these words. First, he assigns a place in the expanse of heaven to the planets and stars; but astronomers make a distinction of spheres and, at the same time, teach that the fixed stars have their proper place in the firmament. Moses makes two great luminaries; but astronomers prove by conclusive reasons that the star of Saturn, which on account of its great distance appears the least of all, is greater than the moon. Here lies the difference: Moses wrote in a popular style things which, without instruction, all ordinary persons endued with common sense are able to understand; but astronomers investigate with great labor whatever the sagacity of the human mind can comprehend. Nevertheless, this study is not to be reprobated, nor this science to be condemned, because some frantic persons are wont boldly to reject whatever is unknown to them. For astronomy is not only pleasant but also very useful to be known; it cannot be denied that this art unfolds the admirable wisdom of God. Wherefore, as ingenious men are to be honored who have expended useful labor on this subject, so they who have leisure and capacity ought not to neglect this kind of exercise. Nor did Moses truly wish to withdraw us from this pursuit in omitting such things as are peculiar to the art; but, because he was ordained a teacher as well of the unlearned and rude as of the learned, he could not otherwise fulfill his office than by descending to this grosser method of instruction. Had he spoken of things generally unknown, the uneducated might have pleaded in excuse that such subjects were beyond their capacity.

Lastly, since the Spirit of God here opens a common school for all, it is not surprising that he should chiefly choose those subjects which would be intelligible to all. If the astronomer inquires respecting the actual dimensions of the stars, he will find the moon to be less than Saturn; but this is something abstruse, for to the sight it appears differently. Moses therefore rather adapts his discourse to common usage. For since the Lord stretches forth, as it were, his hand to us in causing us to enjoy the brightness of the sun and moon, how great would be our ingratitude were we to close our eyes against our own experience. There is therefore no reason why janglers should deride the unskillfulness of Moses in making the moon the second luminary; for he does not call us up into heaven; he only proposes things which lie open before our eyes. Let the astronomers possess their more exalted knowledge; but, in the meantime, they who perceive by the moon the splendor of night are convicted by its use of perverse ingratitude unless they acknowledge the beneficence of God.

Kepler for his part, in the introduction to his Astronomia Nova *(1609), employs similar distinctions between scientific and popular style in order to solve apparent conflicts between heliocentrism and the Bible.*

It must be confessed that there are very many devoted to holiness who dissent from the judgment of Copernicus, fearing to give the lie to the Holy Ghost speaking in the Scriptures, if they should say that the earth moves and the sun stands still. But let such consider that, since we judge of very many and those the most principal things by the sense of seeing, it is impossible that we should alienate our speech from this sense of our eyes. Therefore many things daily occur of which we speak according to the sense of sight, when we certainly know that the things themselves are otherwise, as for example in that verse of Vergil, "We sail forth from the harbor, and lands and cities draw backwards." . . .

Thus we conceive of the rising and setting of the stars, that is to say, of their ascension and descension; we affirm that the sun rises, when at the same time others say that it goes down. . . . So in like manner, the Ptolemaics affirm that the planets stand still, when for some days together they seem to be fixed, although they believe them at that very time to be moved in a direct line either downwards to or upwards from the earth. Thus the writers of all nations use the word "solstice," and yet they deny that the sun does really stand still. . . . And so in other cases of the like nature.

But now the Sacred Scriptures, speaking to men of vulgar matters (in which they were not intended to instruct men) after the manner of men, so that they might be understood by men, do use such expressions as are granted by all, whereby to intimate other things more mysterious and divine. What wonder is it, then, if the Scripture speaks according to man's apprehension at such time when the truth of things dissents from the conception that all men, whether learned or unlearned, have of them. Who knows not that it is a poetical allusion (Psalm 19) where, under the similitude of the sun, the course of the Gospel, as also the peregrination of our Lord Christ in this world, undertaken for our sakes, is described. Thus the sun is said to come forth from his "tabernacle" of the horizon "as a bridegroom coming out of his chamber, and rejoices as a strong man to run a race." . . . The Psalmist knew that the sun went not forth from the horizon as out of its tabernacle, and yet it seems to the eye so to do. Nor did he believe that the sun moved because it appeared to his sight so to do. And yet he says both, because both were so to his seeming. Neither is it to be adjudged false in either sense, for the perception of the eyes has its verity, fit for the more secret purpose of the Psalmist in shadowing forth the current passage of the Gospel, as also the peregrination of the Son of God. . . .

Kepler considers a number of passages from the Psalms and from Job (e.g., chapter 38) and suggests that to give poetic figures a purely literal interpretation is to "meddle with the Holy Spirit" and to bring him into contempt. In reading such texts we should "turn our eyes from natural philosophy, to the scope and intent of Scripture." He continues:

But Psalm 104 is thought by some to contain a discourse altogether physical, in that it concerns natural philosophy. Now God is there said to have "laid the foundations of the earth, that it should not be removed for ever" (v. 5). But here also the Psalmist is far from the speculation of physical causes. For he wholly acquiesces in the greatness of God, who did all these things, and sings a hymn to God the Maker of them, in which he runs over the world in order, as it appeared to his eyes. And if you well consider this psalm, it is a paraphrase upon the six days work of the creation. For in it the three first days were spent in the separation of regions: the first, of light from exterior darkness; the second, of the waters from the waters, by the interposition of the firmament; the third, of the sea from land, when also the earth was clothed with herbage and plants. And the three last days were spent in filling the regions thus distinguished: the fourth, of heaven; the fifth, of the seas and air; the sixth, of the earth. Likewise here in this psalm there are so many distinct parts proportionable to the analogy of the six days works. For in verse 2 he clothes and covers the Creator with light (the first of creatures, and work of the first day) as with a garment. The second part begins at verse 3 and treats of the waters above the heavens, the extent of heaven and of meteors (which the Psalmist seems to intend by the waters above) as namely of clouds, winds, whirl-winds, lightnings. The third part begins at verse 6 and celebrates the earth as the foundation of all those things which he here considers. For he refers all things to the earth, and to those animals which inhabit it, since in the judgment of sight the two principal parts of the world are heaven and earth. He therefore here observes that the earth after so many ages has not faltered, tired, or decayed, whereas no man has yet discovered upon what it is founded. He aims not to teach men what they do not know, but puts them in mind of what they neglect, namely, the greatness and power of God in creating so huge a mass so firm and steadfast.

If an astronomer should teach that the earth is placed among the planets, he overthrows not what the Psalmist here says, nor does he contradict common experience. For it is true notwithstanding that the earth, the structure of God its Architect, does not decay (as our buildings are wont to do) by age, or consume by worms, nor sway and lean to this or that side; that the seats and nests of living creatures are not molested; that the mountains and shores stand immoveable against the violence of the winds and waves, as they were at the beginning. But the Psalmist adds a most elegant hypothesis of the separation of the waters from the continent or mainland, and adorns it with the production of fountains, and the benefits that springs and rocks exhibit to birds and beasts. Nor does he omit the apparelling the earth's surface, mentioned by Moses among the works of the third day, but more sublimely describes it in his case in expressions infused from divine inspiration; and flourishes out the commemoration of the many commodities which redound from that exornation for the nourishment and comfort of man, and covert of beasts.

The fourth part begins at verse 20 celebrating the fourth day's work, namely the sun and moon, but chiefly the commodiousness of those things which in their seasons befall to all living creatures and to man, this being the subject matter of his discourse. So that it plainly appears he acted not the part of an astronomer. For if he had, he would not then have omitted to mention the five planets, than whose motion nothing is more admirable, nothing more excellent, nothing that can more evidently set forth the wisdom of the Creator among the learned. The fifth part begins, verse 25, with the fifth day's work. And it stores the seas with fishes, and covers them with ships. The sixth part is more obscurely hinted at, verse 28, and alludes to the land-creatures that were created the sixth day. And lastly, he declares the goodness of God in general, who daily creates and preserves all things. So that whatever he said of the world is in relation to living creatures. He speaks of nothing but what is granted on all hands, for it was his intent to extol things known, and not to dive into hidden matters, but to invite men to contemplate the benefits that redound unto them from the works of each of these days.

And I also beseech my reader, not forgetting the divine goodness conferred on mankind, the consideration of which the Psalmist chiefly urges, that when he returns from the temple, and enters into the school of astronomy, he would with me praise and admire the wisdom and greatness of the Creator, which I discover to him by a more narrow explication of the world's form, the disquisition of causes, and detection of the errors of sight. And so he will not only extol the bounty of God in the preservation of living creatures of all kinds, and establishing the earth; but even in its motion also, which is so strange, so admirable, he will acknowledge the wisdom of the Creator.

But he who is so stupid as not to comprehend the science of astronomy, or so weak and scrupulous as to think it an offence of piety to adhere to Copernicus, him I advise that, leaving the study of astronomy, and censuring the opinions of philosophers at pleasure, he betake himself to his own concerns, and that, desisting from further pursuit of these intricate studies, he keep at home and manure his own ground; and with those eyes wherewith alone he sees, being elevated towards this to-be-admired heaven, let him pour forth his whole heart in thanks and praises to God the Creator, and assure himself that he shall therein perform as much worship to God, as the astronomer, on whom God has bestowed this gift, that though he sees more clearly with the eye of his understanding, yet whatever he has attained to, he is both able and willing to extol his God above it.

SOURCES: Adapted from *A Commentary of John Calvin, upon the first book of Moses called Genesis*, trans. Thomas Tymme, London, 1578; *Johannes Keplerus, His Reconcilings of Scripture Texts*, in Thomas Salusbury, *Mathematical Collections and Translations*, London, 1661.

21

A Star Never Seen
Before Our Time

Tycho Brahe

Tycho Brahe (1546–1601) often appears as a sort of extended footnote in the history of astronomy, most famously because of his "Tychonic" system, according to which the planets revolve around the sun, which in turn still revolves around a stationary earth. Yet his keen pretelescopic observations of the heavens in fact laid the foundation for subsequent cosmological theory, in particular that of his successor, Johannes Kepler. Moreover, precisely because they were observational rather than principally theoretical, some of Tycho's claims must have had an even more radical imaginative and evidential impact than those of Copernicus. Copernicus's main contribution was to rethink the structure of what we now call the solar system, but he did so within a fundamentally medieval (if expanded) sphere of immutable fixed stars. Tycho, on the other hand, presented ocular evidence of mutability even in those celestial realms. As Annie Jump Cannon puts it, Tycho's "discovery was a death knell to the natural philosophy of Aristotle. A change had taken place in that 'solid crystal sphere of the fixed stars,' which had been assumed, during nearly two thousand years, to be subject neither to growth nor to decay." Or we may put it this way: Contrary to previous beliefs, even the highest heavens themselves are in the grip of Time.

Having already for more than a decade been an avid and knowledgeable sky-watcher, one night in 1572 Tycho noticed, disbelievingly, a new star in the constellation Cassiopeia. In his awed response one can almost hear the cracking of the foundations of medieval cosmology.

Last year [1572], in the month of November, on the eleventh day of that month, in the evening, after sunset, when according to my habit I was contemplating the stars in a clear sky, I noticed that a new and unusual star, surpassing the other stars in brilliancy, was shining almost directly above my head. And since I had almost from boyhood known all the stars of the heavens perfectly . . . it was quite evident to me that there had never before been any star in that place in the sky, even the smallest, to say nothing of a star so conspicuously bright as this. I was so astonished at this sight that I was not ashamed to doubt the trustworthiness of my own eyes.

But when I observed that others, too, on having the place pointed out to them, could see that there was really a star there, I had no further doubts. A miracle indeed, either the greatest of all that have occurred in the whole range of nature since the beginning of the world, or one certainly that is to be classed with those attested by the Holy Oracles, the staying of the sun in its course in answer to the prayers of Joshua, and the darkening of the sun's face at the time of the crucifixion. For all philosophers agree, and facts clearly prove it to be the case, that in the ethereal region of the celestial world no change, in the way either of generation or of corruption, takes place; but that the heavens and the celestial bodies in the heavens are without increase or diminution, and that they undergo no alteration, either in number or in size or in light or in any other respect; that they always remain the same, like unto themselves in all respects, no years wearing them away. Furthermore, the observations of all the founders of science, made some thousands of years ago, testify that all the stars have always retained the same number, position, order, motion, and size as they are found, by careful observation on the part of those who take delight in heavenly phenomena, to preserve even in our own day. Nor do we read that it was ever before noted by any one of the founders that a new star had appeared in the celestial world, except only by Hipparchus [fl. 2nd-century B.C.], if we are to believe Pliny [A.D. 23–79]. For Hipparchus, according to Pliny (Book II of his *Natural History*) noticed a star different from all others previously seen, one born in his own age. . . .

Although more technical than is the norm for this anthology, the following selection shows how crucial it was to establish where the new star was located. If it were sublunary, then the phenomenon would be merely unremarkable. If it were located among the planets, then at least no revision of the doctrine of stellar immutability would be required. However, if the new star were truly among the "fixed" stars, then two thousand-year-old assumptions would be shaken.

The nonmathematician requires only patience to appreciate the simplicity and force of Tycho's argument.

It is a difficult matter, and one that requires a subtle mind, to try to determine the distances of the stars from us, because they are so incredibly far removed from the earth; nor can it be done in any way more conveniently and with greater certainty than by the measure of the [diurnal] parallax, if a star have one. For if a star that is near the horizon is seen in a different place than when it is at its highest point and near the vertex, it is necessarily found in some orbit with respect to which the earth has a sensible size. How far distant the said orbit is, the size of the parallax compared with the semidiameter of the earth will make clear. If, however, a [circumpolar] star, that is as near to the horizon [at lower culmination] as to the vertex [at upper culmination], is seen at the same point of the Primum Mobile, there is no doubt that it is situated either in the eighth sphere or not far below it, in an orbit with respect to which the whole earth is as a point.

In order, therefore, that I might find out in this way whether this star was in the region of the element or among the celestial orbits, and what its distance was from the earth itself, I tried to determine whether it had a parallax, and if so how great a one. And this I did in the following way: I observed the distance between this star and Schedir of Cassiopeia (for the latter and the new star were both nearly in the meridian), when the star was at its nearest point to the vertex, being only 6 degrees removed from the zenith itself (and for that reason, though it were near the earth, would produce no parallax in that place, the visual position of the star and the real position then uniting in one point, since the line from the center of the earth and that from the surface nearly coincide). I made the same observation when the star was farthest from the zenith and at its nearest point to the horizon, and in each case I found that the distance from the above-mentioned fixed star was exactly the same, without the variation of a minute: namely, 7 degrees and 55 minutes. Then I went through the same process, making numerous observations with other stars. Whence I conclude that this new star has no diversity of aspect, even when it is near the horizon. For otherwise in its least altitude it would have been farther away from the above-mentioned star in the breast of Cassiopeia than when in its greatest altitude. Therefore, we shall find it necessary to place this star not in the region of the element, below the moon, but far above, in an orbit with respect to which the earth has no sensible size.

For if it were in the highest region of the air, below the hollow region of the lunar sphere, it would, when nearest the horizon, have produced on the circle a sensible variation of altitude from that which it held when near the vertex.

A simple way of expressing the same argument (though using the Copernican language of earthly rotation) is this: Suppose that, standing on the equator, I view an object X vertically overhead somewhere in the heavens, taking note of the angular distance between X and a "fixed" star also roughly vertically overhead, and then six hours later I view the same object X and the same star both now on the horizon (since I have moved 90° about the earth's axis). If the angular distance between X and the fixed star remains the same, then this lack of parallax justifies the inference that X and the fixed star are at an immeasurable distance from earth.

In the same, way Tycho marshals evidence against the second possibility, that the new star is a planet, and so is left with but one astounding alternative. His conservative interpretation of comets in this context reveals how reluctant he remains to admit evidence of change beyond the sublunary sphere.

Therefore, this new star is neither in the region of the element, below the moon, nor among the orbits of the seven wandering stars, but it is in the eighth sphere, among the other fixed stars, which was what we had to prove. Hence it follows that it is not some peculiar kind of comet or some other kind of fiery meteor become visible. For none of these are generated in the heavens themselves, but they are below the moon, in the upper region of the air, as all philosophers testify, unless one would believe with Albategnius that comets are produced not in the air but in the heavens. For he believes that he has observed a comet above the moon, in the sphere of Venus. That this can be the case is not yet clear to me. But, please God, sometime, if a comet shows itself in our age, I will investigate the truth of the matter. Even should we assume that it can happen (which I, in company with other philosophers, can hardly admit), still it does not follow that this star is a kind of comet; first, by reason of its very form, which is the same as the form of the real stars and different from the form of all the comets hitherto seen, and then because, in such a length of time, it advances neither latitudinally nor longitudinally by any motion of its own, as comets have been observed to do. For, although these sometimes seem to remain in one place several days, still, when the observation is made carefully by exact instruments, they are seen not to keep the same position for so very long or so very exactly.

I conclude, therefore, that this star is not some kind of comet or fiery meteor, whether these be generated beneath the moon or above the moon, but that it is a star shining in the firmament itself—one that has never previously been seen before our time, in any age since the beginning of the world.

SOURCE: Tycho Brahe, *De Nova Stella*, trans. John H. Walden, in *A Source Book in Astronomy*, ed. Harlow Shapley and Helen E. Howarth, New York: McGraw-Hill, 1929.

22

This Little Dark Star
Wherein We Live

Thomas Digges

*Thomas Digges (c. 1546–1595) is notable not only for being the first transla-
tor of Copernicus into English but also for drawing from Copernicus's theory bold inferences concerning the size, and perhaps the infinity, of the
universe. Digges's* A Perfit Description of the Caelestiall Orbes *was published
in 1576 as an appendix to an edition of a largely meteorological work by his
father, Leonard Digges, entitled* A Prognostication *(1553, with subsequent
revisions). The senior Digges prefaces his book with a defense of astronomy
against two standard charges: that it is useless, and that it is impious:*

To avoid tediousness I refer all of that sort which have tasted any learning
(the rest not regarded) to the first part of famous Guido Bonatus' *On the
Common Utility of Astronomy*. . . . Also for brevity I appoint all nice [i.e.,
fastidious] divines, or (as Melanchthon termeth them) "theological Epicureans," to his high commendations touching astronomy, . . . where he showeth
how far wide they allege the Scriptures against the astronomer, which
[agreeth] wholly with the astronomer.

*The controversial themes of astronomy's usefulness and piety become part of
the undercurrent too of Thomas's discussion in* A Perfit Description. *The
younger Digges should not be seen merely as a translator of Copernicus. His
paraphrase is mainly of chapters 10, 7, and 8 (in that order) of the first book
of the* Revolutions, *with some inclusions from other chapters; however, he
both digresses freely from and embellishes Copernicus's text. In his almost*

nervous high praise of the newcomer Copernicus, in his deft use of "our own philosopher Aristotle" against the Aristotelians he knows will oppose the new world picture, and in his charming flattery of "noble English minds," we enjoy a unique taste of the new and inevitably controversial cosmology, which Digges sets forth for the first time in the English language.

Having of late (gentle reader) corrected and reformed sundry faults that by negligence have crept into my father's *General Prognostication*, among other things I found a description or model of the world and situation of spheres celestial and elementary according to the doctrine of Ptolemy, whereunto all universities (led thereto chiefly by the authority of Aristotle) since have consented. But in this our age one rare wit (seeing the continual errors that from time to time more and more have been discovered, besides the infinite absurdities in their theories, which they have been forced to admit that would not confess any mobility in the ball of the earth) hath by long study, painful practice, and rare invention delivered a new theory or model of the world, showing that the earth resteth not in the center of the whole world, but only in the center of this our mortal world or globe of elements which, environed and enclosed in the moon's orb, and together with the whole globe of mortality, is carried yearly round about the sun, which like a king in the midst of all reigneth and giveth laws of motion to the rest, spherically dispersing his glorious beams of light through all this sacred celestial temple. And the earth itself to be one of the planets, having his peculiar and straying courses turning every twenty-four hours round upon his own center, whereby the sun and great globe of fixed stars seem to sway about and turn, albeit indeed they remain fixed.

Digges distances himself from those who would defend Copernicus's model as a mere hypothesis, constructed only to "save the appearances."

So many ways is the sense of mortal men abused, but reason and deep discourse of wit having opened these things to Copernicus, and the same being with demonstrations mathematical most apparently by him to the world delivered, I thought it convenient together with the old theory also to publish this, to the end such noble English minds (as delight to reach above the baser sort of men) might not be altogether defrauded of so noble a part of philosophy. And to the end it might manifestly appear that Copernicus meant not as some have fondly excused him to deliver the grounds of the earth's mobility only as mathematical principles, fained and not as philosophical truly averred, I have also from him delivered both the philosophical reasons by Aristotle and others produced to maintain the earth's stability, and also their solutions and insufficiency, wherein I cannot a little commend the modesty of that grave philosopher Aristotle, who seeing (no doubt) the insufficiency of his own reasons in seeking to confute the earth's motion, useth these words: "We have explained these matters so far as our capacity permits." Howbeit, his disciples have not with like sobriety maintained the same.

Thus much for my own part in this case I will only say: There is no doubt but of a true ground, truer effects may be produced than of principles that are false; and of true principles, falsehood or absurdity cannot be inferred. If therefore the earth be situate immovable in the center of the world, why find we not theories upon that ground to produce effects as true and certain as these of Copernicus? Why cast we not away those "equant circles" and motions irregular, seeing our own Philosopher Aristotle, himself the light of our universities, hath taught us: "A simple body has a simple motion." But if contrary it be found impossible (the earth's stability being granted) but that we must necessarily fall into these absurdities, and cannot by any means avoid them, why shall we so much dote in the appearance of our senses, which many ways may be abused, and not suffer ourselves to be directed by the rule of reason, which the great God hath given us as a lamp to lighten the darkness of our understanding and the perfect guide to lead us to the golden branch of verity amid the forest of errors?

Behold a noble question to be of the philosophers and mathematicians of our universities argued not with childish invectives but with grave reasons philosophical and irreprovable demonstrations mathematical. . . .

The globe of elements enclosed in the orb of the moon I call the globe of mortality, because it is the peculiar empire of death. . . . In the midst of this globe of mortality hangeth this dark star or ball of the earth and water balanced and sustained in the midst of the thin air only with that propriety which the wonderful workman hath given at the creation to the center of this globe with his magnetical force vehemently to draw and hale unto itself all such other elementary things as retain the like nature. This ball every twenty-four hours by natural, uniform, and wonderful sly and smooth motion rolleth round, making with his period our natural day, whereby it seems to us that the huge infinite immovable globe should sway and turn about.

The moon's orb that environeth and containeth this dark star and the other mortal, changeable, corruptible elements and elementary things is also turned round every twenty-nine days, thirty-one minutes, fifty seconds . . . and this period may most aptly be called the month.

Digges's title, repeated in its entirety following the preface and before his paraphrase of chapter 10 of the first book of the Revolutions, *employs a strategy often used by those presenting new ideas. Just as Protestant reform-*

ers insisted that the Reformation was a return to the Scriptures and to the early church, so here Digges depicts Copernicus not as a champion of novelty but as a reviver of "ancient doctrine."

A PERFECT DESCRIPTION OF THE CELESTIAL ORBS ACCORDING TO THE MOST ANCIENT DOCTRINE OF THE PYTHAGOREANS, LATELY REVIVED BY COPERNICUS AND BY GEOMETRICAL DEMONSTRATIONS APPROVED.

Although in this most excellent and difficult part of philosophy in all times have been sundry opinions touching the situation and moving of the bodies celestial, yet in certain principles all philosophers of any account of all ages have agreed and consented: First, that the orb of the fixed stars is of all other the most high, the farthest distant, and comprehendeth all the other spheres of wandering stars. And of these straying bodies called "planets" the old philosophers thought it a good ground in reason that the nighest to the center should swiftliest move, because the circle was least and thereby the soonest overpassed, and the farther distant the more slowly. Therefore as the moon being swiftest in course is found also by measure nighest, so have all agreed that the orb of Saturn, being in moving the slowest of all the planets, is also the biggest; Jupiter the next; and then Mars. But of Venus and Mercury there hath been great controversy, because they stray not every way from the sun as the rest do. And therefore some have placed them above the sun, as Plato in his *Timaeus*; others beneath, as Ptolemy and the greater part of them that followed him. . . .

If we situate the orbs of Saturn, Jupiter, and Mars referring them as it were to the same center, so as their capacity be such as they contain and circulate also the earth, happily we shall nor err, as by evident demonstrations in the residue of Copernicus' *Revolutions* is demonstrated. For it is apparent that these planets nigh the sun are always least, and farthest distant and opposite are much greater in sight and nigher to us; whereby it cannot be but the center of them is rather to the sun than to the earth to be referred, as in the orbs of Venus and Mercury also.

But if all these to the sun as a center in this manner be referred, then must there needs between the convex orb of Venus and the concave of Mars a huge space be left wherein the earth and elementary frame, enclosed with the lunar orb, of duty must be situate. For from the earth the moon may not be far removed, being without controversy of all other nighest in place and nature to it, especially considering between the same orbs of Venus and Mars there is room sufficient. Therefore need we not to be ashamed to confess this whole globe of elements enclosed with the moon's sphere, together with the earth as the center of the same, to be by this great orb, together with the other planets, about the sun turned, making by his revolution our year. And whatsoever seem to us to proceed by the moving of the sun, the same to proceed indeed by the revolution of the earth, the sun still remaining fixed and immovable in the midst. And the distance of the earth from the sun to be such as, being compared with the other planets, maketh evident alterations and diversity of aspects, but if it be referred to the orb of stars fixed, then hath it no proportion sensible, but as a point or a center to a circumference, which I hold far more reasonable to be granted, than to fall unto such an infinity of multitude of absurd imaginations, as they were fain to admit that will needs wilfully maintain the earth's stability in the center of the world. But rather herein to direct ourselves by that wisdom we see in all God's natural works, where we may behold one thing rather endued with many virtues and effects, than any superfluous or unnecessary part admitted.

And all these things although they seem hard, strange, and incredible, yet to any reasonable man that hath his understanding ripened with mathematical demonstration, Copernicus in his *Revolutions*, according to his promise, hath made them more evident and clear than the sunbeams. These grounds therefore admitted, which no man reasonably can impugn, that the greater orb requireth the longer time to run his period, the orderly and most beautiful frame of the heavens doth ensue.

*Continuing to hew closely to Copernicus's own discussion (*Revolutions *1.10), Digges makes clear the mathematical symmetry of the heliocentric system, in particular its elegance in ridding astronomy of the embarrassment of many irregularities in the heavens, irregularities which may now be concluded are merely apparent, and "follow upon the earth's motion."*

However, as one set of problems is solved, another problem comes into view, namely that of annual parallax: If the earth revolves around the sun in an enormous annual orbit, why don't the stars, like the planets, appear to vary their movement as the earth moves?

And that none of these alterations do happen in the fixed stars, it plainly argueth their huge distance and immeasurable altitude, in respect whereof this great orb wherein the earth is carried is but a point, and utterly without sensible proportion being compared to that heaven. For as it is in perspective demonstrated, every quantity hath a certain proportionable distance whereunto it may be discerned and beyond the same it may not be seen, this distance therefore of that immovable heaven is so exceeding great, that the whole "great orbit" vanisheth away, if it be conferred to that heaven.

In concluding thus, Digges agrees exactly with Copernicus. He goes beyond his teacher, however, in expatiating upon the meaning of astronomical magnitude, hinting even at infinitude.

Herein can we never sufficiently admire this wonderful and incomprehensible huge frame of God's work proposed to our senses, seeing first this ball of the earth, wherein we move, to the common sort seemeth great, and yet in respect of the moon's orb is very small, but compared with the great orbit wherein earth is carried, it scarcely retaineth any sensible proportion, so marvelously is that orb of annual motion greater than this little dark star wherein we live. But that great orbit being as is before declared but as a point in respect of the immensity of that immovable heaven, we may easily consider what little portion of God's frame our elementary corruptible world is, but never sufficiently be able to admire the immensity of the rest, especially of that fixed orb garnished with lights innumerable and reaching up in spherical altitude without end.

At this point Digges departs from the Copernican text and draws an inference, namely, there are countless more stars than can be viewed with the naked eye, an inference only confirmed more than three decades later by Galileo with his telescope, though it was theorized earlier by Nicholas Cusanus, whose reasoning Digges in part echoes:

Of these lights celestial it is to be thought that we only behold such as are in the inferior parts of the same orb; and as they are higher, so seem they of less and lesser quantity, even till our sight being not able farther to reach or conceive, the greatest part rest, by reason of their wonderful distance, invisible unto us. And this may well be thought of us to be the glorious court of that great God, whose unsearchable works invisible we may partly, by these his visible, conjecture: to whose infinite power and majesty such an infinite place surmounting all other both in quantity and quality only is convenient.

In this way, although Digges continues to use the traditional term "orb" when referring to the fixed stars, his inference dissolves that term's usual meaning. Hence, in the famous diagram that heads A Perfit Description of the Caelestiall Orbes *(see the figure on the opposite page), the text that accompanies the starry "orb" reads:*

This orb of stars fixed infinitely up extendeth it self in altitude spherically and therefore immovable, the palace of felicity garnished with perpetual shining glorious lights innumerable far excelling our sun both in quantity and quality, the very court of celestial angels devoid of grief and replenished with perfect endless joy, the [habitation] for the elect.

A more sober tone is found in the sentences with which Digges closes his paraphrase of and extrapolations from Copernicus's Revolutions *1.10:*

But because the world hath so long a time been carried with an opinion of the earth's stability, as the contrary cannot but be now very impersuasible, I have thought good out of Copernicus also to give a taste of the reasons philosophical alleged for the earth's stability, and their solutions, that such as are not able with geometrical eyes to behold the secret perfections of Copernicus' theory, may yet by these familiar natural reasons be induced to search farther, and not rashly to condemn for fantastical so ancient doctrine revived, and by Copernicus so demonstratively approved.

Although Digges's version of Revolutions *1.7 and 1.8 is more nearly a straight translation than is his paraphrase of 1.10, near the end of 1.8 he imports a burst of reasoning from 1.9 concerning gravity in which Copernicus adumbrates some implications of the new recognition that the universe contains various orbs and that "these orbs have several centers." For Aristotle, gravity attached to* place, *namely, to the center of the universe—a place, moreover, which the earth occupied simply because it is heavy. But if there are several centers, gravity can no longer be accounted for in terms of a single center.* Body *thus dethrones* place *as the primary gravitational determinant. Copernicus, and with him Digges, dimly anticipating Newton, recur to what Digges has already identified as "that propriety which the wonderful workman hath given at the creation to the center of this globe with his magnetical force vehemently to draw and hale unto itself all such other elementary things as retain the like nature."*

Seeing therefore that these orbs have several centers, it may be doubted whether the center of this earthly gravity be also the center of the world. For gravity is nothing else but a certain proclivity or natural coveting of parts to be coupled with the whole, which by divine providence of the creator of all is given and impressed into the parts, that they should restore themselves into

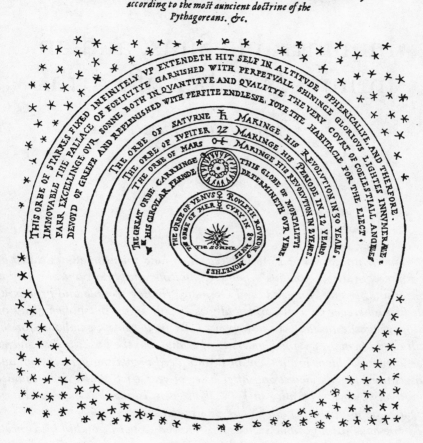

A perfit defcription of the Cæleftiall Orbes,
according to the moft auncient doctrine of the Pythagoreans. &c.

their unity and integrity concurring in spherical form, which kind of propriety or affection it is likely also that the moon and other glorious bodies want
not to knit and combine their parts together, and to maintain them in their
round shape, which bodies notwithstanding are by sundry motions sundry
ways conveyed.

SOURCE: Thomas Digges, *A Perfit Description of the Caelestiall Orbes*; appended to
Leonard Digges, *A Prognostication Everlasting*, London, 1576.

23

Innumerable Suns, and an Infinite Number of Earths

Giordano Bruno

Giordano Bruno, born in 1548 in Nola, Italy (and thus sometimes called by his contemporaries "Nolanus"), pursued an itinerant philosophical career in Switzerland, France, England, and Germany, finding trouble and provoking polemics wherever he went, principally as a result of his outspoken views on various topics, especially cosmology. In 1591 he became a candidate for the chair of mathematics at the University of Padua, but the position was given a year later to Galileo. From then on Bruno faced increasing persecution, and prosecution, for his views; and after a seven-year trial, the Roman Inquisition finally burned him alive in 1600. This event is one that impressed itself indelibly, no doubt, on Galileo's consciousness.

At one level Bruno can be seen as a convinced Copernican, but his conception of the universe goes far beyond finite heliocentrism and, with its emphasis on infinity, forms a bridge between the atomism of Lucretius and the absolute space of Newton. What seems most to have rankled the authorities of his day is Bruno's anti-Aristotelianism, which is conspicuous from the start of his 1584 dialogue On the Infinite Universe and Worlds.

The speakers here are Philotheo (Bruno's main spokesperson), Fracastoro, and Elpino. Philotheo attempts to deconstruct the very notions of finite place and space upon which Aristotle's physics is based.

Philotheo. If the world is finite, and if there is nothing beyond the world, then I ask you: *Where* is the world? *Where* is the universe? Aristotle's reply is: The world is in itself. The convex surface of the primordial heaven is universal space, and as the primordial container it is not contained by anything else; for location is merely the containing body's surfaces and limit, so that he who has no containing body has no location. But, dear Aristotle, what do you mean by "the location is in itself"? What will you tell us about that which is beyond the world? If you say there is nothing, then the heavens and the world will surely not be anywhere at all.

Fracastoro. Therefore the world will be nowhere. Everything will be in nothing.

Philotheo. The world will then be something impossible to find. It certainly seems to me that you are trying somehow to avoid the terms *vacuum* and *nothing*. But if you say that beyond the world there is a divine intellectual being, so that God becomes the location of all things, then you are going to have a hard job explaining how something at once incorporeal, intelligible, and dimensionless can be the location of something dimensional. And if you say that this location as it were contains a form, as the spirit contains the body, then you are not answering the question about the "beyond," or about what exists beyond the universe. And if you wish to excuse yourself by declaring that where there is nothing, and where not anything is, there is no such thing as location or beyond or outside, I shall not be at all satisfied. For such are mere words and excuses.

Having thus unsettled the Aristotelian conception of space and place, Philotheo is able to hypothesize "countless other spaces like this one." Elpino objects that the infinite—in particular, infinite Good—"certainly exists, but is incorporeal." Philotheo responds:

We agree concerning the incorporeal infinite. But what prevents our similarly accepting the existence of the good *corporeal* infinite? Why should that infinite which is implicit in the absolutely simple and indivisible primordial First Cause not wish to make itself explicit in his own infinite and boundless image, instead of within such narrow bounds? Surely it would be a shame for us to continue to believe that the body of this world, which to us seems so great and vast, may from God's perspective appear a mere point, even as a nothing.

Elpino. Since the greatness of God does not by any means consist in corporeal size (nor does this world add anything to him), we ought not to think of the greatness of his image as consisting in its greater or lesser size.

Philotheo. True enough; but you are not addressing the heart of my argument. I am not insisting on infinite space—and nature does not possess infinite space—for the sake of sheer magnitude or physical size, but for the sake of corporeal natures and species themselves. For infinite excellence is incomparably better expressed in things innumerable than in things merely numerable and finite. Indeed, it is fitting that an inaccessible divine countenance should have an infinite likeness with infinite parts—such as those countless worlds I have postulated. Moreover, since innumerable degrees of perfection must unfold God's incorporeal excellence by corporeal means, it follows that there must be innumerable individuals such as those great creatures are (our earth being one of them—the divine mother who gave birth to us, nourishes us, and will finally receive us again into herself). To encompass these innumerable creatures requires an infinite space. Yet it is good that this should be so. It means that there can be innumerable worlds like ours, which achieved and continues to achieve existence. And existence is good.

Various influences are traceable in Bruno's cosmology. There is a good dose of Platonism, and Bruno's repeated reference to the physical universe as a mirror naturally recalls Timaeus's description of the cosmos as a moving image of eternity. Philotheo also cites Nicholas Cusanus, whose assertion of the oneness of the unbounded universe seems to undergird Bruno's position, which is successfully imparted to the receptive Elpino, whose conventional views slowly succumb to Philotheo's combination of reason and imagination. It is worth emphasizing that the theory Bruno unfolds in his dialogue precedes Galileo's telescopic "expansion" of the universe.

Philotheo. All things then are one: the heavens, the immensity of space, our mother earth, the encompassing universe, the ethereal region through which all things move and continue on their way. Herein our senses may perceive innumerable heavenly bodies, stars, spheres, suns, and earths; and reason may deduce an infinitude of them. The universe, immense and infinite, is the sum total of all that space and all the bodies it contains.
Elpino. So there are no orbs with surfaces concave or convex, no deferent circles. Instead, all is one field, a single common envelope.
Philotheo. That is right.
Elpino. Thus the opinion concerning diverse heavens has come about as a result of diverse astral motions, along with what appears to be a star-filled heaven circling about the earth. These lights can by no means be seen to move relative to each other; rather, they always maintain the same distance and mutual relation, and the same fixed arrangement. And they seem to revolve about the earth, just as a wheel on which are attached countless mirrors turns about its own axis. So it is considered obvious from ocular evidence that the luminous bodies have no motion of their own, nor can they

move about like birds in flight. Instead, they move only with the turning of the circles to which they are affixed, and governed by the divine impulse of a higher intelligence.

Philotheo. Yes, that is the commonly accepted opinion. But once that motion is attributed to this earthly star that we inhabit, affixed to no such deferent circle but impelled through the open fields of space by means of its own inherent principle, spirit, and nature, revolving about the sun and rotating about its own center—then this illusion will be dispelled. Our understanding of the true principles of nature will be set free, and with great strides we shall advance along the road of truth, which has been obscured by such sordid and bestial imaginations, hidden until now by the injuries of time and the vicissitudes of things—ever since the daylight of ancient wisdom was overtaken by the dark night of rash sophistry. . . .

Elpino. There is no doubt that this entire fantasy of star- and fire-bearing orbs, of axes, of deferent circles, of cranking epicycles—along with plenty of other monstrous notions—is founded merely on the illusory notion that, as it appears, the earth is in the midpoint and center of the universe, while everything else circles about this fixed stationary earth.

Philotheo. This appearance is the same for those who dwell on the moon and on the other stars sharing the same space, be they earths or suns.

Elpino. For now, then, let us suppose the earth's own motion is the reason for what appears to be the daily rotation of the universe, and thus that the diversity of earth's motion accounts for the apparent motions of the countless stars. If so, then we would still affirm that the moon (which is another earth) moves by her own force through the air about the sun. Likewise, Venus, Mars, and the others—which are still other earths—pursue their journeys about the same source of life.

Philotheo. That is right.

Elpino. The respective motions of these bodies are those they appear to have *apart from* those attributable to the presumed rotation of the universe. And the actual motions of those known as the fixed stars (though their apparent fixity, along with the universal rotation, is only relative to the earth) are more various and plentiful than are those bodies themselves. For if we could observe the motion of each one, we would see that no two stars ever exhibit the same pattern or measure, and what keeps us from realizing the variations is the stars' huge distance from us. However much these stars revolve about the solar fire or rotate about their own centers in order to partake of the vital heat of the sun, we simply cannot apprehend their various decreasing or increasing distances from us.

Philotheo. That is right.

Elpino. And so there are innumerable suns, and likewise an infinite number of earths circling about those suns, just like the seven near to us which we see circling about the sun.
Philotheo. That is right.

SOURCE: Translated (with kind advice from Arielle Saiber) from *De l'infinito universo et Mondi,* 1584; in *Le opere italiane di Giordano Bruno,* Göttingen, 1888.

24

Neither Known Nor Observed by Anyone Before

Galileo Galilei

In 1898 Agnes Clerke wrote:

No one could at first have divined the momentous character of the accident by which Hans Lippershey, a spectacle-maker at Middleburg in Holland, hit upon an arrangement of lenses serving virtually to abridge distance. It happened in 1608; and Galileo Galilei (1564–1642), hearing of it shortly afterwards at Venice, prepared on the hint a "glazed optic tube," and viewed with it, early in 1610, the satellites of Jupiter, the mountains of the moon, the starstreams of the Milky Way, and in 1611, the phases of Venus, the spots on the sun, and the strange appendages of Saturn. Thus, amid a tumult of applause, the telescopic revelation of the heavens began.

It does not diminish Galileo's accomplishment—though perhaps it enhances our sense of his humanity—that he in fact openly invited the applause, especially from the box seat occupied by the Grand-Duke of Tuscany, Cosimo de' Medici II, to whom he dedicated his 1610 pamphlet, Sidereus Nuncius (The Starry Messenger), *and after whom he christened the four moons of Jupiter "the Medicean Stars." Galileo begins the text of his potent little work with a blurb summarizing, excitedly but truly without exaggeration, the astonishing list of discoveries he is about to unfold.*

In the present small treatise I set forth some matters of great interest for all observers of Nature to look at and consider. They are of great interest, I think, first, from their intrinsic excellence; secondly, from their utter novelty;

and lastly, also on account of the instrument by whose aid they have been presented to my sight.

To this day the fixed stars which observers have been able to view without artificial powers of sight can be counted. Therefore it is certainly a great thing to add to their number and to expose to our view myriads of other stars never seen before and outnumbering the old, previously known stars more than ten to one.

Again, it is a most beautiful and delightful thing to behold the body of the moon—which is actually distant from us nearly sixty semidiameters of the earth—as if it were only two such semidiameters distant from us; so that the diameter of the same moon appears about thirty times larger, its surface about nine hundred times, and its solid mass nearly 27,000 times larger than when it is viewed only with the naked eye. Consequently, anyone may know with palpable certainty that the surface of the moon really is not smooth and polished but instead rough, uneven and, just like the face of the earth itself, everywhere full of vast protuberances, deep chasms, and sinuosities.

Then to have got rid of disputes about the galaxy or Milky Way, and to have made its nature clear to the very senses, not to say to the understanding—this too would seem a matter of no small importance. In addition to this, to trace out with one's finger the nature of those stars which all astronomers until now have called *nebulous*, and to demonstrate that it is very different from what has hitherto been believed—this also will be thrilling and beautiful. But what is most exciting and astonishing by far, and what particularly moved me to address myself to all astronomers and philosophers, is this: I have discovered four planets, neither known nor observed by anyone before my time.

Maintaining his almost conversational narrative style, Galileo recounts his construction of a telescope, as Clerke says, from a hint; and he carries on to describe, step by step, his observations of the surface of the moon. What he sees, amazingly, is phenomena that are conspicuously earth-like.

About ten months ago a rumor reached my ears that a certain Dutchman had constructed a telescope, by means of which visible objects, although at a great distance from the eye of the observer, were seen distinctly as if nearby. Testimonies concerning its amazing powers were reported, but some believed these and others denied them. A few days later I received confirmation of the report in a letter written from Paris by a noble Frenchman, Jacques Badovere. This finally caused me to devote myself first to working out the principle of the telescope, and then to considering how I might achieve the invention of a similar instrument, which in a short while I succeeded in doing through a study of the theory of refraction. At first I prepared a tube of lead into whose ends I fitted two glass lenses, both plane on one side, but on the other side one spherically convex, and the other concave. Then bringing my eye to the concave lens I saw objects gratifyingly large and near. In fact they appeared one-third as distant and nine times as large as when seen with the natural eye alone. I shortly afterwards constructed another, superior telescope, which magnified objects more than sixty times. Finally, by sparing neither labor nor expense, I succeeded in constructing for myself such an excellent instrument that it rendered objects seen through it nearly a thousand times larger and more than thirty times nearer than they appeared when viewed by unaided natural powers of sight.

It would be a complete waste of time to list here the many great advantages this instrument will afford when used by land or sea. But leaving earthly things behind, I devoted myself to observing the heavens. First of all, I viewed the moon as if it were distant scarcely two semidiameters of the earth. And afterwards I frequently observed other heavenly bodies, both fixed stars and planets, with incredible delight. . . .

Now let me review my observations during the two months just past, again calling upon all who love true philosophy to undertake the contemplation of truly great things.

Let me speak first of the surface of the moon turned towards us, which for the sake of clarity I divide into two parts, the brighter and the darker. The brighter part seems to surround and pervade the whole hemisphere; but the darker part, like a sort of cloud, discolors the moon's surface and makes it appear covered with spots. Now these spots, as they are somewhat dark and fairly large, are plain to everyone, and every age has seen them. Accordingly I shall call them *great* or *ancient* spots, to distinguish them from other spots, smaller in size, but so thickly scattered that they sprinkle the whole surface of the moon, especially the brighter portion of it. These spots have never been observed by anyone before me; and from my repeated observations I have arrived at the following conclusion: that we undoubtedly do not perceive the surface of the moon to be perfectly smooth, free from inequalities and exactly spherical (as a large school of philosophers believes concerning both the moon and other heavenly bodies), but on the contrary to be full of inequalities, uneven, full of hollows and protuberances. It is like the surface of the earth itself, which is everywhere varied with lofty mountains and deep valleys. The appearances from which we may derive these conclusions are as follows.

On the fourth or fifth day after new moon, when the moon presents herself to us with bright horns, the boundary which divides the part in shadow from the enlightened part does not appear as a consistent curve, as it would on a

perfectly spherical body, but traces an irregular, uneven, and very wavy line.
. . . Several bright excrescences, so to speak, extend beyond the boundary of
light and shadow into the dark part, and on the other hand bits of shadow
encroach upon the light. Indeed, quite a lot of small blackish spots, quite separated from the dark part, are sprinkled across almost the entire area now
otherwise flooded with sunlight, except only for the region of those great and
ancient spots. I noticed that the small spots just mentioned consistently display this characteristic: they always have the dark part towards the sun,
while on the side away from the sun they have brighter boundaries like shining ridges. Indeed on earth about sunrise we have a very similar phenomenon: we behold the valleys not yet flooded with light while the mountains
surrounding them on the side opposite to the sun are already ablaze with the
splendor of his beams. And just as the shadows in the hollows of the earth diminish in size as the sun rises higher, so also these spots on the moon lose
their blackness as the illuminated part grows larger and larger. Again, not
only are the boundaries of light and shadow in the moon seen to be uneven
and sinuous, but—even more astoundingly—there appear many bright
points within the darkened portion of the moon quite divided and broken off
from the illuminated region and located no small distance away. Gradually,
as time passes, these grow larger and brighter, and after an hour or two are
joined on to the rest of the bright part, which is now somewhat larger. In the
meantime others within the shaded part, one here and another there, sprout
up and are ignited, increase in size, and at last are joined up with the rest of
the luminous surface, which has now spread even further. . . .

And is it not the case on earth before sunrise, that while the level plain is
still in shadow, the peaks of the most lofty mountains are illuminated by the
sun's rays? Then, a little later, does not the light spread further, while the
middle and larger parts of those mountains are gradually illuminated; and finally, when the sun has risen, do not the illuminated parts of the plains and
hills join together? The grandeur, however, of such prominences and depres-
sions in the moon seems to surpass both in magnitude and extent the rugged-
ness of the earth's surface, as I hope to demonstrate later. However, let me
mention here what a remarkable spectacle I observed while the moon was
rapidly approaching her first quarter. . . . A large bulge of the shadow pro-
truded into the illuminated part near the lower horn of the crescent. As I
studied this indentation longer, I observed that it was dark throughout. Finally, after about two hours, a bright peak began to arise a little below the
middle of the depression. This gradually increased and formed a triangular
shape but was still quite detached and separate from the illuminated surface.
Soon, three other small bright points emerged round about it, until, when the
moon was just about to set, that triangular shape, having now extended and

widened, became connected with the rest of the illuminated part and, still surrounded with the three bright peaks, suddenly burst into the dark bulge like a vast promontory of light.

Galileo's observations of the play of light upon the surface of the moon reveal not only a great deal about the nature of that surface but also something of the reciprocal relationship between moon and earth. The importance of this finding can hardly be overstated for a world previously thought to be qualitatively set apart—virtually quarantined—within the "sublunary sphere." As expounded by Thomas Edgerton in the next chapter, Galileo's demonstration of "earthshine" established that heavenly influence could be a two-way street. Galileo states explicitly that he introduces the discussion of earthshine "chiefly in order that the connection and resemblance between the moon and the earth may appear more plainly." Poetically and tellingly, he describes the phenomenon as constituting a kind of commerce.

The earth, with fair and grateful exchange, pays back to the moon an illumination like that which it receives from the moon nearly the whole time during the darkest gloom of night. Let me explain the matter more clearly. At new moon, when the moon occupies a position between the sun and the earth, the moon is illuminated by the sun's rays on the hemisphere facing the sun but turned away from the earth. The other hemisphere, which faces the earth, is covered with darkness and thus in no way illumines the earth's surface. As the moon slightly recedes from the sun, she is at once partly illumined on the half facing us. She turns towards us a slender silvery crescent, and slightly illumines the earth. The sun's illumination increases upon the moon as she approaches her first quarter, and the reflection of that light increases on the earth. The brightness in the moon then extends beyond the semicircle, and our nights grow brighter. Finally, the entire surface of the moon facing the earth is irradiated with the most intense brightness by the sun. This happens when the sun and moon are on opposite sides of the earth, and far and wide the surface of the earth shines with the flood of moonlight. Afterward, when the moon is waning, it sends out less powerful beams and so the earth is illumined less powerfully. Finally, the moon draws near her first position of conjunction with the sun and again black night invades the earth.

Thus in its cycle each month, the moon gives us alternations of brighter and fainter illumination. But the benefit of her light to the earth is balanced and repaid by the benefit of the light of the earth to her. For while the moon approaches the sun about the time of the new moon, she has in front of her the entire surface of that hemisphere of the earth which is exposed to the sun and vividly illumined with his beams, and so receives light reflected from the earth. Because of this reflection, the hemisphere of the moon nearer to us, though deprived of sunlight, appears of considerable brightness. When the moon is removed from the sun through a quadrant, she sees only one half of the earth's hemisphere illuminated, namely, the western half, for the other, the eastern, is covered with the shades of night. The moon is therefore less brightly enlightened by the earth, and accordingly that secondary light appears fainter to us. But if you imagine the moon to be positioned on the opposite side of the earth to the sun, she will see the hemisphere of the earth, now between the moon and the sun, quite dark, and steeped in the gloom of night. If, therefore, while the moon is in this position an eclipse should occur, she will receive no light at all, being altogether deprived of the illumination of the sun or earth. In any other position with respect to the earth and the sun, the moon receives more or less light by reflection from the earth in proportion to how much or how little of the earthly hemisphere illuminated by the sun she beholds. This is the law observed between these two orbs: whenever the earth is most brightly enlightened by the moon, that is when the moon is least enlightened by the earth, and vice versa.

That is all I need say for now on this subject, which I will consider more fully in my *System of the Universe*, where many arguments and experimental proofs will be provided to demonstrate a very strong reflection of the sun's light from the earth—this for the benefit of those who assert, principally on the grounds that it has neither motion nor light, that the earth must be excluded from the dance of the stars. For I will prove that the earth does have motion, that it surpasses the moon in brightness, and that it is not the sump where the universe's filth and ephemera collect.

This last paragraph is highly instructive in view of the persistent myth that ancient and medieval geocentrism placed the earth and humankind in a position of supreme importance in the universe. Galileo's comment indicates exactly the contrary: that the center, as a place where heavy, gross things settle, was seen as a place of disrepute. There is much evidence in the writings of Galileo and of Kepler (see Chapter 26) that their version of heliocentrism was in fact motivated by a desire to reconstruct the place of humankind as a position of prominence within the universe.

Galileo next moves from the moon to the stars; indeed, at a stroke he lays stellar astronomy's observational foundation and takes the first steps towards answers to some of its central questions: the nature and number of the stars, the nature of the Milky Way, and the composition of nebulae.

In what follows it is useful to remember that stellar magnitude, though often used loosely as a unit of apparent stellar size, is technically a measure of brightness, and that the smaller the number, the brighter the star. Thus first magnitude is brightest, down to sixth magnitude, which is the dimmest degree of brightness visible to the naked eye. Galileo extends this scale further downwards.

Now I will briefly announce the phenomena which I have observed so far concerning fixed stars. First, it is worth noting the following: the stars, fixed as well as erratic [i.e., the planets], when seen through a telescope, do not appear enlarged in the same proportion as are other objects, including the moon. On the contrary, the stars appear much less magnified. Accordingly, if for example a telescope magnifies other objects a hundred times, you will find it magnifies the stars scarcely four or five times. But the reason for this is that, when stars are viewed with natural sight they do not, as regards their size, present themselves to us so to speak naked and unadorned but radiant with a certain splendor and fringed with a sparkling aura, especially as night wears on. Thus they seem much larger than they would if stripped of these inessential fringes, for apparent size within one's field of vision is determined not by the actual body of a star but by the luster with which it is circumfused.

Perhaps you will grasp this most readily from a well-known phenomenon: stars emerging at sunset in the first coming on of twilight, even stars of the first magnitude, appear very small. Indeed, even Venus, whenever she is visible in broad daylight, seems so small as scarcely to equal a little star of the least magnitude. It is different for other objects, including the moon, which always appears the same size whether viewed in noonday brightness or in darkest night. Thus stars seen at midnight in uncurtailed glory can be shorn of their fringes by the light of day—indeed not only by the light but also by any little cloud which comes between a star and the eye of an observer. A dark veil or colored glass has the same effect: placed between eye and object, these banish from the stars their radiant halos. And a telescope does the same thing, removing from the stars their inessential and extraneous splendors but then enlarging their actual spheres (if this is truly their shape). The result is that the stars seem proportionately less magnified than other objects, with one of the fifth or sixth magnitude appearing through a telescope merely like one of first magnitude. . . .

But below stars of the sixth magnitude you will see through the telescope a host of other stars that escape natural sight—so many that it is almost unbelievable—more than six other degrees of magnitude, and the largest of them, which we can call seventh magnitude stars, or first magnitude invisible stars, appearing through the telescope larger and brighter than second magnitude stars seen with the naked eye. . . .

The next thing I observed is the essence, or substance, of the Milky Way. With a telescope this can be perceived so palpably that all the disputes that have tormented philosophers for so many centuries are quashed by sheer ocular proof, and we are released from all those wordy arguments. For the galaxy is nothing else but a mass of innumerable stars planted together in clusters. Point your telescope in any direction within the galaxy and at once a great mass of stars comes into view. Of these, many are fairly large and robustly conspicuous, but the number of small ones is utterly unfathomable.

Now this milky brightness, like that of whitish clouds, is seen not only in the Milky Way; several disks of a similar color shine faintly here and there throughout the ether. Turn your telescope upon any one of them and what you will discover is a cluster of stars packed close together. Furthermore—and even more amazingly—the stars which absolutely all astronomers until now have called *nebulous* are swarms of small stars astonishingly packed together. Although on account of smallness or immense distance from us each such star eludes our sight, nevertheless from the commingling of their rays there arises that brightness which until now was thought to be the denser part of the heavens capable of reflecting rays from the stars or the sun.

Although at a stroke Galileo had changed forever how human beings would view the moon, the stars, and the galaxy, he had still to serve up what he regarded his pièce de résistance: the discovery of the moons of Jupiter, "the Medicean Stars."

I have yet to present what I consider most important in this undertaking, namely the announcement and exposition of my discovery and observation of four *planets* never seen from the beginning of the world to the present age, together with their positions and the notes I have made on them over the past two months. . . .

On January 7th of this year, 1610, in the first hour after midnight, while I was viewing the celestial constellations through a telescope, Jupiter appeared before my sight. Because I had made for myself an exceptional instrument, I noticed . . . that three little stars, small but very bright, were near the planet. Although I believed them to belong to the number of the fixed stars, they did make me wonder, for they seemed arranged exactly in a straight line parallel to the ecliptic. They were brighter than other stars of the same size. And their positions relative to one another and to Jupiter was as follows:

On the east side there were two stars, and a single one towards the west. The star which was furthest towards the east, and the western star, appeared rather larger than the third.

I was not at all bothered about the distance between them and Jupiter, for as already mentioned I started out thinking they were fixed stars. However, when on January 8th I returned to the same observation, led by I know not what fate, I discovered a very different state of affairs: three little stars all of them west of Jupiter, nearer together than on the previous night, and spaced at equal intervals. . . .

At this point, although I had not given any thought to the mutual configuration of these stars, yet it piqued my curiosity how Jupiter could one day appear to the east of all those fixed stars when the day before it had been to the west of two of them. And I started to worry that perhaps the planet, violating the predictions of astronomers, passed those stars by means of its own motion. Therefore I waited for the next night with intense longing, but my hopes were frustrated, for the sky was covered with clouds in every direction.

But on January 10th the stars appeared in the following position relative to Jupiter: there were only two, both of them on the east side of Jupiter, with the third, I thought, hidden behind. As before, they were exactly in the same straight line with Jupiter and along the Zodiac.

Given these observations, and knowing that such changes of position could not be explained by reference to Jupiter—and also recognizing that the stars I saw were consistently the same ones, for there were no others in Jupiter's path in front or behind for a great distance in the Zodiac—finally moving from doubt to amazement, I saw that the change of positions was attributable not to Jupiter but to those stars. And therefore I reckoned they had better be observed more accurately and attentively.

Thus on January 11th I saw an arrangement as follows: only two stars to the east of Jupiter, the nearer of which was three times as far from the planet as from the star that stood further to the east. And the star furthest to the east was nearly twice as large as the other one, whereas on the previous night they had appeared nearly the same size. I therefore concluded beyond doubt that there are three stars in the heavens circling about Jupiter as do Venus and Mercury around the sun. Afterwards, many subsequent observations showed the same thing as plain as day, and also that there are not only three, but four wandering stars performing revolutions around Jupiter.

Galileo's discovery, as he well knew, was no mere curiosity, and certainly more than just an opportunity for an ambitious but struggling academic approaching middle age to flatter the equivalent of his granting agency, the Medici family. Its greatest significance as far as cosmology is concerned was that it observationally confirmed what seemed one of the most incredible tenets of Copernicanism: in Copernicus's words, that "there is more than one center." In short, it demonstrated an orbit upon an orbit. In Galileo's words,

"No one can doubt that [the Medicean stars] make their revolutions about [Jupiter] while at the same time together completing twelve-year orbits about the center of the universe."

Besides, we have a notable and splendid argument with which to remove the scruples of people who serenely tolerate the revolution of the planets round the sun in the Copernican system, yet are so perturbed by the motion of one moon about the earth while the two together accomplish an annual orbit about the sun that they think this theory of the universe's constitution must be rejected as impossible. For now we have not just one planet revolving about another while both traverse a vast orbit about the sun, but we actually see four satellites circling about Jupiter, like the moon about the earth, while the whole system makes a great orbital journey about the sun.

So long as it is not read as diminishing the immense beauty and wonder of Galileo's achievement, Agnes Clerke's comment remains a worthy if incomplete assessment: "The problem of the heavens, stripped . . . of metaphysical obscurities, was laid bare to the reason as one of pure mechanics: The planets came to be treated as ordinary projectiles, and distinct reasoning about the nature of their paths was rendered possible. Newton's great task was thus prepared and defined by Galileo."

SOURCES: Galileo Galilei, *Sidereus Nuncius*, Venice, 1610; translation adapted from *The Sidereal Messenger of Galileo Galilei*, trans. E. S. Carlos, London, 1880; Agnes M. Clerke et al., *The Concise Knowledge Astronomy*, London, 1898.

25

Galileo and the Geometrization of Astronomical Space

Samuel Edgerton

While no one doubts that science influences art, historian Samuel Edgerton (b. 1926) presents a complementary picture of Galileo and late Renaissance Italy in which influence flows the other way—in which a particular technique of painting provides the foundation for a monumental advance in astronomical knowledge.

Everyone knows about Galileo's extraordinary contributions to astronomy—his discoveries, for instance, of the earthlike topography of the moon, the moonlike phases of Venus, and the four satellites of Jupiter—but few historians, even modern Galileo scholars, have paid serious heed to the famous Florentine's interest in the fine arts. As his contemporaries often remarked, he knew something of painting and was particularly skilled in the specialized Florentine practice of *disegno*. This activity was much more than just a casual pastime for Galileo. It contributed crucially to at least one of his revolutionary astronomical discoveries: the true physical appearance of the surface of the moon. . . .

By the sixteenth century the study of linear perspective in general and of chiaroscuro in particular appealed not only to artists but even more to professional scholars, especially in Italy and Germany, who otherwise had no interest in the visual arts. Highly technical perspective books were published with this audience in mind.

Edgerton summarizes the published material on perspective available to Galileo and provides examples of drawings from, for example, Daniel

Barbaro's Pratica della perspettiva *(1568), which includes the illustration shown above.*

In sum, how could Galileo, lover of geometry and living in the most competitive art center of western Europe, have missed these chiaroscuro spheroid exercises that so challenged the mind's eye?

Let us for a moment take leave of Florence and look in on Jacobean London during the summer of 1609, where we find Galileo's scientific contemporary Thomas Harriot (1560–1621) turning his attention from mapping the Virginia colonies . . . to a study of the moon; in fact, observing it through a six-power telescope that he managed to procure from its Flemish inventors. Oddly, Harriot's primacy in this matter, preceding Galileo as he did by some six months, goes unmentioned in most modern astronomy textbooks. Harriot even made an extant drawing of the moon as seen through his "perspective tube" (as the English called the new device). Unfortunately, he added no explanation save the Julian date and time of his observation: "1609, July 26, hor. 9 p.m." . . . In any case—and this is why he is so seldom recorded in books on modern astronomy—Harriot's crude sketch reveals nothing new.

Europeans of this time still had no reason to doubt Aristotle's description of the moon as a perfect sphere, the prototypical form of all planets and stars

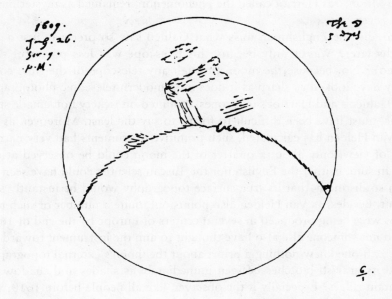

in the cosmos. Christian dogma added to this euphoric image by having the moon symbolize the Virgin's Immaculate Conception. "Pure as the moon" became a commonplace expression for Mary, implying that the universe, like herself, was incorruptible, that God would not have created the moon or any other heavenly body in another shape. Renaissance artists, especially those who served zealous Catholic patrons, frequently depicted the Virgin standing on such as moon, as did Bartolomé Esteban Murillo (1617–1682) as late as the mid-seventeenth century in Spain . . . —[a moon] marbled like translucent alabaster but with a highly polished, utterly smooth surface.

In Thomas Harriot's England, the anti-Aristotelian Francis Bacon had concluded that the lunar body was not solid at all, but rather composed of some unexplained "vapour." Harriot's own opinion about the moon's composition remains unrecorded. Nonetheless, he drew the terminator—that is, the demarcation line between the illuminated and shaded portions of the moon—with short, ragged strokes as if it fell over a roughened surface. On the upper half of the sphere Harriot indicated the configurations of what we now know as the great lunar "seas," the Maria Tranquilitatis, Crisium, and Serenitatis, which do seem to have appeared to him as surface markings rather than internal, vaporous discolorations. Nevertheless, he was unable to recognize the significance of these observations. The telescope only confirmed more or less what the ancients had always said he would see. The "strange spottednesse of the Moon," as Harriot called the phenomenon, remained as mysterious to him as ever.

Why did the Englishman miss what Galileo saw so precisely just a few months later? Was it only because his telescope was less powerful than Galileo's? No, because the moon through any telescope of the time could hardly have looked as sharp as it does in [a modern telescopic photograph]. Both Galileo's and Harriot's telescopes, mounted on rickety homemade stanchions, must have been difficult to focus, to say the least. Moreover, as Albert Van Helden has calculated, such primitive instruments had very narrow fields of view; only about a quarter of the moon could be observed at one time. In sum, neither the English nor the Tuscan scientist could have seen the moon so distinctly that its true surface topography would be instantly self-evident. Besides, as Van Helden also points out, quite a number of such telescopes were being produced in several centers of Europe by the end of 1609. Would not someone else also have thought to aim the instrument toward the sky? . . . If one knew nothing a priori about the moon's external topography, would its grayish blotches be seen immediately as shades and shadows of mountain ridges? Especially if the observer, like all people before 1610, was already certain that such blotches had something to do with the moon's translucent internal composition? . . .

[In Harriot's time] no serious study of geometric perspective . . . existed in England at all. Demand in Britain for perspective training was so slight that no indigenous book on the subject was published until 1635, when John Wells edited a crude manual titled *Sciographia, or The Art of Shadows*, too late of course to have been much use to Harriot.

In the meantime back in Padua, where Galileo was living and teaching, the Tuscan scientist heard nothing of Harriot's lunar observations. In fact, he learned of the recent Flemish invention of the telescope only in May 1609. Immediately he sent for instructions. With remarkable ingenuity, not to say alacrity, he applied his considerable perspective experience to the optical problems and managed by the end of the year to build a number of the instruments with magnification improved to twenty power and with the addition even of aperture stops. . . .

Galileo's recordings of the moon's phases date from November and December 1609. Since his observations during these two months could be affirmed only when the moon appeared in partial shadow, his viewing nights were limited to about twenty-four, not all of which would conveniently be free of clouds.

Perhaps Galileo made some illustrations from the beginning, right there on the spot as he stared at the moon from the San Giorgio Maggiore campanile. No such drawings have survived, but we are in possession of seven finished sepia studies, obviously done later but probably based on firsthand ad hoc sketches. These small wash drawings, four of the waxing and three of the

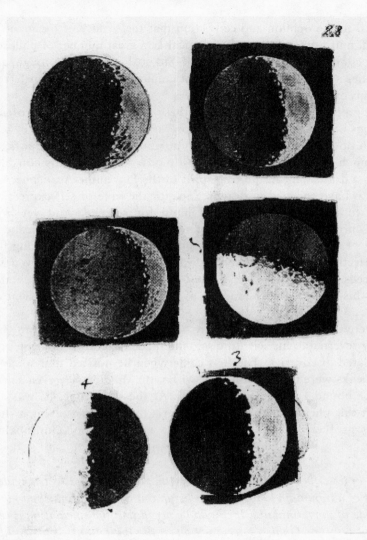

waning moon were certainly done by someone well practiced in the manipulation of ink washes, especially the rendering of chiaroscuro effects. They are by an experienced artist, and we have no reason to believe it was anyone other than Galileo himself. The astronomer no doubt prepared these washes as models for the engraver who would illustrate his book *Sidereus nuncius,* . . . which he rushed to publication in March 1610, barely five months after he began looking at the skies through his telescope.

Only five engravings of the moon's phases were printed in *Sidereus nuncius,* none exactly replicating the wash drawings. . . . Galileo's accompanying matter-of-fact description of these engravings belies both his own excitement and the stupendous impression they made on an unsuspecting world: "I have been led to the opinion and conviction that the surface of the moon is not smooth, uniform, and precisely spherical as a great number of philosophers believe it (and the other heavenly bodies) to be, but is uneven, rough, and full of cavities and prominences, being not unlike the face of the Earth, relieved by chains of mountains and deep valleys." . . .

Is it preposterous to claim that these simple yet highly professional paintings belong as much to the history of art as to the history of science? Though no comparable artwork also attributable to Galileo exists, we do have much contemporary verbal testimony concerning his considerable skill as a draftsman. In the true spirit of the Florentine Academia, Galileo seems to have engaged in drawing not for the sake of self-expression but rather to discipline his eye and hand for science. And yet in these chiaroscuro washes he has anticipated the independent landscape in the history of art. His almost impressionistic technique for rendering fleeting light effects reminds us of Constable and Turner, and perhaps even Monet. One needs only to read on in *Sidereus nuncius* to appreciate his wonder, as well as his rational understanding, as he first gazed at the transient moonscape. . . .

Moreover, after [marvelling] at the picturesque lunar terrain, Galileo quickly reverted to his scientific self and made two other amazing perspective-related discoveries. The first came when he noticed that some of the lunar peaks were tipped with light within the shadow side even as the terminator boundary lay a long way off. At the same time, he was able to convert this phenomenon into a geometric diagram for solving a shadow-casting problem such as he may have recalled from Guidobaldo del Monte.

Edgerton describes the simple geometry that Galileo used to triangulate the heights of a mountain whose top peeks up out of the moon's shadow on the dark side of the terminator. It should be noted that the calculation is remarkable not only for Galileo's accuracy given the limitations of his equipment, but also for the very fact that geometry—literally "earth measurement"—was applied extraterrestrially. The implication—an anti-Aristotelian one—is that space is qualitatively the same up there as it is down here. Hence the geometrization of astronomical space.

Since the moon's diameter was known to be two-sevenths of the earth's diameter, or about 2,000 miles, Galileo . . . revealed by Pythagorean calculation that . . . the mountain's height on center from its base reached more than four miles into the lunar sky! By applying a problem well known to students of Renaissance perspective, Galileo added yet another fact to his already wondrous revelations, that the mountains on the moon were more spectacular than the Alps here on earth.

His next observation had to do with what is today referred to as "earthshine," described thus in *Sidereus nuncius*:

> When the Moon is not far from the Sun, just before or after a new Moon, its globe offers itself to view not only on the side where it is adorned with shining horns, but a certain faint light is also seen to mark out the periphery of the dark part which faces away from the Sun, separating this from the aether. Now if we examine the matter more closely, we shall see that not only does the extreme limb of the shaded side glow with this uncertain light, but the entire face of the Moon (including the side which does not receive the glare of the Sun) is whitened by a not inconsiderable gleam. . . . It is then found that this region of the Moon, though deprived of sunlight, also shines not a little. The effect is heightened if the gloom of night has already deepened through departure of the Sun, for in a darker field a given light appears brighter. . . . This remarkable gleam has afforded no small perplexity to philosophers. . . . Some would say it is an inherent and natural light of the Moon's own; others that it is imparted by Venus; others yet, by all the stars together; and still others derive it from the Sun, whose rays they would have permeate the thick solidity of the Moon. But statements of this sort are refuted and their falsity evinced with little difficulty. For if this kind of light were the Moon's own, or were contributed by the stars, the Moon would retain it particularly during eclipses. . . . Now since the secondary light does not inherently belong to the Moon, and is not received from any star or from the Sun, and since in the whole universe there is no other body left but the Earth, what must we conclude? What is to be proposed? Surely we must assert that the lunar body (or any other dark and sunless orb) is illuminated by the Earth. Yet what is so remarkable about this? The Earth, in fair and grateful exchange, pays back to the Moon an illumination similar to that which it receives from her throughout nearly all the darkest gloom of night.

How was Galileo able to make such a discovery? What led him to raise this issue in the first place? The fact is, as any seventeenth-century Florentine connoisseur of art would have known, the ability to depict reflected light was one of the outstanding achievements of Renaissance painting. . . . While growing up in Tuscany, the young scientist may have seen many unforgettable examples. . . . Moreover, Galileo, through association with Cigoli and the Florentine Accademia del Disegno, is likely to have known the relevant instructions in Leon Battista Alberti's treatise *On Painting*, available since 1568 in a popular Italian-language edition:

> A shadow is made when rays of light are intercepted. Rays that are intercepted are either reflected elsewhere or return upon themselves. They are reflected, for instance, when they rebound off the surface of water onto the ceiling; as mathematicians prove, reflection of rays always takes place at equal angles. . . . Reflected rays assume the color they find on the surface from which they are reflected. We see this happen when the faces of people walking about in the meadows appear to have a greenish tinge.

Any would-be artist since the quattrocento had to learn to draw this optical phenomenon just as Alberti described it—but of course only in relation to terrestrial experience. . . . By   applying the same painterly logic to the moon, Galileo discovered what had eluded professional astronomers for centuries.

SOURCE: Samuel Y. Edgerton, Jr., *The Heritage of Giotto's Geometry: Art and Science on the Eve of the Scientific Revolution*, Ithaca: Cornell UP, 1991.

26

This Boat Which Is Our Earth

Johannes Kepler

Kepler (1571–1630) and Galileo form a delightful diptych: the German and the Italian, the profound theorizer and the penetrating observer, the two greatest among second-generation Copernicans—and, inevitably, rivals. Moreover, we find Kepler at perhaps his most charming, exasperating, and revealing even as, in 1610, he confronts the stunning discoveries Galileo has just published in his little book Sidereus Nuncius. *Kepler asks himself, it seems, "Why didn't I think of that?"—and answers, in most cases, that he had thought of it.*

In 1609, Kepler had published his Astronomia Nova (The New Astronomy), *most famous for his first two laws of planetary motion. Entertaining an idea that even Copernicus had not considered, namely, that the orbits of the planets might be other than circular, Kepler explained those orbits as ellipses (his "first law") which nevertheless displayed regularity: The changing velocity of a planet within its elliptical orbit renders areas of an ellipse "swept out" in equal intervals of time equal. (Picture an elliptical pizza whose wedge-shaped slices are all actually equal in weight and calories, even though the width of each piece at the wide end varies; this is the "second law.") One of the assumptions of these laws is that the planets are physical earth-like objects—a claim Galileo in 1610 makes regarding the moon. One may perhaps be more patient in reading Kepler if one reflects on the fact that Kepler at least acknowledges Galileo's accomplishments even when Galileo does not return the favor. Kepler even salutes Galileo's writing ability.*

I may perhaps seem rash in accepting your claims so readily with no support from my own experience. But why should I not believe a most learned mathematician, whose very style attests the soundness of his judgment? He has no intention of practicing deception in a bid for vulgar publicity, nor does he pretend to have seen what he has not seen. Because he loves the truth, he does not hesitate to oppose even the most familiar opinions, and to bear the jeers of the crowd with equanimity. Does he not make his writings public, and could he possibly hide any villainy that might be perpetrated? Shall I disparage him, a gentleman of Florence, for the things he has seen? Shall I with my poor vision disparage him with his keen sight?

Kepler recognizes at once that the success of Galileo's observations has implications for the very nature of space or its contents. He explains that, though he had the theoretical knowledge of optics necessary for the invention of the telescope, he did not push ahead with the project because he assumed that the cumulative opacity of the "aether" would prevent accurate vision at enormous distances.

I believed that the air is dense and blue in color, so that the minute parts of visible things at a distance are obscured and distorted. Since this proposition is intrinsically certain, it was vain, I understood, to hope that a lens would remove this substance of the intervening air from visible things. Also with regard to the celestial essence, I surmised some such property as could prevent us, supposing that we enormously magnified the body of the moon to immense proportions, from being able to differentiate its tiny particles in their purity from the lowest celestial matter.

For these reasons, reinforced by other obstacles, I refrained from attempting to construct the device.

But now, most accomplished Galileo, you deserve my praise for your tireless energy. Putting aside all misgivings, you turned directly to visual experimentation. And indeed by your discoveries you caused the sun of truth to rise, you routed all the ghosts of perplexity together with their mother, the night, and by your achievement you showed what could be done.

Under your guidance I recognize that the celestial substance is incredibly tenuous. To be sure, this property is made known on page 127 of my "Optics." If the relative densities of air and water are compared with the relative densities of the aether and air, the latter ratio undoubtedly shows a much greater disparity. As a result, not even the tiniest particle of the sphere of the stars (still less of the body of the moon, which is the lowest of the heavenly bodies) escapes our eyes, when they are aided by your instrument. A single fragment of the lens interposes much more matter (or opacity) between the eye and the object viewed than does the entire vast region of the aether. For a slight indistinctness arises from the lens, but from the aether none at all. Hence we must virtually concede, it seems, that that whole immense space is a vacuum.

Like most scientists who make major breakthroughs, Galileo and Kepler are naturally best known as thinkers who are forward-looking and before their time. However, even aspects of their discussion that strike us as antiquated are engaging and imaginative, as in the following section in which Kepler acknowledges the role of Plutarch in the ongoing discussion on the nature of the moon, and divulges his belief that somehow nature had sought out both him and Galileo as conduits of astronomical revelation.

What shall I say now [Galileo] about your very acute analysis of the ancient spots on the moon? On page 251 of my book I cited the opinion of Plutarch, who regarded those ancient spots on the moon as lakes or seas, and the bright areas as continents. I did not hesitate to oppose him and to reverse his interpretation, by attributing the spots to continents, and the purity of the bright region to the effects of a liquid. Wackher used to give strong approval to my stand on this question. We were deeply engaged in these discussions last summer (I suppose, because nature was seeking the same results through us as it achieved a little later through Galileo). To please Wackher, I even founded a new astronomy for the inhabitants of the moon, as it were; in plain language, a sort of lunar geography. Among its basic propositions was this thesis, that the spots are continents, while the bright areas are seas. . . . Suppose that the moon, like the island of Crete, is composed of a white soil (as Lucian said that the moon is a cheese-like land). We shall have to admit that the soil shines by sunlight more vividly than the seas, however little they may be tinged with black.

My book, consequently, does not prevent me from agreeing with you, as you adduce mathematical arguments against me in favor of Plutarch with brilliant and irrefutable logic. Certainly the bright areas are broken up by many cavities; the bright areas are bounded by an irregular line; the bright areas contain great peaks, on account of which they light up sooner than the neighboring region. Where they face the sun, they are bright; where they face away from the sun, they are dark. All these characteristics suit a dry, solid, and high material, but not a fluid. On the other hand, the dark spots, known since antiquity, are flat. The dark spots light up later—a fact which proves their low elevation—when the surrounding peaks are already aglow far and wide. When the dark spots are illumined, a certain shadow-like black effect differentiates them from the peaks. The boundary of the illumination in the dark area is a straight line at half-moon. These characteristics, in turn, belong to a liquid, which seeks the lowest levels and on account of its weight settles in a horizontal position.

By these arguments, I say, you have proved your point completely. I admit that the spots are seas, I admit that the bright areas are land.

Kepler seems irresistibly drawn beyond observation to speculation—speculation that nurtured what later generations would call science fiction.

I cannot help wondering about the meaning of that large circular cavity in what I usually call the left corner of the mouth. Is it a work of nature, or of a trained hand? Suppose that there are living beings on the moon (following in the footsteps of Pythagoras and Plutarch, I enjoyed toying with this idea, long ago in a disputation written at Tübingen in the year 1593, later on in my "Optics" on page 250, and most recently in my aforementioned lunar geography). It surely stands to reason that the inhabitants express the character of their dwelling place, which has much bigger mountains and valleys than our earth has. Consequently, being endowed with very massive bodies, they also construct gigantic projects. Their day is as long as 15 of our days, and they feel insufferable heat. Perhaps they lack stone for erecting shelters against the sun. On the other hand, maybe they have a soil as sticky as clay. Their usual building plan, accordingly, is as follows. Digging up huge fields, they carry out the earth and heap it in a circle, perhaps for the purpose of drawing out the moisture down below. In this way they may hide deep in the shade behind their excavated mounds and, in keeping with the sun's motion, shift about inside, clinging to the shadow. They have, as it were, a sort of underground city. They make their homes in numerous caves hewn out of that circular embankment. They place their fields and pastures in the middle, to avoid being forced to go too far away from their farms in their flight from the sun.

But let us follow the thread of your discourse still further. You ask why the moon's outermost circle does not also appear irregular. I do not know how carefully you have thought about this subject, or whether your query, as is more likely, is based on the popular impression. For in my book . . . I stated that there was surely some imperfection in this outermost circle during full moon. Study the matter, and once again tell us how it looks to you, for I shall have confidence in your telescopes.

Assuming the fact to be established, you answer the question in two ways. The first way is not incompatible with my findings. For the multitude of peaks, crowded one behind another, presents the appearance of a perfect circle at the outermost limb of the visible hemisphere. This can happen only if the peaks have been smoothed and polished on a lathe so that any tiny crevices or bumps fail to show up. This situation would be consistent with my observations.

Your second way of answering the question is to wrap a sphere of air around the moon. Where this sphere curves back to the recesses of the lunar globe, it presents some depth to the rays from the sun and the earth, and thus to our eyes also. Hence the limb gleams pure and spotless, while the entire interior of the face, where this air does not obstruct our vision so deeply, abounds with numerous spots.

Pages 252 and 302 of my book could have told you about this air on the moon. These passages in my book are splendidly confirmed by your pertinent observations.

When he turns to Galileo's findings regarding the number of the stars and the telescopic appearance of their light, Kepler reveals how profoundly anthropocentric his Copernicanism is and how radically his conception of the sun differs from more modern (and in Bruno's case, earlier) views of the sun as one among countless stars.

Your second highly welcome observation concerns the sparkling appearance of the fixed stars, in contrast with the circular appearance of the planets. What other conclusion shall we draw from this difference, Galileo, than that the fixed stars generate their light from within, whereas the planets, being opaque, are illuminated from without; that is, to use Bruno's terms, the former are suns, the latter, moons or earths?

Nevertheless, let him not lead us on to his belief in infinite worlds, as numerous as the fixed stars and all similar to our own. Your third observation comes to our support: the countless host of fixed stars exceeds what was known in antiquity. You do not hesitate to declare that there are visible over 10,000 stars. The more there are, and the more crowded they are, the stronger becomes my argument against the infinity of the universe, as set forth in my book on the "New Star." . . . This argument proved that where we mortals dwell, in the company of the sun and the planets, is the primary bosom of the universe; from none of the fixed stars can such a view of the universe be obtained as is possible from our earth or even from the sun. For the sake of brevity, I forbear to summarize the passage. Whoever reads it in its entirety will be inclined to assent.

Let me add this consideration to buttress my case. To my weak eyes, any of the larger stars, such as Sirius, if I take its flashing rays into account, seems to be only a little smaller than the diameter of the moon. But persons with unimpaired vision, using astronomical instruments that are not deceived by these wavy crowns, as is the naked eye, ascertain the dimensions of the stars' diameters in terms of minutes and fractions of minutes. Suppose that we took only 1000 fixed stars, none of them larger than one minute. . . . If these were all merged in a single round surface, they would equal (and even surpass) the diameter of the sun. If the little disks of 10,000 stars are fused into one, how much more will their visible size exceed the apparent disk of the sun? If this is true, and if they are suns having the same nature as our sun, why do not these suns collectively outdistance our sun in brilliance? Why do they all together transmit so dim a light to the most accessible places? . . . Will my opponent tell me that the stars are very far away from us? This does not help his cause at all. For the greater their distance, the more does every single one of them outstrip the sun in diameter. But maybe the intervening aether obscures them? Not in the least. For we see them with their sparkling, with their various shapes and colors. This could not happen if the density of the aether offered any obstacle.

Hence it is quite clear that the body of our sun is brighter beyond measure than all the fixed stars together, and therefore this world of ours does not belong to an undifferentiated swarm of countless others.

Kepler goes on in his slightly deflating way to compliment Galileo on his resolution of the Milky Way, but from this discovery he draws an implication that betrays his still mystical and essentially medieval notion of the immutability of the fixed stars.

You have conferred a blessing on astronomers and physicists by revealing the true character of the Milky Way, the nebulae, and the nebulous spirals. You have upheld those writers who long ago reached the same conclusion as you: they are nothing but a mass of stars, whose luminosities blend on account of the dullness of our eyes.

Accordingly, scientists will henceforth cease to create comets and new stars out of the Milky Way, after the manner of Brahe, lest they irrationally assert the passing away of perfect and eternal celestial bodies.

When he turns to Galileo's visual discovery of the planets of Jupiter, Kepler gives himself credit for having predicted the same thing at the theoretical level. More interesting than this specific claim, however, is Kepler's engagement of the fact—and the ongoing mystery—that often, in science, human beings have indeed achieved a priori knowledge or conceptions of things that have only later been proven experimentally.

Finally, I move on with you to the new planets, the most wonderful topic in your little book. . . . I rejoice that I am to some extent restored to life by your work. If you had discovered any planets revolving around one of the fixed stars, there would now be waiting for me chains and a prison amid Bruno's innumerabilities, I should rather say, exile to his infinite space. Therefore, by reporting that these four planets revolve, not around one of the fixed stars, but around the planet Jupiter, you have for the present freed me from the great fear which gripped me as soon as I had heard about your book from my opponent's triumphal shout.

Wackher of course had once more been seized by deep admiration of that dreadful philosophy [of Bruno's]. What Galileo recently saw with his own eyes, I had many years before not only proposed as a surmise, but thoroughly established by reasoning. It is doubtless with perfect justice that those men attain fame whose intellect anticipates the senses in closely related branches of philosophy. Theoretical astronomy, at a time when it had never set foot outside Greece, nevertheless disclosed the characteristics of the Arctic Zone. . . . Who does not honor Plato's myth of Atlantis, Plutarch's legend of the gold-colored islands beyond Thule, and Seneca's prophetic verses about the forthcoming discovery of a New World, now that the evidence for such a place has finally been furnished by that Argonaut from Florence [Vespucci]? Columbus himself keeps his readers uncertain whether to admire his intellect in divining the New World from the direction of the winds, more than his courage in facing unknown seas and the boundless ocean, and his good luck in gaining his objective. . . . Surely those thinkers who intellectually grasp the causes of phenomena, before these are revealed to the senses, resemble the Creator more closely than the others, who speculate about the causes after the phenomena have been seen.

In this way Kepler at once compliments Galileo and puts him in his place. And, as he did in connection with his discussion of the moon, Kepler unleashes his imagination with regard to extraterrestrial inhabitants. In so doing, he reveals much concerning his view of the teleology of the cosmos and of humans' unfolding role within it. He begins with what was to become the persistent analogy between voyages of discovery and colonization to America and those to other "new worlds" in space.

I cannot refrain from contributing this additional feature to the unorthodox aspects of your findings. It is not improbable, I must point out, that there are inhabitants not only on the moon but on Jupiter too or . . . that those areas are now being unveiled for the first time. But as soon as somebody demonstrates the art of flying, settlers from our species of man will not be lacking. Who would once have thought that the crossing of the wide ocean was calmer and safer than of the narrow Adriatic Sea, Baltic Sea, or English Channel? Given ships or sails adapted to the breezes of heaven, there will be those who will not shrink from even that vast expanse. Therefore, for the sake of those who, as it were, will presently be on hand to attempt this voyage, let us establish the astronomy, Galileo, you of Jupiter, and me of the moon.

Let the foregoing pleasantries be inserted on account of the miracle of human courage, which is evident in the men of the present age especially. For the revered mysteries of sacred history are not a laughing matter for me.

I have also thought it worth while, in passing, to tweak the ear of the higher philosophy. Let it ponder the questions whether the almighty and provident Guardian of the human race permits anything useless and why, like an experienced steward, he opens the inner chambers of his building to us at this particular time. Such was the opinion put forward by my good friend Thomas Seget, a man of wide learning. Or does God the creator, as I replied, lead mankind, like some growing youngster gradually approaching maturity, step by step from one stage of knowledge to another? (For example, there was a period when the distinction between the planets and the fixed stars was unknown; it was quite some time before Pythagoras or Parmenides perceived that the evening star and the morning star are the same body; the planets are not mentioned in Moses, Job, or the Psalms). Let the higher philosophy reflect, I repeat, and glance backward to some extent. How far has the knowledge of nature progressed, how much is left, and what may the men of the future expect?

But let us return to humbler thoughts, and finish what we began. There are in fact four planets revolving around Jupiter at different distances with unequal periods. For whose sake, the question arises, if there are no people on Jupiter to behold this wonderfully varied display with their own eyes? For, as far as we on the earth are concerned, I do not know by what arguments I may be persuaded to believe that these planets minister chiefly to us, who never see them. We should not anticipate that all of us, equipped with your telescopes, Galileo, will observe them hereafter as a matter of course. . . . It becomes evident that these four new planets were ordained not primarily for us who live on the earth but undoubtedly for the Jovian beings who dwell around Jupiter.

With this daring suggestion Kepler's imagination opens onto a still more arresting scene in which inhabitants of spaceship earth—"this boat, which is our earth"—exercise their planetary patriotism by asserting the superiority of their location within the universe.

Well, then, someone may say, if there are globes in the heaven similar to our earth, do we vie with them over who occupies the better portion of the universe? For if their globes are nobler, we are not the noblest of rational creatures. Then how can all things be for man's sake? How can we be the masters of God's handiwork?

It is difficult to unravel this knot, because we have not yet acquired all the relevant information. We shall hardly escape being labeled foolish if we expatiate at length on this subject.

Yet I shall not pass over in silence those philosophical arguments which, it seems to me, can be brought to bear. They will establish not merely in general . . . that this system of planets, on one of which we humans dwell, is located in the very bosom of the world, around the heart of the universe, that is, the sun. These arguments will also establish in particular that we humans live on the globe, which by right belongs to the primary rational creature, the noblest of the (corporeal) creatures.

In support of the former proposition concerning the inmost bosom of the world, . . . the evidence . . . was based, first, on the fixed stars, which by their vast numbers truly enclose this area like a wall and, secondly, on our sun, which is more splendid than the fixed stars. . . .

Let us now also indicate why the earth surpasses Jupiter and better deserves to be the abode of the predominant creature.

In the center of the world is the sun, heart of the universe, fountain of light, source of heat, origin of life and cosmic motion. But it seems that man ought quietly to shun that royal throne. Heaven was assigned to the lord of heaven, the sun of righteousness, but earth, to the children of man. God has no body, of course, and requires no dwelling place. Yet more of the force which rules the world is revealed in the sun . . . than in all the other globes. Because man's house is otherwise, therefore, let him recognize his own wretchedness and the opulence of God. Let him acknowledge that it is not the source and origin of the world's splendor, but that he is dependent on the true source and origin thereof. Moreover, as I said in the "Optics," in the interests of that contemplation for which man was created, and adorned and equipped with eyes, he could not remain at rest in the center. On the contrary, he must make an annual journey on this boat, which is our earth, to perform his observations. So surveyors, in measuring inaccessible objects, move from place to place for the purpose of obtaining from the distance between their positions an accurate base line for the triangulation.

After the sun, however, there is no globe nobler or more suitable for man than the earth. For, in the first place, it is exactly in the middle of the principal globes (if we exclude, as we should, Jupiter's satellites and the moon revolving around the earth). Above it are Mars, Jupiter, and Saturn. Within the embrace of its orbit run Venus and Mercury, while at the center the sun rotates, instigator of all the motions, truly an Apollo, the term frequently used by Bruno.

Thus Kepler redefines centrality itself in a most dynamic manner based on the idea that geometrical comprehension, triangulation, requires variation of place. Earlier Kepler had commented that "geometry . . . shines in the mind of God. The share of it which has been granted to man is one of the reasons why he is the image of God." To exercise or actualize this image properly, humans must be able to observe the universe from a "central" but changing point of view. On the other hand, Kepler generously theorizes that God—to mitigate interplanetary envy—has granted the Jovians a few extra moons by way of compensation.

We on the earth have difficulty in seeing Mercury, the last of the principal planets, on account of the nearby, overpowering brilliance of the sun. From Jupiter or Saturn, how much less distinct will Mercury be? Hence this globe seems assigned to man with the express intent of enabling him to view all the planets. Will anyone then deny that, to make up for the planets concealed from the Jovians but visible to us earth-dwellers, four others are allocated to Jupiter, to match the four inferior planets, Mars, Earth, Venus, and Mercury, which revolve around the sun within Jupiter's orbit?

    Let the Jovian creatures, therefore, have something with which to console themselves. Let them . . . have their own planets [i.e., their moons]. We humans who inhabit the earth can with good reason (in my view) feel proud of the pre-eminent lodging place of our bodies, and we should be grateful to God the creator.

SOURCE: *Kepler's Conversation with Galileo's Sidereal Messenger*, trans. Edward Rosen, New York and London: Johnson Reprint Corporation, 1965.

27

The Two Books of God Agree with Each Other

Tommaso Campanella

Tommaso Campanella (1568–1639) provides an example of one who was willing to mount a defense of the beleaguered Galileo from within Roman Catholicism. So much has been written about the warfare between science and theology, with Galileo as prime exhibit of victimization by the latter, that it is stimulating to hear arguments arising from scriptural hermeneutics and theology in defense of the scientist. As in most interpretation, the largest issue is the nature of truth itself. But close behind it is that of how the human subject, immersed in the world's contingencies, perceives the truth. Such is the first point made by Tobias Adami, publisher of Campanella's Apologia Pro Galileo.

GREETINGS TO THE BENEVOLENT READER
from the Publisher

For insignificant creatures like us, who live in this world surrounded on all sides like worms in a cheese, it is no small matter to engage in grave disputes about the structure of the world, such as whether our abode or house, which we call the earth, rotates on high around the sun together with the other globes similar to it, or whether the sun rotates around the earth. We are indeed such small creatures that we are very ignorant of such matters. We are like a mouse in a ship who, when asked by a fellow mouse about the ship being at rest on the sea, would never be able to say whether the ship, their common home, is in motion or whether it remains fixed in one and the same place.

As a result many judge these investigations to be more complex than is commonly thought, especially after so many new things have been detected in the celestial globes by means of the optical instrument which the Lycean philosophers in Rome call a telescope. On the other hand, although certain arrogant people, who wish to pass themselves off as philosophers, have seen most of these things, still their spontaneous amazement did not prevent them from turning others away from a more careful investigation of the truth. And indeed many theologians, both Catholic and Protestant, are especially eager to suppress this investigation by appealing to the unchanging authority of the Sacred Scriptures. But whoever loves the truth must give special consideration to what is right or wrong in this matter. Both in our day and in times past, many famous people who were well informed about both profane and sacred studies, beginning with the Pythagoreans, have defended and still defend this view; and they ought not to be accused rashly of either impiety or ignorance.

Campanella's defense of Galileo itself is really a principled theological defense of the kind of investigation Galileo has engaged in, rather than a defense of the rightness of the heliocentric system as such. The work has a quite medieval flavor, with a palpable concern for citation of biblical and patristic authorities even while arguing that truth itself supersedes authority.

I will never be sufficiently astonished at those potbellied theologians who locate the limits of human genius in the writings of Aristotle. The fact that not even Ptolemy reached the truth is shown by the new phenomena which his theory cannot explain, and thus he does not remove disorder from the heavens. I will also pass over the errors which Copernicus introduced into astronomy, for example, that there is a regular motion of a sphere around a center other than its own. . . . But Copernicus . . . has returned to the teachings of the ancient Pythagoreans, which provide a better account of the appearances. In addition to this, Galileo has discovered new planets and new worlds and previously unknown changes in the heavens.

Therefore, anyone is insane and most ignorant to think that an adequate knowledge of the heavens is to be found in Aristotle, who contributed nothing on his own and who encouraged others to investigate such matters. And those who came after him are uncertain and are still fighting with each other.

Galileo, in a letter to the Grand Duchess of Tuscany written a year before Campanella composed his defense, cites Tertullian on the "parallel texts" of Nature and Scripture: "God is known . . . by Nature in his works, and by doctrine in his revealed Word." Upon this foundation Galileo defends him-

self, arguing that the two "books" must be interpreted consistently with each other, so that "having arrived at any certainties in physics, we ought to utilize these as the most appropriate aids in the true interpretation of the Bible" (Discoveries and Opinions of Galileo, ed. Stillman Drake [1957], 183). Similarly, Campanella defends Galileo on the universal grounds that God is the author of both books—of Scripture and of Nature—and that "one truth does not contradict another truth."

Hence human science does not contradict divine science, nor do the works of God contradict God. . . . Therefore, although theology in itself does not need proofs taken from the human sciences, nevertheless, for our sake theology does need to do this so that we can strengthen our convictions by understanding the supernatural in terms of the sensible and the natural. . . .

Now it is clear that the sciences exist in the human race as a whole, and not only in this or that individual person. For God made man to know God, and by knowing to love Him, and by loving to please Him; and for this man has senses and reason. But if the purpose of reason is to attain knowledge, then humans would act contrary to the divine natural order, just like a man who would not wish to use his feet to walk, unless one uses this gift of God according to the divine plan, as Chrysostom has regularly argued. As Aristotle has said, "All men by nature desire to know," and as Moses said in Genesis 1, "God put man in paradise to cultivate and take care of it." But this was not manual labor or the caring for animals. . . . Rather man's work was to know things, and to observe the heavens and the natural world out of curiosity, so that he would as a result investigate everything to meet his obligation to venerate God, which cannot be done without first having knowledge, for "The invisible things of God are known through what He has made," as the Apostle said [Rom. 1:20].

Even if it be granted that all the sciences were infused into Adam, still he lacked experiential knowledge. Further, this command to learn was given to him, not as an individual person but as the head of the human race, and hence it has been also given to us, his descendants, as the Fathers testify. . . .

As a result, from the beginning the world has been called the "Wisdom of God" (as was revealed to St. Brigid) and a "Book" in which we can read about all things. Hence . . . St. Leo says, "We understand the meaning of God's will from these very elements of the world, as from the pages of an open book." . . .

Since the more wonderful and more extraordinary things in the world are better images of God, their author, they should be investigated for this reason with greater care. And by this study divinity is shown to the human soul. Such are the heavens and the stars and the great system of the world. Thus

Anaxagoras has said that man was made to contemplate the heavens. And Ovid was much praised by all the theologians, and especially by Lactantius, for having said of God, "While the other animals look down towards the earth, he created man to face upwards, and he ordered him to see the heavens and to stand erect, turning his gaze to the stars."

David reveals the reason for this when he sings in Psalm 18, "The heavens proclaim the glory of God, and the firmament speaks of the work of his hands," and in Psalm 8, "For I will look at your heavens, the work of your fingers, and at the moon and the stars, which you have made." Moreover Plato . . . proves the dignity and the deification of man and the immortality of his soul from his knowledge of the heavens. . . . Ovid also confirms this when he says to the astronomers, "Yours is a happy lot because your primary role is to know about these matters and to rise up to the celestial houses; you bring the distant stars closer to our eyes, and you subject the heavens to your genius."

These praises belong to Galileo more than to anyone else.

Campanella engages in a balancing act: praising Galileo, but refusing to idolize him or to see him—or anyone else either—as the epitome of knowledge.

The wisdom of God is exceedingly vast and cannot be confined to the genius of any one human. The more it is sought, the more it is found to contain, and we then realize that we know nothing in comparison to the numerous and marvelous things of which we are ignorant. This is the knowledge which Solomon envisioned in Ecclesiastes, and which the Apostle praises, and which Socrates found in himself. Those who think that they know because they know Aristotle or because, like Galileo, they know something new about the world, the book of God, do not know the method required for knowledge. They are not truly wise unless they know that there are many more things of which they are ignorant, and that they should not stop their investigations as if they already knew everything. . . . For what we know is only a glimmer.

Therefore, wisdom is to be read in the immense book of God, which is the world, and there is always more to be discovered. Hence the sacred writers refer us to that book and not to the small books of humans.

Campanella's defense of Galileo from within the faith is motivated not only by high principle but also by raw apprehension of the embarrassment to be suffered should heliocentrism prove true.

For . . . if Galileo wins out, our theologians of the Roman faith will be the cause of a great deal of ridicule among the heretics, for his theory and the use of the telescope have by now been enthusiastically accepted in Germany, France, England, Poland, Denmark, Sweden, etc. . . . Therefore I think that this philosophical theory should not be condemned. One reason for this is that it will be embraced even more enthusiastically by the heretics and they will laugh at us. For we know how greatly those who live north of the Alps complained about some of the decrees adopted at the Council of Trent. What will they do when they hear that we have attacked the physicists and astronomers? Will they not immediately proclaim that we have done violence to both nature and the Scriptures?

The book containing these words was published not in 1616, when it was written, but only six years later, in 1622, and not in Italy but in Germany. Moreover, the anxious questions of Campanella, a Dominican, appear prophetic when we consider John Milton's defense of free enquiry after truth (his Areopagitica*), written some two decades later in another northern country—England—in which he cites his firsthand experience of Italy and the Roman establishment's most famous prisoner:*

Lords and Commons, . . . I could recount what I have seen and heard in other countries, where this kind of inquisition tyrannizes; when I have sat among their learned men, for that honor I had, and been counted happy to be born in such a place of philosophic freedom, as they supposed England was, while themselves did nothing but bemoan the servile condition into which learning amongst them was brought; that this was it which had damped the glory of Italian wits; that nothing had been there written now these many years but flattery and fustian. There it was that I found and visited the famous Galileo, grown old a prisoner to the Inquisition, for thinking in astronomy otherwise than the Franciscan and Dominican licensers thought.

And having made this reference, Milton goes on to endorse once more the principle upon which Campanella's defense of Galileo is founded: the consistency of truth with truth.

To be still searching what we know not by what we know, still closing up truth to truth as we find it (for all her body is homogeneal and proportional), this is the golden rule in theology as well as in arithmetic.

SOURCES: Thomas Campanella, *A Defense of Galileo the Mathematician from Florence*, trans. Richard J. Blackwell, Notre Dame and London: U of Notre Dame P, 1994; John Milton, *Areopagitica*, London, 1644.

28

They Hoist the Earth Up and Down Like a Ball

Robert Burton

Robert Burton's monumentally digressive Anatomy of Melancholy *(1638) offers a glimpse of how the astronomical and cosmological debates in the century after Copernicus may have appeared to a learned non-scientist. Burton's copious prose conveys a sense of the mental and psychological readjustment which the ordinary seventeenth-century observer was forced to undergo in confronting the new cosmology with its denial of impenetrable spheres and annihilation indeed of entire elements. The readjustment was expressed perhaps most famously and succinctly by John Donne, in 1611, in "An Anatomy of the World":*

> And new philosophy calls all in doubt,
> The element of fire is quite put out,
> The sun is lost, and th'earth, and no man's wit
> Can well direct him where to look for it.
> And freely men confess that this world's spent,
> When in the planets and the firmament
> They seek so many new; they see that this
> Is crumbled out again to his atomies.
> 'Tis all in pieces, all coherence gone . . .

Although Burton in 1638 seems somewhat less grudging than Donne in 1611, he wavers between comprehension of the main reasons for the Coper-

nican system and bewilderment at the plethora of contrary positions taken
by astronomers.

In the following section Burton moves from a dry cataloguing of positions
against the old notion of impenetrable crystalline orbs to an imaginative stir-
ring at the prospect that comes into view "if the heavens then be penetrable."

Saluciensis and Kepler take upon them to demonstrate that no meteors,
clouds, fogs, vapors, arise higher than fifty or eighty miles, and all the rest to
be purer air or element of fire. . . . Cardan, Tycho, and John Pena manifestly
confute by refractions, and many other arguments, there is no such element
of fire at all. If, as Tycho proves, the moon be distant from us fifty and sixty
semidiameters of the earth, and, as Peter Nonius will have it, the air be so august, what proportion is there betwixt the other three elements and it? To
what use serves it? Is it full of spirits which inhabit it, as the Paracelsians and
Platonists hold, the higher the more noble, full of birds, or a mere vacuum to
no purpose? It is much controverted between Tycho Brahe and Christopher
Rotman, the Landgrave of Hesse's mathematician, in their astronomical epistles, whether it be the same *Diaphanum*, clearness, matter of air and heavens,
or two distinct essences. Christopher Rotman, John Pena, Jordanus Brunus,
with many other late mathematicians contend it is the same and one matter
throughout, saving that the higher still the purer it is, and more subtle, as
they find by experience in the top of some hills in America: if a man ascend,
he faints instantly for want of thicker air to refrigerate the heart. . . . Tycho
will have two distinct matters of heaven and air. But to say truth, with some
small qualification they have one and the selfsame opinion about the essence
and matter of the heavens: that it is not hard and impenetrable, as Peripatet-
ics hold, transparent, of a fifth essence; "but that it is penetrable and soft as
the air itself is, and that the planets move in it, as birds in the air, fishes in the
sea."

This they prove by motion of comets . . . and as Tycho, Roeslin, Hagge-
sius, Pena, Rotman, Fracastorius demonstrate by their progress, parallaxes,
refractions, motions of the planets, which interfere and cut one another's
orbs, now higher, and then lower, as Mars amongst the rest, which sometimes, as Kepler confirms by his own and Tycho's accurate observations,
comes nearer the earth than the sun, and is again eftsoons aloft in Jupiter's
orb; and other sufficient reasons, far above the moon; exploding in the
meantime that element of fire, those fictitious first watery movers, those
heavens I mean above the firmament, which Delrio, Lodovicus Imola, Patri-
cius, and many of the fathers affirm; those monstrous orbs of eccentrics, and
epicycles departing from the eccentric, which—howsoever Ptolemy, Al-
hasen, Vitellio, Purbachius, Maginus, Clavius, and many of their associates

stiffly maintain to be real orbs, eccentric, concentric, circles equant, etc.—
are absurd and ridiculous. For who is so mad to think that there should be
so many circles, like subordinate wheels in a clock, all impenetrable and
hard, as they feign, add, and subtract at their pleasure? . . .

If the heavens then be penetrable, as these men deliver, and no lets, it were
not amiss in this aerial progress to make wings and fly up . . . or if that may
not be, yet with a Galileo's glass, or Icaromenippus's wings in Lucian, command the spheres and heavens, and see what is done amongst them.

*If the penetrability of the heavens encourages one optimistically to imagine
exploration of them by means of telescope or by winged travel, the new astronomy also introduces pessimism or at least anxiety (see Pascal, chapter
31) when we consider the relative magnitudes of earth and other heavenly
bodies. In other words, Donne's image of earth "crumbled out again to his
atomies" can be read quite literally as implying earth's atomic size relative to
the rest of what is out there. This, as Burton indicates, is only one of a tor-
rent of questions the ongoing controversies generate.*

Examine . . . whether the stars be of that bigness, distance, as astronomers
relate, so many in number, 1026, or 1725, as J. Bayerus; or as some Rabbins,
29,000 myriads; or as Galileo discovers by his glasses, infinite, and that
Milky Way a confused light of small stars, like so many nails in a door; or all
in a row, like those 12,000 isles of the Maldives in the Indian Ocean.
Whether the least visible star in the eighth sphere be eighteen times bigger
than the earth; and as Tycho calculates, 14,000 semidiameters distant from
it. Whether they be thicker parts of the orbs, as Aristotle delivers; or so many
habitable worlds, as Democritus. Whether they have light of their own, or
from the sun, or give light round, as Patritius discourseth. Whether they be
equally distant from the center of the world. Whether light be of their
essence; and that light be a substance or an accident. Whether they be hot by
themselves, or by accident cause heat. Whether there be such a precession of
the equinoxes as Copernicus holds, or that the eighth sphere move.

*Burton repeats the claim that for Copernicus heliocentrism was a hypothe-
sis. Yet, he recognizes that it has become more than merely hypothetical,
even if it is apparently (in 1638) still a minority opinion.*

To omit all smaller controversies as matters of less moment and examine that
main paradox, of the earth's motion, now so much in question: Aristarchus
Samius, Pythagoras maintained it of old, Democritus and many of their
scholars, Didacus Astunica, Anthony Fascarinus, a Carmelite, and some

other commentators, will have Job to insinuate as much (Job 9:6)—"Which shaketh the earth out of her place", etc.—and that this one place of Scripture makes more for the earth's motion than all the other prove against it; whom Pineda confutes, most contradict. Howsoever, it is revived since by Copernicus, not as a truth, but as a supposition, as he himself confesseth in the preface to Pope Nicholas, but now maintained in good earnest by Calcagninus, Telesius, Kepler, Rotman, Gilbert, Digges, Galileo, Campanella, and especially Lansbergius, as comporting with nature, reason, and truth. . . . For if the earth be the center of the world, stand still, and the heavens move, as the most received opinion is, which they call "a disordered arrangement of the heaven," though stiffly maintained by Tycho, Ptolemeus, and their adherents, . . . that shall drive the heavens about with such incomprehensible celerity in twenty-four hours, when as every point of the firmament and in the equator must needs move (so Clavius calculates) 176,660 in one 246th part of an hour; and an arrow out of a bow must go seven times about the earth whilst a man can say an Ave Maria, if it keep the same space, or compass the earth 1884 times in an hour, which is beyond human conceit.

Wavering again between the poles of comprehension and even assent, at one extreme, and bewilderment or incredulity at the other, Burton segues from a neutral summary of the Copernican system to a charged recognition of the magnitudes that it implies, and again to renewed speculation about extraterrestrial life.

They ascribe a triple motion to the earth, the sun immovable in the center of the whole world, the earth center of the moon alone, above Venus and Mercury [and] beneath Saturn, Jupiter, Mars, . . . a single motion to the firmament, which moves in thirty or twenty-six thousand years; . . . and so solve all appearances better than any way whatsoever, calculate all motions . . . without epicycles, intricate eccentrics, etc., "more accurately and fittingly by means of a single motion of the earth," says Lansbergius, much more certain than by those Alphonsine or any such tables, which are grounded from those other suppositions. And 'tis true they say, according to optic principles, the visible appearances of the planets do so indeed answer to their magnitudes and orbs, and come nearest to mathematical observations and precedent calculations, there is no repugnancy to physical axioms, because no penetration of orbs. But then between the sphere of Saturn and the firmament there is such an incredible and vast space or distance (7,000,000 semidiameters of the earth, as Tycho calculates) void of stars; and besides, they do so enhance the bigness of the stars, enlarge their circuit, to solve those ordinary objections or parallaxes and retrogradations of the fixed stars, that alteration of the poles, elevation in several places or latitude of cities here on earth (for, they say, if a man's eye were in the firmament, he should not at all discern that great annual motion of the earth, but it would still appear "an indivisible point" and seem to be fixed in one place, of the same bigness) that it is quite opposite to reason, to natural philosophy, and all out as absurd as disproportional (so some will) as prodigious, as that of the sun's swift motion of heavens.

But to grant this their tenet of the earth's motion: if the earth move, it is a planet, and shines to them in the moon, and to the other planetary inhabitants, as the moon and they do to us upon the earth. But shine she doth, as Galileo, Kepler, and others prove, and then it follows that the rest of the planets are inhabited, as well as the moon, which he grants in his dissertation with Galileo's *Nuncius Sidereus* "that there be Jovial and Saturn inhabitants," etc. . . .

We may likewise insert with Campanella and Brunus . . . there be infinite worlds, and infinite earths or systems, in an infinite ether. . . . For if the firmament be of such an incomparable bigness as these Copernical giants will have it, infinite, or approaching infinity, so vast and full of innumerable stars, as being infinite in extent, one above another, some higher, some lower, some nearer, some farther off, and so far asunder, and those so huge and great, insomuch that if the whole sphere of Saturn and all that is included in it—"if the whole entirety," as Fromundus argues, "were carried off among the stars, we would not even be able to see it, it would be like a mere point, so enormous is the distance between earth and the fixed stars." If our world be small in respect, why may we not suppose a plurality of worlds, those infinite stars visible in the firmament to be so many suns with particular fixed centers, to have likewise their subordinate planets, as the sun hath his dancing still round him? Which Cardinal Cusanus, Walkarinus, Brunus, and some others have held, and some still maintain (albeit spirits fed on Aristotle and educated in minute speculations may think otherwise). Though they seem close to us, they are infinitely distant, and so it follows that they are infinite habitable worlds. What hinders? Why should not an infinite cause (as God is) produce infinite effects? . . .

Kepler (I confess) will by no means admit of Brunus's infinite worlds, or that the fixed stars should be so many suns, with their compassing planets, yet the said Kepler between jest and earnest . . . seems in part to agree with this, and partly to contradict; . . . and so doth Tycho in his astronomical epistles . . . break into some such like speeches, that he will never believe those great and huge bodies were made to no other use than this that we perceive, to illuminate the earth, a point insensible in respect of the whole. But who shall dwell in these vast bodies, earths, worlds, "if they be inhabited? rational creatures?" as Kepler demands, "or have they souls to be saved? or do they inhabit a better part of the world than we do? Are we or they lords of the world? And how are all things made for man?" . . . This only he proves, that we are in the best place, best world, nearest the heart of the sun. . . .

The anxiety that emerges from Burton's account has less to do with the movement of the earth or with the incoherence of the universe than with the jumble of theories that astronomers and mathematicians propose. It is still a familiar lay person's complaint against scientists that, if they know so much, how can they disagree so radically with each other? Something like this (potentially) ad hominem *response causes Burton's prose to take a delightfully satirical turn.*

But to avoid these paradoxes of the earth's motion (which the Church of Rome hath lately condemned as heretical . . .) our later mathematicians have rolled all the stones that may be stirred, and to solve all appearances and objections have invented new hypotheses, and fabricated new systems of the world, out of their own Daedalian heads. Fracastorius will have the earth stand still, as before; and to avoid that supposition of eccentrics and epicycles, he hath coined seventy-two homocentrics, to solve all appearances. Nicholas Ramerus will have the earth the center of the world, but movable. . . . Tycho Brahe puts the earth [at] the center, the stars immovable, the rest with Ramerus, the planets without orbs to wander in the air, keep time and distance, true motion, according to that virtue which God hath given them. Roeslin censureth both, with Copernicus, whose hypothesis about the earth's movement Lansbergius hath lately vindicated. . . . The said Lansbergius, 1633, hath since defended his assertion against all the cavils and calumnies of Fromundus . . . , Morinus, and Bartholinus. Fromundus, 1634, hath written against him again, J. Rosseus of Aberdeen, etc. (sound drums and trumpets) whilst Roeslin (I say) censures all and Ptolemeus himself as insufficient. . . . In his own hypothesis he makes the earth as before the universal center, the sun to the five upper planets; to the eighth sphere he ascribes diurnal motion, eccentrics, and epicycles to the seven planets, which hath been formerly exploded. And so . . . as a tinker stops one hole and makes two, he corrects them, and doth worse himself; reforms some, and mars all. In the meantime, the world is tossed in a blanket amongst them; they hoist the earth up and down like a ball, [and] make it stand and go at their pleasure.

SOURCE: Robert Burton, *The Anatomy of Melancholy*, London, 1638; rpt. London, 1886.

29

A World in the Moon

John Wilkins

John Wilkins (1614–1672) had a great career in university, church, and scientific affairs in the England of the Interregnum and the Restoration. Among other things, he served as Master of Trinity College, Cambridge, and as Bishop of Chester. He was also one of the founders of the Royal Society. However, part of the charm of the writing presented here, excerpted from The Discovery of a World in the Moon *(1638), is that it is the work of a young person, only 24 years old, but powerfully aware that he stands on the brink of a brave new universe, one that is yet only dimly understood by those whose reading is limited to the English language.*

Though appearing more than sixty years after Digges's presentation of the outlines of the Copernican system in English, Wilkins's Discovery *clearly anticipates an audience that will find some of its suggestions shocking. The heliocentric "hypothesis" itself may at first, he says, seem "horrid." Having appealed to the authority of writers from Aristarchus to Copernicus and Lansbergius, he cites Campanella to the effect that*

Very many others both English and French . . . affirmed our earth to be one of the planets, and the sun to be the center of all, about which the heavenly bodies move. And how horrid soever this may seem at the first, yet is it likely enough to be true, nor is there any maxim or observation in optics . . . that can disprove it.

Now if our earth were one of the planets (as it is according to them), then why may not another of the planets be an earth?

It is this pivotal inference—in Kepler, Wilkins, and others—that unleashed the entire genre of science fiction and began the modern history of specula-tion concerning extraterrestrial life—both that genre and that history, of course, themselves providing new ways of examining our world and its in-habitants. Such a reciprocal process engages the imaginative potential of what Hans Blumenberg calls "reflexive telescopics": No sooner did Galileo observe the moon through his telescope than the human race started to won-der how the earth and its inhabitants would look to someone observing them through a telescope on the moon. At a moral level we may see claims con-cerning extraterrestrials as involving a planetary reciprocity akin to "doing unto others as we would have them do unto us." If physical reality is homo-geneous, then the planets represent a kind of society in which it is only fair to grant similar privileges to all.

In any case, Wilkins prefaces his main hypothesis with an examination of the hurdles that true science must overcome. Sounding a little like Francis Bacon a generation earlier, he acknowledges how hard it can be to transcend mere appearance and settled opinion.

Many evident truths seem incredible to such who know not the causes of things. You may as soon persuade some country peasants that the moon is made of green cheese (as we say) as that 'tis bigger than his cart-wheel, since both seem equally to contradict his sight, and he has not reason enough to lead him farther than his senses. Nay, suppose (saith Plutarch) a philosopher should be educated in such a secret place where he might not see either sea or river, and afterwards should be brought out where one might show him the great ocean, telling him the quality of that water, that it is blackish, salt, and not potable, and yet there were many vast creatures of all forms living in it, which make use of the water as we do of the air. Questionless, he would laugh at all this, as being monstrous lies and fables, without any color of truth.

Just so will this truth which I now deliver appear unto others. Because we never dreamt of any such matter as a world in the moon, because the state of that place hath as yet been veiled from our knowledge, therefore we can scarcely assent to any such matter. Things are very hardly received which are altogether strange to our thoughts and our senses. The soul may with less dif-ficulty be brought to believe any absurdity, when it has formerly been ac-quainted with some colors and probabilities for it. But when a new and an unheard of truth shall come before it, though it have good grounds and rea-sons, yet the understanding is afraid of it as a stranger and dares not admit it into its belief without a great deal of reluctance and trial. And besides, things that are not manifested to the senses are not assented unto without some labor of mind, some travail and discourse of the understanding. And many lazy souls had rather quietly repose themselves in an easy error than take pains to search out the truth.

The strangeness then of this opinion which I now deliver will be a great hindrance to its belief. . . . I have stood the longer in the preface because that prejudice which the mere title of the book may beget cannot easily be removed without a great deal of preparation. . . .

I must needs confess, though I had often thought with myself that it was possible there might be a world in the moon, yet it seemed such an uncouth opinion that I never durst discover it for fear of being counted singular and ridiculous. But afterward, having read Plutarch, Galileo, Kepler, with some others, and finding many of my own thoughts confirmed by such strong authority, I then concluded that it was not only possible there might be, but probable that there was another habitable world in that planet.

Wilkins's speculations do not concern the nature of the lunar inhabitants themselves but only what their world is like. In hindsight, we may find his thesis amusing, but this is a risk he knows he is taking. Moreover, his speculations, like most good science fiction (though Wilkins would not consider it fiction), are built upon a foundation of scientific plausibility. Part of Wilkins's foundation, as provided largely by Galileo, is the advantageous light that the moon receives from the earth. His eleventh proposition is "that as their world is our moon, so our world is their moon."

If there be such a world in the moon, 'tis requisite that their seasons should be some way correspondent unto ours, that they should have winter and summer, night and day, as we have.

Now that in this planet there is some similitude of winter and summer is affirmed by Aristotle himself, since there is one hemisphere that hath always heat and light, and the other that hath darkness and cold. True indeed, their days and years are always of one and the same length, but 'tis so with us also under the poles, and therefore that great difference is not sufficient to make it altogether unlike ours, nor can we expect that everything there should be in the same manner as it is here below, as if nature had no way but one to bring about her purposes. . . .

However, it may be questioned whether it doth not seem to be against the wisdom of providence to make the night of so great length, when they have such a long time unfit for work. I answer no, since 'tis so, and more with us also under the poles; and besides, the general length of their night is somewhat abated in the bigness of their moon, which is our earth. For this returns as great a light unto that planet as it receives from it.

Wilkins spends more time supporting this opinion than we might think necessary, simply because it was indeed a new and as yet undigested idea. Only once space itself was firmly conceived of as homogeneous ("but one region"), and the two "planets" thought of (in Wilkins's words) as "loving friends," could their reciprocal relationship clearly be seen as involving what Galileo had called a "grateful exchange" of light.

'Tis the general consent of philosophers that the reflection of the sunbeams from the earth doth not reach much above half a mile high, where they terminate the first region, so that to affirm they might ascend to the moon were to say there were but one region of air, which contradicts the proved and received opinion.

Unto this it may be answered: that it is indeed the common consent that the reflection of the sunbeams reach only to the second region; but yet some there are, and those too philosophers of good note, who thought otherwise. Thus Plotinus is cited by Calius, "If you did conceive yourself to be in some such high place, where you might discern the whole globe of the earth and water when it was enlightened by the sun's rays, 'tis probable it would then appear to you in the same shape as the moon doth now unto us." Thus also Carolus Malapertius, whose words are these: "If we were placed in the moon, and from thence beheld this our earth, it would appear unto us very bright, like one of the nobler planets." Unto these doth Fromondus assent, when he says, "I believe that this globe of earth and water would appear like some great star to any one who should look upon it from the moon." Now this could not be, nor could it shine so remarkably, unless the beams of light were reflected from it. . . .

If you behold the moon a little before or after the conjunction, when she is in a sextile with the sun, you may discern not only the part which is enlightened but the rest also to have in it a kind of duskish light. But if you choose out such a situation where some house or chimney (being some 70 or 80 paces distant from you) may hide from your eye the enlightened horns, you may then discern a greater and more remarkable shining in those parts unto which the sunbeams cannot reach. Nay, there is so great a light that by the help of a good perspective [i.e., a telescope] you may discern its spots. . . .

But now this light is not proper to the moon. It doth not proceed from the rays of the sun which doth penetrate her body, nor is it caused by any other of the planets or stars. Therefore it must necessarily follow that it comes from the earth. . . . This light must necessarily be caused by that which with a just gratitude repays to the moon such illumination as it receives from her.

And as loving friends equally participate of the same joy and grief, so do these mutually partake of the same light from the sun, and the same darkness from the eclipses, being also severally helped by one another in their greatest wants. For when the moon is in conjunction with the sun, and her upper part receives all the light, then her lower hemisphere (which would otherwise be altogether dark) is enlightened by the reflection of the sunbeams from the earth. When these two planets are in opposition, then that part of the earth which could not receive any light from the sunbeams is most enlightened by the moon, being then in her full. And as she doth most illuminate the earth when the sunbeams cannot, so the grateful earth returns to her as great, nay greater, light when she most wants it. So that always that visible part of the moon which receives nothing from the sun is enlightened by the earth, as is proved by Galileo, with many more arguments. . . .

The manner of this mutual illumination betwixt these two you may plainly discern in this figure [on the opposite page], where A represents the sun, B the earth, and C the moon.

Now suppose the moon C to be in a sextile of increase, when there is only one small part of her body enlightened, then the earth B will have such a part of its visible hemisphere darkened as is proportionable to that part of the moon which is enlightened. And as for so much of the moon as the sunbeams cannot reach unto, it receives light from a proportional part of the earth which shines upon it, as you may plainly perceive by the figure.

You see then that agreement and similitude which there is betwixt our earth and the moon. Now the greatest difference which makes them unlike is this: that the moon enlightens our earth round about, whereas our earth gives light only to that hemisphere of the moon which is visible unto us, as may be certainly gathered from the constant appearance of the same spots, which could not thus come to pass if the moon had such a diurnal motion about its own axis as perhaps the earth hath. And though some suppose her to move in an epicycle, yet this doth not so turn her body round that we may discern both hemispheres.

At the end of his book, Wilkins returns to the question of the progress of knowledge—and, we would say, of technology. He won't speculate on the nature of the moon's inhabitants, "because I know not any ground whereon to build any probable opinion." But he does expound an enthusiastic dream of the expansion of human knowledge, a dream some parts of which have in-

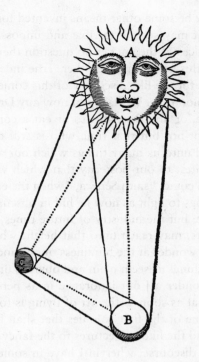

deed come true, even if the progress of time has not turned up evidence of lunar inhabitants.

I think that future ages will discover more; and our posterity, perhaps, may invent some means for our better acquaintance with these inhabitants. 'Tis the method of providence not presently to show us all, but to lead us along from the knowledge of one thing to another. 'Twas a great while ere the planets were distinguished from the fixed stars, and sometime after that ere the morning and evening star were found to be the same. And in greater space I doubt not but this also, and far greater mysteries, will be discovered.

In the first ages of the world, the islanders either thought themselves to be the only dwellers upon the earth, or else if there were any other, yet they could not possibly conceive how they might have any commerce with them, being severed by the deep and broad sea. But the after-times found out the invention of ships, in which notwithstanding none but some bold daring men durst venture, there being few so resolute as to commit themselves unto the vast ocean. And yet now how easy a thing is this, even to a timorous and cowardly nature?

So, perhaps, there may be some other means invented for a conveyance to the moon. And though it may seem a terrible and impossible thing ever to pass through the vast spaces of the air, yet no question there would be some men who durst venture this as well as the other. True indeed, I cannot conceive any possible means for the like discovery of this conjecture, since there can be no sailing to the moon. . . . We have not now any Drake or Columbus to undertake this voyage, or any Daedalus to invent a conveyance through the air. However, I doubt not but that time who is still the father of new truths, and hath revealed unto us many things which our ancestors were ignorant of, will also manifest to our posterity that which we now desire but cannot know. "Time will come," saith Seneca, "when the endeavors of afterages shall bring such things to light as now lie hid in obscurity." Arts are not yet come to their solstice; but the industry of future times, assisted with the labors of their forefathers, may reach unto that height which we could not attain to. . . . As we now wonder at the blindness of our ancestors, who were not able to discern such things as seem plain and obvious unto us, so will our posterity admire [i.e., wonder at] our ignorance in as perspicuous matters. Kepler doubts not but that as soon as the art of flying is found out, some of their nation will make one of the first colonies that shall inhabit that other world. But I leave this and the like conjectures to the fancy of the reader, desiring now to finish this discourse, wherein I have in some measure proved what at the first I promised, a world in the moon.

SOURCE: John Wilkins, *The Discovery of a World in the Moon: or, A Discourse tending to prove that 'tis probable there may be another habitable World in that Planet,* London, 1638.

30

A Very Liquid Heaven

René Descartes

The ideas of René Descartes (1596–1650) figure prominently in histories of philosophy but in some histories of cosmology receive only a footnote. His notion of vortices, or whirlpools of matter that account for the origin and structure of planetary systems, was decisively displaced by Newtonian conceptions and was subsequently, by most, remembered no more. In the seventeenth century, however, it was an influential idea. To us it can perhaps seem somewhat ridiculous simply because our common conceptions of space and motion are themselves still doggedly Newtonian. Yet if we bracket the assumptions that motion is naturally rectilinear and that space is empty, absolute, and devoid of any structure of its own, then we can better enter into the process of thought which Descartes exhibits in the following selection from his Principles of Philosophy *(published in Latin in 1644, in French in 1647). Once Descartes has rejected vacuous space and made the fundamental assumption that the heavens are liquid, the analogy between celestial motions and the familiar phenomenon of whirlpools within a current of water becomes almost inevitable. In this way too he is able to offer a hypothesis that accounts for the dynamism of the heavens while avoiding (he says, somewhat equivocally) any imputation of motion to the earth.*

I will be more careful than Copernicus not to attribute motion to the earth, and will attempt to show how my thoughts on this subject might be truer than Tycho's: I will propose here a hypothesis that seems to me to be the simplest and most convenient of all, in order to explicate the planets as well as to research their natural causes. . . .

First, because we do not yet know for sure what the distance is between the earth and the fixed stars, and because we cannot imagine them so distant as to be at odds with experience, let us not be content merely to put them above Saturn, where all the astronomers acknowledge they are; but let us take the liberty of supposing them to be as distant from it as will be useful for our goal. If we wish to judge their height in comparison with the distances we see between objects on earth, our estimations of them will have as little credibility as the greatest distance we could imagine. On the other hand, if we consider the omnipotence of God, who created them, the greatest distance we can conceive is no less believable than the smallest. And . . . we can no better explain the appearance of comets or planets unless we suppose a very great space between Saturn's sphere and the fixed stars. . . .

Secondly, because the sun is similar both to a flame and to the fixed stars, in that light departs from it which is not borrowed from elsewhere, let us also imagine that its motion is similar to a flame and that its position is similar to the stars. . . . We may assume that the sun is composed of a very liquid material, the particles of which are so extremely agitated that they carry away heavenly particles which are their neighbors and which surround them. Moreover, the sun is likewise similar to the fixed stars in that it does not travel from one part of the heavens to another. . . .

Furthermore, it must be noted here that, if the sun and fixed stars resemble one another in their positions, we may not assume they are all on the surface of the same sphere (as several suppose them to be), since the sun is not able to be with them on the surface of this sphere. Instead, we must assume that, just as it is surrounded by a vast space in which there are no fixed stars, so too each fixed star is very remote from all the others, and some of these stars are more remote from us and from the sun than others are. . . .

Thirdly, let us assume that the material of the heavens, like that which composes the sun and fixed stars, is liquid. This is an opinion that is now commonly held by astronomers, because it is almost impossible to explain well the phenomena without it. . . .

But, wishing to grant that the heavens are liquid, some make the mistake of imagining them to be a totally empty space that not only does not resist the movements of other bodies, but also has no force to move them or carry them along. For one thing, there can be no such vacuum in nature. For another, all liquids share this property, namely that they do not resist the movement of other bodies—not because they have less agitation than particles of solid matter but because they have the same or more. For this reason their small particles can easily be made to move every which way. But when they do all move together in the same direction, they necessarily carry along any body which they embrace and surround (and which no external force prevents from being thus carried along)—even though such bodies are hard, solid, and completely at rest. . . .

Fourthly, since we see that the earth is neither supported by columns nor suspended in the air by cables, but rather surrounded on all sides by a very liquid heaven, let us assume that it is at rest and that it has no propensity to movement, for we observe no such thing. However, let us not assume that this stops it from being carried along by the heaven's current, just as a boat that is neither driven along by wind or oar nor held back by anchor remains at rest in the midst of the sea, though the ebb and flow of this great mass of water imperceptibly carries it along with itself. . . .

Just as the other planets resemble earth (opaque, and reflecting the sun's rays), so too we have reason to believe that they likewise resemble it in remaining at rest in the part of heaven where each is located, and that every change we observe in their positions arises from the fact of their submitting to the motion of the heaven which contains them. . . .

In this regard, let us recall . . . the nature of movement. *Properly speaking* it is the transference of a body from the vicinity of those bodies which are in immediate contact with it and which we think of as at rest. Yet *in common usage* "movement" is predicated of every action that makes a body pass from one place to another. In this sense one can say that the same thing at the same time both is and is not in motion, depending on the different ways of determining its position. However, neither in the earth nor in the planets is there any movement properly speaking. They are not transferred from the vicinity of the heavenly particles they are in contact with. And thus they are at rest. . . .

If we consider movement according to common usage, we may well say that all the other planets move, and so do the sun and fixed stars. But the same thing cannot be said of the earth except very improperly. We determine the positions of the stars according to what we take to be certain fixed places on earth, and we consider them to have moved when they deviate from the positions we have thus determined. . . . But if a philosopher, who makes a profession out of seeking truth, takes heed of the fact that the earth is a globe floating in a liquid heaven whose particles are extremely agitated, and that the fixed stars always keep the same positions among themselves—if a philosopher wishes to conceive of these stars as stable in order to determine the earth's position, and consequently intends to conclude that it moves, then he is making a mistake, and his discourse lacks rational support. For . . . in its true sense, position is necessarily determined by the bodies in direct contact with whatever is said to be in motion, not by those very distant, as are the fixed stars in relation to the earth. . . . Whoever considers God's grandeur and the fallibility of our senses will judge that . . . perhaps beyond all the visible stars there are still other bodies in relation to which . . . the earth is at rest and the stars are in motion. . . . Therefore, even if, in order to accommodate ourselves to custom, we attribute some movement to the earth, we must acknowledge we are speaking unphilosophically, just as when we sometimes say of those who are asleep on a boat that they nevertheless move from Calais to Dover because the boat is carrying them there.

Having by these arguments removed all our misgivings concerning the earth's movement, let us assume that the heavenly material in which the planets are located revolves incessantly like a vortex with the sun at its center, that its particles close to the sun move faster than those (to a certain limit) that are more distant, and that all the planets (among which we include the earth) remain continually suspended amid the same segments of this heavenly material. By this hypothesis alone, without the help of any other machinery, we will easily understand everything we observe in the heavens.

Picture the bend of a river, where the water coils itself, turning in circles, some major, some quite minor. Things floating in this current, we notice, are carried along by it and turned round and round, even heavy objects, some of which revolve about their own centers. Those objects nearer the center of the eddy containing them complete their revolution sooner than do those farther from the center. Finally, though these eddies always turn in a circle, they hardly ever describe a perfect circle . . . so that not all the particles on the circumference are equidistant from the center. Let us likewise imagine the same things happening to the planets. And this is all we need in order to explain all of their phenomena.

SOURCE: Adapted from René Descartes, *Principles of Philosophy*, trans. Blair Reynolds, *Studies in the History of Philosophy*, vol. 6, Lewiston, N.Y.: Edwin Mellen, 1988.

31

The Eternal Silence of These Infinite Spaces

Blaise Pascal

Blaise Pascal (1623–1662) possessed a remarkable combination of talents. He invented what he called "a machine for making arithmetical calculations without pen or counters," considered by some to be the world's first digital calculator. Fittingly, then, his name reappears in computer programming languages as well as in laws and units of fluid pressure, in the physics of which he did groundbreaking experimental *work. Indeed, he said that geometry, arithmetic, music, physics, medicine, and architecture are all "sciences that are subject to experiment and reasoning"; and he criticized "the blindness of those who bring authority alone as proof in physical matters." When he wrote these words, Pascal particularly had in mind the age-old question regarding the possibility of a vacuum, which contemporaries such as Descartes denied for quite nonempirical reasons. Descartes and Pascal in fact met with each other for two days in September 1647 to discuss the issue, a meeting whose lack of harmonious outcome is attested by Descartes' snide remark to Huygens that Pascal "has too much vacuum in his head."*

Pascal was also a philosopher and a Christian apologist who engaged with imagination and meditative awe questions of humankind's place in the post-Copernican universe.

Let man then contemplate the whole of nature in her full and grand majesty, and turn his vision from the low objects which surround him. Let him gaze on that brilliant light, set like an eternal lamp to illumine the universe; let the earth appear to him a point in comparison with the vast circle described by the sun; and let him wonder at the fact that this vast circle is itself but a very fine point in comparison with that described by the stars in their revolution round the firmament. But if our view be arrested there, let our imagination pass beyond; it will sooner exhaust the power of conception than nature that of supplying material for conception. The whole visible world is only an imperceptible atom in the ample bosom of nature. No idea approaches it. We may enlarge our conceptions beyond an imaginable space; we only produce atoms in comparison with the reality of things. It is an infinite sphere, the center of which is everywhere, the circumference nowhere. In short, it is the greatest sensible mark of the almighty power of God that imagination loses itself in that thought.

Returning to himself, let man consider what he is in comparison with all existence; let him regard himself as lost in this remote corner of nature; and from the little cell in which he finds himself lodged, I mean the universe, let him estimate at their true value the earth, kingdoms, cities, and himself. What is a man in the Infinite?

But to show him another prodigy equally astonishing, let him examine the most delicate things he knows. Let a mite be given him, with its minute body and parts incomparably more minute, limbs with their joints, veins in the limbs, blood in the veins, humors in the blood, drops in the humors, vapors in the drops. Dividing these last things again, let him exhaust his powers of conception, and let the last object at which he can arrive be now that of our discourse. Perhaps he will think that here is the smallest point in nature. I will let him see therein a new abyss. I will paint for him not only the visible universe, but all that he can conceive of nature's immensity in the womb of this abridged atom. Let him see therein an infinity of universes, each of which has its firmament, its planets, its earth, in the same proportion as in the visible world; in each earth animals, and in the last mites, in which he will find again all that the first had, finding still in these others the same thing without end and without cessation. Let him lose himself in wonders as amazing in their littleness as the others in their vastness. For who will not be astounded at the fact that our body, which a little while ago was imperceptible in the universe, itself imperceptible in the bosom of the whole, is now a colossus, a world, or rather a whole, in respect of the nothingness which we cannot reach? He who regards himself in this light will be afraid of himself, and observing himself sustained in the body given him by nature between those two abysses of the Infinite and Nothing, will tremble at the sight of these marvels; and I think that, as his curiosity changes into admiration, he will be more disposed to contemplate them in silence than to examine them with presumption.

For, in fact, what is man in nature? A Nothing in comparison with the Infinite, an All in comparison with the Nothing, a mean between nothing and everything. Since he is infinitely removed from comprehending the extremes, the end of things and their beginning are hopelessly hidden from him in an impenetrable secret; he is equally incapable of seeing the Nothing from which he was made, and the Infinite in which he is swallowed up.

What will he do then, but perceive the appearance of the middle of things, in an eternal despair of knowing either their beginning or their end. All things proceed from the Nothing, and are borne towards the Infinite. Who will follow these marvelous processes? The Author of these wonders understands them. None other can do so.

· · ·

When I consider the short duration of my life, swallowed up in the eternity before and after, the little space which I fill, and even can see, engulfed in the infinite immensity of spaces of which I am ignorant, and which know me not, I am frightened, and am astonished at being here rather than there; for there is no reason why here rather than there, why now rather than then. Who has put me here? By whose order and direction have this place and time been allotted to me? . . .

The eternal silence of these infinite spaces frightens me.

SOURCE: Blaise Pascal, *Thoughts*, trans. W. F. Trotter, New York: Collier, 1910.

32

This Pendent World

John Milton

John Milton (1608–1674) is best known not as a cosmologist but as a poet, and his great poem Paradise Lost *is most famous as a magnificent re-telling of the biblical story of the fall of humankind, packaged as a classical epic. It should not be surprising, however, that a work dealing with the grand themes of God and human destiny would engage in a serious way questions about what the world is like and how it came to be. Although Milton's universe is eclectic and not easy to categorize, we can see its relation to various developments in cosmological history by charting the experiences or responses of three different characters in* Paradise Lost, *a devil, a human, and an angel.*

Following the journey of Satan allows us to "zoom in" on earth in three distinct stages. First, Satan sets out on a voyage from Hell, which is located at one "end" or "side" of what seems to be a boundless, infinite uncreated realm called Chaos. Milton avoids dualism by saying that the substance of Chaos came originally from God, but the picture we are given of it has much in common with the state described by the early atomists. Satan, together with the personified figures of Sin and Death, looks out from the gates of Hell and views the disordered, precreational state of matter.

> Before their eyes in sudden view appear
> The secrets of the hoary deep—a dark
> Illimitable ocean, without bound,
> Without dimension, where length, breadth, and height,
> And time, and place, are lost; where eldest Night
> And Chaos, ancestors of Nature, hold
> Eternal anarchy, amidst the noise

Of endless wars, and by confusion stand.
For Hot, Cold, Moist, and Dry, four champions fierce,
Strive here for mastery, and to battle bring
Their embryon atoms; they around the flag
Of each his faction, in their several clans,
Light-armed or heavy, sharp, smooth, swift, or slow,
Swarm populous, unnumbered as the sands
Of Barca or Cyrene's torrid soil,
Levied to side with warring winds, and poise
Their lighter wings. To whom these most adhere
He rules a moment; Chaos umpire sits,
And by decision more embroils the fray
By which he reigns. Next him, high arbiter
Chance governs all. Into this wild abyss,
The womb of nature and perhaps her grave,
Of neither sea, nor shore, nor air, nor fire,
But all these in their pregnant causes mixed
Confusedly, and which thus must ever fight,
Unless the Almighty Maker them ordain
His dark materials to create more worlds—
Into this wild abyss the wary fiend
Stood on the brink of Hell and looked a while,
Pondering his voyage.

(2.890–919)

On the analogy of an explorer voyaging to discover the New World (and also to corrupt it!), Satan crosses the "ocean" of Chaos and approaches what we readers know is our place in the cosmos. It is a scene that causes us to reflect on the astonishing scale of the physical reality Milton presents. Ptolemy had inferred the enormous size of the sphere of the fixed stars when he realized that the earth was but a point by comparison with it. Copernicus's system demanded a further expansion of that sphere, because even the entire orbit of the earth around the sun was immeasurably small in relation to the sphere of the fixed stars. But in Milton, that immensely large sphere is itself rendered not much larger than a point by comparison with the size of Heaven and Chaos. Our spherical universe—not the earth, but the entire solar and stellar system—appears as small as a faint, sixth magnitude star symbolically hanging down from heaven.

But now at last the sacred influence
Of light appears, and from the walls of Heaven

Shoots far into the bosom of dim Night
A glimmering dawn. Here Nature first begins
Her farthest verge, and Chaos to retire,
As from her outmost works, a broken foe,
With tumult less and with less hostile din;
That Satan with less toil, and now with ease,
Wafts on the calmer wave by dubious light,
And, like a weather-beaten vessel, holds
Gladly the port, though shrouds and tackle torn;
Or in the emptier waste, resembling air,
Weighs his spread wings, at leisure to behold
Far off the empyreal Heaven, extended wide
In circuit, undetermined square or round,
With opal towers and battlements adorned
Of living sapphire, once his native seat;
And, fast by, hanging in a golden chain,
This pendent world, in bigness as a star
Of smallest magnitude close by the moon.
                                            (2.1034–53)

*Having landed on the outer surface of our universe and gaining entrance
through a kind of hatch intended for the use of heavenly beings, Satan makes
the second leg of his journey, this time to the sun. Although the geometry of
Milton's universe is Ptolemaic in the sense that earth is at the center, virtually
everything else about it is related to some aspect of the new astronomy. There
are no crystalline spheres; the stars are not all carried, equidistant from the
center, by a single sphere (their locomotion is instead "magnetic," an idea Milton may have borrowed from Kepler); space is homogeneous and permits
"space travel," as Satan illustrates; the stars may be "other worlds"; and the
earth itself is a "star" reflecting sunlight as do the moon and other planets. The
following passage ironically refers to Satan, after his arrival on the sun, as a
"spot" (i.e. blemish) unlike any sunspot seen by Galileo through his telescope.*

Down right into the world's first region [Satan] throws
His flight precipitant, and winds with ease
Through the pure marble air his oblique way
Amongst innumerable stars, that shone
Stars distant, but nigh hand seemed other worlds;
Or other worlds they seemed, or happy isles,

Like those Hesperian gardens famed of old,
Fortunate fields, and groves, and flowery vales,
Thrice happy isles; but who dwelt happy there
He staid not to inquire. Above them all
The golden sun, in splendor likest Heaven,
Allured his eye; thither his course he bends
Through the calm firmament (but up or down,
By center, or eccentric, hard to tell,
Or longitude) where the great luminary [i.e. the sun]
Aloof the vulgar constellations thick,
That from his lordly eye keep distance due,
Dispenses light from far; they, as they move
Their starry dance in numbers that compute
Days, months, and years, towards his all-cheering lamp
Turn swift their various motions, or are turned
By his magnetic beam, that gently warms
The universe, and to each inward part
With gentle penetration, though unseen,
Shoots invisible virtue even to the deep;
So wondrously was set his station bright.
There lands the fiend, a spot like which perhaps
Astronomer in the sun's lucent orb
Through his glazed optic tube yet never saw.

(3.562–90)

The last stage of Satan's journey is from the sun to the earth. However, the reason he stops on the sun is to request directions. It is not obvious to him which "orb" humans live on—a further indication of the homogeneity of the universe. Earth is not conspicuously unique, and Satan's question echoes other seventeenth-century speculation about interplanetary travel and settlement. He asks an angel named Uriel, "In which of all these shining orbs hath man / His fixed seat, or fixed seat hath none, / But all these shining orbs his choice to dwell[?]" Uriel's answer gives not only Satan but also the reader a glimpse of earth as seen from space.

Look downward on that globe, whose hither side
With light from hence, though but reflected, shines;
That place is earth, the seat of man; that light
His day, which else, as the other hemisphere,

Night would invade; but there the neighboring moon
(So call that opposite fair star) her aid
Timely interposes, and her monthly round
Still ending, still renewing, through mid heaven,
With borrowed light her countenance triform
Hence fills and empties to enlighten the earth,
And in her pale dominion checks the night.
That spot, to which I point, is Paradise,
Adam's abode; those lofty shades, his bower.
Thy way thou canst not miss.

                                        (3.722–35)

*The human perspective on the universe is given by Adam in conversation
with the angel Raphael, who has been sent from heaven to instruct him on
various items, particularly on the importance of watching out for temptation. But the conversation includes discussion of astronomy. Adam's view is
essentially Ptolemaic, but not in any stereotypically naive sense. Like
Oresme, for example, Adam recognizes the lack of "economy" in assuming
that the immense heavens make a daily revolution around this tiny earth:*

When I behold this goodly frame, this world,
Of heaven and earth consisting, and compute
Their magnitudes, this Earth a spot, a grain,
An atom, with the firmament compared
And all her numbered stars, that seem to roll
Spaces incomprehensible (for such
Their distance argues, and their swift return
Diurnal) merely to officiate light
Round this opaque earth, this punctual spot,
One day and night, in all her vast survey
Useless besides; reasoning I oft admire,
How Nature wise and frugal could commit
Such disproportions, with superfluous hand
So many nobler bodies to create,
Greater so manifold, to this one use,
For aught appears, and on their orbs impose
Such restless revolution day by day
Repeated; while the sedentary earth,
That better might with far less compass move,
Served by more noble than herself, attains

Her end without least motion, and receives
As tribute such a sumless journey brought
Of incorporeal speed, her warmth and light;
Speed, to describe whose swiftness number fails.

                                                    (8.15–38)

Raphael's answer does not explicitly take sides in any simplistically Ptole-maic-versus-Copernican debate. Like Robert Burton's account of astronomy ("They hoist the earth up and down like a ball"), Raphael's takes the subject seriously but adopts a somewhat satirical posture towards human efforts to "wield" the universe according to various theories. Raphael's reply to Adam ranges across issues from geocentrism to heliocentrism, extraterrestrial life, the size and purpose of the heavens, and limits (or lack of limits) to the speed of space travel.

To ask or search, I blame thee not; for heaven
Is as the book of God before thee set,
Wherein to read his wondrous works, and learn
His seasons, hours, or days, or months, or years:
This to attain, whether heaven move or earth,
Imports not, if thou reckon right; the rest
From man or angel the great Architect
Did wisely to conceal, and not divulge
His secrets to be scanned by them who ought
Rather admire; or, if they list to try
Conjecture, he his fabric of the heavens
Hath left to their disputes, perhaps to move
His laughter at their quaint opinions wide
Hereafter, when they come to model heaven
And calculate the stars, how they will wield
The mighty frame, how build, unbuild, contrive
To save appearances, how gird the sphere
With centric and eccentric scribbled o'er,
Cycle and epicycle, orb in orb:
Already by thy reasoning this I guess,
Who art to lead thy offspring, and supposest
That bodies bright and greater should not serve
The less not bright, nor heaven such journeys run,
Earth sitting still, when she alone receives
The benefit. Consider, first, that great

Or bright infers not excellence: the earth,
Though in comparison of heaven so small,
Nor glistering, may of solid good contain
More plenty than the sun that barren shines,
Whose virtue on itself works no effect,
But in the fruitful earth; there first received,
His beams, unactive else, their vigor find.
Yet not to earth are those bright luminaries
Officious, but to thee, Earth's habitant.
And for the heaven's wide circuit, let it speak
The Maker's high magnificence, who built
So spacious, and his line stretched out so far;
That man may know he dwells not in his own;
An edifice too large for him to fill,
Lodged in a small partition, and the rest
Ordained for uses to his Lord best known.
The swiftness of those circles attribute,
Though numberless, to his omnipotence,
That to corporeal substances could add
Speed almost spiritual; me thou thinkest not slow,
Who since the morning hour set out from heaven
Where God resides, and ere mid-day arrived
In Eden, distance inexpressible
By numbers that have name. But this I urge,
Admitting motion in the heavens, to show
Invalid that which thee to doubt it moved;
Not that I so affirm, though so it seem
To thee who hast thy dwelling here on earth.
God, to remove his ways from human sense,
Placed heaven from earth so far, that earthly sight,
If it presume, might err in things too high,
And no advantage gain. What if the sun
Be center to the world, and other stars,
By his attractive virtue and their own
Incited, dance about him various rounds?
Their wandering course now high, now low, then hid,
Progressive, retrograde, or standing still,
In six thou seest; and what if seventh to these
The planet earth, so stedfast though she seem,
Insensibly three different motions move?
Which else to several spheres thou must ascribe,

Moved contrary with thwart obliquities,
Or save the sun his labor, and that swift
Nocturnal and diurnal rhomb supposed,
Invisible else above all stars, the wheel
Of day and night; which needs not thy belief,
If earth industrious of herself fetch day
Travelling east, and with her part averse
From the sun's beam meet night, her other part
Still luminous by his ray. What if that light
Sent from her through the wide transpicuous air
To the terrestrial moon be as a star,
Enlightening her by day, as she by night
This earth? reciprocal, if land be there,
Fields and inhabitants: her spots thou seest
As clouds, and clouds may rain, and rain produce
Fruits in her softened soil for some to eat
Allotted there; and other suns perhaps,
With their attendant moons thou wilt descry
Communicating male and female light,
Which two great sexes animate the world,
Stored in each orb perhaps with some that live.
For such vast room in nature unpossessed
By living soul, desert and desolate,
Only to shine, yet scarce to contribute
Each orb a glimpse of light, conveyed so far
Down to this habitable, which returns
Light back to them, is obvious to dispute.
But whether thus these things, or whether not;
Whether the sun, predominant in Heaven,
Rise on the earth, or earth rise on the sun,
He from the east his flaming road begin;
Or she from west her silent course advance,
With inoffensive pace that spinning sleeps
On her soft axle, while she paces even,
And bears thee soft with the smooth air along;
Solicit not thy thoughts with matters hid;
Leave them to God above; him serve, and fear!

(8.66–168)

Source: John Milton, *Paradise Lost: A Poem in Twelve Books* (2nd. ed.), London, 1674.

33

But One Little Family of the Universe

Bernard le Bouvier de Fontenelle and Aphra Behn

A highly engaging and imaginative encapsulation of ideas about the universe and "other worlds" in the late seventeenth century—one influenced both by Descartes and by John Wilkins—appears in the writings of Bernard le Bouvier de Fontenelle (1657–1757). His dialogue A Discovery of New Worlds *was translated into English by (in the words of Sam Briscoe, one of her later publishers) "the Sappho of our nation, the incomparable Mrs. [Aphra] Behn" (1640–1689). Thus the first professional female author in the English language became party to the popular dissemination of Copernican ideas.*

Behn's own preface to Fontenelle's dialogue is interesting for its engagement of the two issues just alluded to: the role of women in scientific discourse, and the desirability and danger of popularizing science. Behn admits she was motivated to translate the work by "the novelty of the subject in vulgar [i.e., vernacular] languages," and by "the author's introducing a woman as one of the speakers." At the same time, she is openly doubtful about whether Fontenelle succeeds—either in popularizing scientific ideas with appropriate dignity, or in creating a female character with sufficient credibility:

The design of the author is to treat of this part of natural philosophy in a more familiar way than any other hath done, and to make everybody understand him. For this end he introduceth a woman of quality . . . whom he feigns never to have heard of any such thing as philosophy before. How well he hath performed his undertaking you will best judge when you have perused the book. But if you would know beforehand my thoughts, I must tell you freely, he hath failed in his design; for endeavoring to render this part of natural philosophy familiar, he hath turned it into ridicule. He hath pushed his wild notion of the plurality of worlds to that height of extravagancy that he most certainly will confound those readers who have not judgment and wit to distinguish between what is truly solid (or at least probable) and what is trifling and airy. . . .

And for his Lady Marquiese, he makes her say a great many very silly things, though sometimes she makes observations so learned that the greatest philosophers in Europe could make no better. . . .

[The author] endeavors chiefly two things. One is that there are thousands of worlds inhabited by animals, besides our earth, and hath urged this fancy too far. I shall not presume to defend his opinion, but one may make a very good use of many things he hath expressed very finely, in endeavoring to assist his wild fancy. For he gives a magnificent idea of the vastness of the universe, and of the almighty and infinite power of the Creator, to be comprehended by the meanest capacity. This he proves judiciously by the appearances and distances of the planets and fixed stars. And if he had let alone his learned men, philosophical transactions, and telescopes in the planet Jupiter and his inhabitants not only there but in all the fixed stars, and even in the Milky Way, and only stuck to the greatness of the universe, he had deserved much more praise.

The other thing he endeavors to defend and assert is the system of Copernicus. As to this, I cannot but take his part as far as a woman's reasoning can go.

In the dialogue itself, the reader is to imagine five successive nights' conversation between the author and a beautiful noblewoman, a marchioness. Fontenelle sets the scene:

We went one evening after supper to walk in the park. The air was cool and refreshing. . . . The moon was about an hour high, which shining through the boughs of the trees made a most agreeable mixture, and checkered the paths beneath with a most resplendent white upon the green, which appeared to be black by that light. There was no cloud to be seen that could hide from us or obscure the smallest of the stars, which looked all like pure polished gold, whose lustre was extremely heightened by the deep azure field on which they were placed. These pleasant objects set me a-thinking, and had it not been for Madam the Marchioness, I might have continued longer in that silent contemplation. But the presence of a person of her wit and beauty hindered me from giving up my thoughts entirely to the moon and stars.

After a brief exchange in which the glories of the night are compared favorably to those of the day, the author summarizes his complaint regarding sunlight:

"The scene of the universe by day-light appears too uniform, we beholding but one great luminary in an arched vault of azure, of a vast extent, while all the stars appear confusedly dispersed, and disposed as it were by chance in a thousand different figures, which assists our roving fancies to fall agreeably into silent thoughts."

"Sir," replied Madam the Marchioness, "I have always felt those effects of night you tell me of. I love the stars, and could be heartily angry with the sun for taking them from my sight."

"Ah," cried I, "I cannot forgive his taking from me the sight of all those worlds that are there."

"Worlds!" said she. "What worlds?" And looking earnestly upon me, asked me again what I meant.

"I ask your pardon, Madam," said I. "I was insensibly led to this fond discovery of my weakness."

"What weakness?" said she, more earnestly than before.

"Alas," said I, "I am sorry that I must confess I have imagined to myself that every star may perchance be another world, yet I would not swear that it is so. But I will believe it to be true, because that opinion is so pleasant to me, and gives me very diverting ideas, which have fixed themselves delightfully in my imaginations. And 'tis necessary that every solid truth should have its agreeableness."

"Well," said she, "Since your folly is so pleasing to you, give me a share of it. I will believe whatever you please concerning the stars, if I find it pleasant."

"Ah, Madam," said I, hastily, "'Tis not such a pleasure as you find in one of Molière's plays. 'Tis a pleasure that is—I know not where, in our reason, and which only transports the mind."

"What?" replied she, "Do you think me then incapable of all those pleasures which entertain our reason, and only treat the mind? I will instantly show you the contrary, at least as soon as you have told me what you know of your stars."

"Ah, Madam," cried I, "I shall never endure to be reproached with that neglect of my one happiness, that in a grove, at ten o'clock of the night, I talked of nothing but philosophy to the greatest beauty in the world. No, Madam, search for philosophy somewhere else."

In a subtly ironic reversal, the author describes the marchioness's persuasion as a kind of seduction to which he, in the end, cannot help but submit—on the condition that she be "secret for the saving of my honor." He then begins his discourse on celestial mechanics by way of analogy with an opera played out in a theater.

"Pray, Madam, imagine to yourself the ancient philosophers beholding one of our operas, such a one as Pythagoras, Plato, Aristotle, and many more, whose names and reputations make so great a noise in the world. And suppose they were to behold the flying of Phaeton [the chariot driver], who is carried aloft by the winds, and that they could not discern the ropes and pulleys, but were altogether ignorant of the contrivance of the machine behind the scenes. One of them would be apt to say, 'It is a certain secret virtue that carries up Phaeton.' Another, that 'Phaeton is composed of certain numbers, which make him mount upwards.' The third, that 'Phaeton has a certain kindness [or natural inclination] for the highest part of the theater, and is uneasy when he is not there.' And a fourth, that 'Phaeton was not made for flying, but he had rather fly than leave the upper part of the stage void.' Besides a hundred other notions which I wonder have not entirely ruined the reputation of the ancients.

"In our age Descartes and some other moderns would say that 'Phaeton's flight upwards is because he is hoisted by ropes, and while he ascends, a greater weight than he descends.' And now men do not believe that any corporeal being moves itself, unless it be set in motion, or pushed by another body, or drawn by ropes, nor that any heavy thing ascends or descends, without a counterpoise equal with it in weight to balance it, or that it is guided by springs. And could we see nature as it is, we should see nothing but the hinder part of the theater at the opera."

"By what you say," said Madam the Marchioness, "philosophy is become very mechanical."

"So very mechanical," said I, "that I am afraid men will quickly be ashamed of it. For some would have the universe no other thing in great, than a watch is in little; and that all things in it are ordered by regular motion, which depends upon the just and equal disposal of its parts. Confess the truth, Madam, have not you had heretofore a more sublime idea of the universe, and have not you honored it with a better opinion than it deserved? I have known several esteem it less since they believed they knew it better."

"And for my part," said she, "I esteem it more since I knew it is so like a watch. And 'tis most surprising to me, that the course and order of nature, however admirable it appears to be, moves upon principles and things that are so very easy and simple."

"I know not," replied I, "who has given you so just ideas of it, but 'tis not ordinary to have such. Most people retain in their minds some false principle or other of admiration, wrapped up in obscurity, which they adore. They admire nature only because they look on it as a kind of miracle which they do not understand; and 'tis certain that those sort of people never despise anything but from the moment they begin to understand it. But, Madam, I find you so well disposed to comprehend all I have to say to you that, without further preface, I need only draw the curtain and show you the world."

The author, thus continuing in his role as impresario, now briefly reviews the scenes of ancient astronomy from the Chaldeans and Egyptians to Ptolemy, at last arriving at an admission of difficulties in the stagecraft of the Ptolemaic system.

"Madam," said I without vanity, "I have very much softened and explained this system. Should I expose it to you such as it was first invented by its author Ptolemy, or by those that have followed his principles, it would frighten you. The motion of the planets being irregular, they move sometimes fast, sometimes slow; sometimes towards one side, sometimes to another; at one time near the earth, at another far from it. The ancients did imagine I know not how many circles, differently interwoven one with another, by which they fancied to themselves they understood all the irregular phenomena or appearances in nature. And the confusion of these circles was so great that at that time, when men knew no better, a King of Arragon, a great mathematician (not over devout), said that if God had called him to his council, when he formed the universe, he could have given him good advice. The thought was impious, yet 'tis odd to reflect that the confusion of Ptolemy's system gave occasion for the sin of that King. The good advice he would have given was, no doubt, for surpassing these different circles, which had so embarrassed the celestial motions, and it may be also with regard to the two or three superfluous spheres which they had placed above the fixed stars. The philosophers, to explain one kind of motion of the heavenly bodies, did fancy a sphere of crystal above that heaven which we see, which set the inferior heaven on motion; and if anyone made a new discovery of any other motion, they immediately made a new sphere of crystal. In short, these crystalline heavens cost them nothing."

"But why spheres of crystal?" said Madam the Marchioness. "Would no other substance serve?"

"No," said I. "For there was a necessity of their being transparent, that the light might penetrate, as it was requisite for them to be solid beams. Aristotle had found out that solidity was inherent in the excellency of their nature. And because he said it, nobody would adventure to question the truth of it. But there have appeared comets, which we know to have been vastly higher from the earth than was believed by the ancients. These in their course would have broken all those crystal spheres, and indeed must have ruined the universe; so that there was an absolute necessity to believe the heavens to be made of fluid substance. At least 'tis not to be doubted from the observation of this and the last age, that Venus and Mercury move round the sun and not round the earth. So that the ancient system is not to be defended as to this particular. But I will propose one to you which solves all objections, and which will put the King of Arragon out of a condition of advising; and which is so surprisingly simple and easy that that good quality alone ought to make it preferable to all others."

"Methinks," said Madam the Marchioness, "that your philosophy is a kind of sale, or farm, where those that offer to do the affair at the smallest expense are preferred."

"'Tis very true," said I, "and 'tis only by that that we are able to guess at the scheme upon which Nature hath framed her work. She is very saving, and will take the shortest and cheapest way. Yet notwithstanding, this frugality is accompanied with a most surprising magnificence which shines in all she has done. But the magnificence is in the design, and the economy in the execution. And indeed there is nothing finer than a great design carried on with a little expense. But we are very apt to overturn all these operations of nature by contrary ideas. We put economy in the design, and magnificence in the execution. . . ."

"I shall be very glad," said she, "that this system you are to speak of will imitate nature so exactly. For this good husbandry will turn to the advantage of my understanding, since by it I shall have less trouble to comprehend what you have to say."

"There is in this system no more unnecessary difficulties. Know then that a certain German [sic] named Copernicus does at one blow cut off all these different circles and crystalline spheres invented by the ancients, destroying the one and breaking the other in pieces; and being inspired with a noble astronomical fury, takes the earth and hangs it at a vast distance from the center of the world, and sets the sun in its place, to whom that honor does more properly belong. The planets do no longer turn round the earth, nor do they any longer contain it in the circle they describe. And if they enlighten us, it is by chance, and because they find us in their way. All things now turn round the sun, among which the globe itself, to punish it for the long rest so falsely attributed to it before. And Copernicus has loaded the earth with all those motions formerly attributed to the other planets, having left this little globe none of all the celestial train, save only the moon, whose natural course it is to turn round the earth."

"Soft and fair!" said Madam the Marchioness. "You are in so great a rapture, and express yourself with so much pomp and eloquence, I hardly understand what you mean. You place the sun unmovable in the center of the universe. Pray, what follows next?"

In his reply the author sketches the solar system according to Copernicus. Moreover—contrary to most medieval interpretations but in keeping with the popular modern notion—he interprets Copernicus's displacement of the earth as if rejection of geocentrism implied rejection of anthropocentrism.

"Ah, but," said Madam the Marchioness, interrupting me, "you forget the moon."

"Do not fear," said I. "Madam, I shall soon find her again. The moon turns round the earth and never leaves it. And as the earth moves in the circle it describes round the sun, the moon follows the earth in turning round it. And if the moon do move round the sun, it is only because she will not abandon the earth."

"I understand you," said she. "I love the moon for staying with us when all the other planets have left us. And you must confess that your German, Copernicus, would have taken her from us too, had it been in his power. For I perceive by his procedure, he had no great kindness for the earth."

"I am extremely pleased with him," said I, "for having humbled the vanity of mankind, who had usurped the first and best situation in the universe. And I am glad to see the earth under the same circumstances with the other planets."

"That's very fine," said Madam the Marchioness. "Do you believe that the vanity of man places itself in astronomy, or that I am any way humbled, because you tell me the earth turns round the sun? I'll swear, I do not esteem myself one whit the less."

"Good Lord, Madam!" said I. "Do you think I can imagine you can be as zealous for a precedency in the universe as you would be for that in a chamber? No, Madam. The rank of places between two planets will never make such a bustle in the world as that of two ambassadors. Nevertheless, the same inclination that makes us endeavor to have the first place in a ceremony prevails with a philosopher, in composing his system, to place himself in the center of the world if he can. He is proud to fancy all things made for himself, and without reflection flatters his senses with this opinion, which consists purely in speculation."

"Oh," said Madam the Marchioness, "this is a calumny of your own invention against mankind, which ought never to have received Copernicus's opinion, since so easy and so humble."

"Copernicus, Madam" said I, "himself was the most diffident of his own system, so that it was a long time before he would venture to publish it, and at last resolved to do it at the earnest entreaty of people of the first quality. But do you know what he did, the day they brought him the first printed copy of his book? That he might not be troubled to answer all the objections and contradictions he was sure to meet with, he wisely left the world, and died."

The author explains the usual objections concerning the imperceptibility of the movement of the earth by recourse to the usual analogy of a ship whose inhabitants do not perceive its absolute movement from place to place. His next task is to explain how the earth moves both around the sun and around its own axis.

"Pray, sir," said she, ". . . I understand very well how we imagine the sun describes that circle which indeed we ourselves describe. But this requires a whole year's time, when one would think the sun passes over our heads every day. How comes that to pass?"

"Have you not observed," said I, "that a bowl thrown on the earth has two different motions. It runs toward the jack to which it is thrown. And at the same time it turns over and over several times before it comes that length, so that you will see the mark that is on the bowl sometimes above, sometimes below. 'Tis just so with the earth. In the time it advances on the circle it makes round the sun in its yearly course, it turns over once every four and twenty hours upon its own axis; so that in space of time, which is one natural day, every point of the earth (which is not near the South or North Poles) loses and recovers the sight of the sun. And as we turn towards the sun we imagine the sun is rising upon us, so when we turn from it we believe she is setting."

"This is very pleasant," said Madam the Marchioness. "You make the circle to do all, and the sun to stand idle. And when we see the moon, planets, and fixed stars turn round us in four and twenty hours, all is but bare imagination."

"Nothing else," said I, "but pure fancy."

In the second night of conversation, the author moves very quickly to speculation concerning the moon. Building on the likeness of the moon to the earth, and upon their reciprocity as concerns reflected light, he unfolds the idea that the moon too is inhabited.

"Well," said I, "since the sun . . . is now immoveable, and no longer a planet, and that the earth that moves round the Sun is now one, be not surprised if I tell you, the moon is another earth, and is by all appearances inhabited."

Said she, "I never heard of the moon's being inhabited but as a fable."

"So it may be still," said I. ". . . I will tell you my reasons that make me take part with the inhabitants of the moon. Suppose, then, there had never been any commerce between Paris and St. Denis, and that a citizen of Paris, who had never been out of that city, should go up to the top of the steeple of Our Lady and should view St. Denis at a distance, and one should ask him if he believed St. Denis to be inhabited. He would answer boldly, 'Not at all, for I see the inhabitants of Paris, but I do not see those of St. Denis, nor ever heard of 'em.' It may be somebody standing by would represent to him that it was true, one could not see the inhabitants of St. Denis from Our Lady's church, but that the distance was the cause of it; yet that all we could see of St. Denis was very like to Paris. For St. Denis had steeples, houses, and walls; and that it might resemble Paris in everything else, and be inhabited as well as it. All these arguments would not prevail upon my citizen, who would continue still obstinate in maintaining that St. Denis was not inhabited, because he saw none of the people. The moon is our St. Denis and we the citizens of Paris that never went out of our town."

"Ah," interrupted Madam the Marchioness, "you do us wrong. We are not so foolish as your citizen of Paris. Since he sees that St. Denis is so like to Paris in everything, he must have lost his reason if he did not think it was inhabited. But for the moon, that's nothing like the earth."

"Have a care, Madam," said I, "what you say. For if I make it appear that the moon is in everything like the earth, you are obliged to believe that the moon is inhabited."

"I acknowledge," said she, "if you do that, I must yield. And your looks are so assured that you frighten me already. The two different motions of the earth, which would never have entered into my thoughts, make me very apprehensive of all you say. But is it possible that the earth can be an enlightened body as the moon is? For to resemble it, it must be so."

"Alas, Madam," said I, "to be enlightened is not so great a matter as you imagine, and the sun only is remarkable for that quality: 'Tis he alone that is enlightened of himself by virtue of his particular essence. But the other planets shine only as being enlightened by the sun. The sun communicates his light to the moon, which reflects it upon the earth, as the earth, without doubt, reflects it back again to the moon, since the distance from the moon to the earth is the same as from the earth to the moon."

What does mark a difference between the earth and the moon is the fact that the rotation of the moon is exactly coincident with the period of its revolution about the earth. This is why from earth we see but one side of the moon.

"I shall never be satisfied," said Madam the Marchioness, "with the injury we do the earth in being too favorably engaged for the inhabitants of the moon unless you can assure me that they are as ignorant of their advantages as we are of ours, and that they take our earth for a star, without knowing that the globe they inhabit is one also."

"Be assured of that, Madam," said I, "that the earth appears to them to perform all the functions of a star. 'Tis true, they do not see the earth describe a circle round 'em, but that's all one. I'll explain to you what it is. That side of the moon which was turned towards the earth at the beginning of the world has continued towards the earth ever since, which still represents to us these same eyes, nose, and mouth, which our imaginations fancy we see composed of these spots, lights, and shadows which are the surface of the moon. Could we see the other half of the moon, 'tis possible our fancy would represent to us some other figure. This does not argue, but the moon turns however upon her own axis, and takes as much time to perform that revolution as she does to go round the earth in a month. But then, when the moon performs a part of her revolutions on her own axis, and that she ought to hide from us (for example) one cheek of this imaginary face, and appear to us in another position, she does at the same time perform as much of the circle she describes in turning round the earth. And though she is in a new point of sight or opposition as to us, yet she represents to us still the same cheek. So that the moon, in regard to the sun and the other planets, turns upon her own axis, but does not so as to the earth. The inhabitants of the moon see all the other planets rise and set in the space of fifteen days, but they see our earth always hanging in the same point of the heavens. This seeming immovability does not very well agree with a body that ought to pass for a planet. But the truth is, the earth is not in such perfection. Besides, the moon has a certain trembling quality which does sometimes hide a little of her imaginary face, and at other times shows a little of her opposite side. And no doubt but the inhabitants of the moon attribute this shaking to the earth, and believe we make a certain swinging in the heavens, like the pendulum of a clock."

"All these planets," said Madam the Marchioness, "are like us mortals, who always cast our own faults upon others: Says the earth, 'It is not I that turn round, 'tis the sun.' Says the moon, 'It is not I that tremble, 'tis the earth.' There are errors and mistakes everywhere."

Having learned to speak familiarly about the inhabitants of the moon, the author and the marchioness find it but a small step farther to imagine inhabitants of the other planets. In this way the by-now familiar speculations of extraterrestrial science fiction beckon.

"We, the inhabitants of the earth, are but one little family of the universe; we resemble one another. The inhabitants of another planet are another family, whose faces have another air peculiar to themselves. By all appearance, the difference increases with the distance, for could one see an inhabitant of the earth and one of the moon together, he would perceive less difference between them than between an inhabitant of the earth and an inhabitant of Saturn. Here (for example) we have the use of the tongue and voice, and in another planet it may be they only speak by signs. In another the inhabitants speak not at all. Here our reason is formed and made perfect by experience. In another place experience adds little or nothing to reason. Further off, the old know no more than the young. Here we trouble ourselves more to know what's to come than to know what's past. In another planet, they neither afflict themselves with the one nor the other; and 'tis likely they are not the less happy for that. Some say we want a sixth sense by which we should know a great many things we are now ignorant of. It may be the inhabitants of some other planet have this advantage, but want some of those other five we enjoy. It may be also that there are a great many more natural senses in other worlds, but we are satisfied with the five that are fallen to our share, because we know no better. . . ."

"Truly," said she, "I find not so much difficulty to comprehend these differences of worlds. My imagination is working upon the model you have given me. And I am representing to my own mind odd characters and customs for these inhabitants of the other planets. Nay more, I am forming extravagant shapes and figures for 'em."

By the fifth night, speculation has indeed taken wings, and the conversation between the author and the beautiful marchioness soars from the moon and the planets to the stars themselves. As with so much science fiction, this dialogue has a gentle undercurrent of satire directed at the inhabitants of our planet. And at the same time it registers in the respective responses of the marchioness and the author the two contrary emotions one may feel upon contemplating the immensity of the universe: a sense of philosophical humility at the smallness of oneself and of earthly things, and exhilaration at the grandeur of the cosmic prospect. Neither of these emotions eclipses the other emotion that has energized the author's efforts: love.

My lady Marchioness was very impatient to know what should become of the fixed stars.

"Can they be inhabited as the planets are?" said she to me. "Or are they not inhabited? What shall we make of 'em?"

"If you would take the pains, you could not fail to guess," said I. "Madam, the fixed stars cannot be less distant from the earth than fifty millions of leagues. Nay, some astronomers make the distance yet greater. That between the sun and the remotest planet is nothing if compared to the distance between the sun or earth and the fixed stars. We do not trouble ourselves to number 'em. Their lustre as you see is both clear and bright. If the fixed stars receive their light from the sun, it must certainly be very weak and faint before it comes to 'em, having passed through a hundred and fifty millions of miles of the celestial substance I spoke of before. Then consider, the fixed stars are obliged to reflect this borrowed light upon us at the same distance, which in reason must make that light yet paler and more faint. It is impossible that this light, if it were borrowed from the sun and not only suffered a reflection but passed through twice the distance of an hundred and fifty millions of miles, could have the force and vivacity that we observe in the fixed stars. Therefore I conclude they are enlightened of themselves and are, by consequence, so many suns."

"Do not I deceive myself?" cried out the Marchioness. "Do not I see whither you are going to lead me? Are you not about to tell me the fixed stars are so many suns, and that our sun is the center of a great *tourbillion* [or vortex], which turns round him? What hinders but a fixed star may be the center of a *tourbillion*, whirling or turning round it? Our sun has planets, which he enlightens. Why may not every fixed star have planets also?"

"I have nothing to answer, but what Phaedra said to Oenone, ''Tis you that have hit it.'"

"But," said she, "I see the universe to be so vast that I lose myself. I know not where I am, and having conceived nothing all this while. What is the universe thus divided into *tourbillions*, confusedly cast together? Is every fixed star the center of a *tourbillion*, and it may be full as big as our sun? Is it possible that all this immense space wherein our sun and planets have their revolution is nothing but an inconsiderable part of the universe? And that every fixed star must comprehend and govern an equal space with our sun? This confounds, afflicts, and frightens me."

"And for my part," said I, "it pleases and rejoices me. When I believed the universe to be nothing but this great azure vault of the heavens wherein the stars are placed, as it were so many golden nails or studs, the universe seemed to me too little and straight. I fancied myself to be confined and oppressed. But now when I am persuaded that this azure vault has a greater depth and vaster extent, and that 'tis divided into a thousand and a thousand different *tourbillions* or whirlings, I imagine I am at more liberty, and breathe a freer air; and the universe appears to me to be infinitely more magnificent. Nature has spared nothing in her production, and hath profusely bestowed her treasures upon a glorious work worthy of her. You can represent nothing so august to yourself as this prodigious number of *tourbillions*, whose center is possessed by a sun, and that makes the planets turn round him. The inhabitants of the planets of any of these infinite *tourbillions* see from all sides the enlightened center of the *tourbillion* with which they are environed, but cannot discover the planets of another, who enjoy but a faint light, borrowed from their own sun, which it does not dart further than its own sphere or activity."

"You show me," said Madam the Marchioness, "so vast a prospect that my sight cannot reach to the end of it. I see clearly the inhabitants of our world. And you have plainly presented to my reason the inhabitants of the moon, and other planets of our *tourbillion* or whirlings. After this you tell me of the inhabitants of the planets of all the other *tourbillions*. I confess they seem to me to be sunk in so boundless a depth that whatever force I put upon my fancy, I cannot comprehend 'em by the expressions you made use of in speaking of 'em and their inhabitants. You must certainly call 'em the inhabitants of one of the planets of one of these infinite *tourbillions*. And what shall become of us in the middle of so many worlds, since the title you give to the rest agrees to this of ours? And for my part, I see the earth so dreadfully little that hereafter I shall scorn to be concerned for any part of it. And I admire why mankind are so very fond of power, so earnest after grandeur, laying design upon design, circumventing, betraying, flattering, and poorly lying, and are at all this mighty pains to grasp a part of the world they neither know nor understand, nor anything of these mighty *tourbillions*. For me, I'll lazily contemn it, and my carelessness shall have this advantage by my knowledge, that when anybody shall reproach me with my poverty, I will with vanity reply, 'Oh! You do not know what the fixed stars are.'" . . .

"As for me who know 'em," [said I,] "I am very sorry I can draw no advantage from that knowledge, which can cure nothing but ambition and disquiet—and none of these diseases trouble me. I confess a kind of weakness in love, a kind of frailty for what is delicate and handsome. This is my distemper, wherein the *tourbillions* are not concerned at all. The infinite multitude of other worlds may render this little in your esteem, but they do not spoil fine eyes, a pretty mouth, or make the charms of wit ever the less. These will still have their true value, still are a price in spite of all the worlds in the universe."

Thus love conquers all, even in a possibly infinite universe. Although the author goes on to discourse on the Milky Way—"an infinity of little fixed stars," an "ant-hill of stars, . . . seeds of worlds"—the dialogue concludes by returning (as befits its participants) to the gentle analogy between astronomical instruction and the persuasions of a lover.

"It is a strange thing," said Madam the Marchioness, laughing, "that love saves himself from all dangers, and there is no system or opinion can hurt him." . . .

"Madam," said I, "since we are always in the humor of mixing some little gallantries with our most serious discourses, give me leave to tell you that mathematical reasoning is in some things near akin to love. And you cannot allow the smallest favor to a lover, but he will soon persuade you to yield another, and after that a little more; and in the end [he] prevails entirely. So if you grant the least principle to a mathematician, he will instantly draw a consequence from it which you must yield also, and from that another, and then a third; and maugre your resistance, in a short time he will lead you so far that you cannot retreat. These two sorts of men, the lover and philosopher, always take more than is given 'em."

. . .

"Let us return yet more," said I, "and, if you please, make this subject no longer that of our discourse. Besides, you are arrived at the utmost bounds of heaven. . . . 'Tis sufficient for me to have carried your understanding as far as your sight can penetrate."

"What," cried out Madam the Marchioness, "have I the systems of all the universe in my head? Am I become so learned?"

"Yes, Madam, you know enough, and with this advantage, that you may believe all or nothing of what I have said, as you please. I only beg this as a recompense for my pains: that you will never look on the heavens, sun, moon, or stars, without thinking of me."

SOURCE: Bernard le Bouvier de Fontenelle, *A Discovery of New Worlds*, trans. Aphra Behn, London, 1688.

34

Into the Celestial Spaces

Isaac Newton

Isaac Newton (1642–1727) ranks with Copernicus and Einstein among principal shapers of our physical worldview. However, whereas we can appreciate Copernicus against the backdrop of Ptolemy, and Einstein against the backdrop of Newton, Newton himself may strike the average educated non-scientist as somewhat unremarkable. For our everyday physical conceptions are themselves so thoroughly Newtonian that, to us, Newton may sound merely as if he is expounding common sense. Today perhaps only a small proportion of people are aware, when they use terms such as inertia, mass, gravity, and centripetal or centrifugal force, that it was Newton who either coined these terms or else worked them into the definitions they still generally retain.

Newton indeed begins his Mathematical Principles of Natural Philosophy—*his* Principia—*with definitions, and in them we may glimpse something both of the spirit of the whole work, whose essence is mathematical and quantitative, and of its magisterial lucidity. It is precisely the abstract character of mathematics that allows Newton to move so deftly, almost imperceptibly, from air, snow, and dust, to planets; and from a bullet fired from a mountaintop, to the critical orbit of the moon.*

DEFINITION I

The quantity of matter is the measure of the same, arising from its density and bulk conjointly.

Thus air of a double density, in a double space, is quadruple in quantity; in a triple space, sextuple in quantity. The same thing is to be understood of snow, and fine dust or powders, that are condensed by compression or liquefaction, and of all bodies that are by any causes whatever differently condensed. I have no regard in this place to a medium, if any such there is, that freely pervades the interstices between the parts of bodies. It is this quantity that I mean hereafter everywhere under the name of body or mass. And the same is known by the weight of each body, for it is proportional to the weight, as I have found by experiments on pendulums, very accurately made. . . .

DEFINITION II

The quantity of motion is the measure of the same, arising from the velocity and quantity of matter conjointly.

The motion of the whole is the sum of the motions of all the parts; and therefore in a body double in quantity, with equal velocity, the motion is double; with twice the velocity, it is quadruple.

DEFINITION III

The vis insita, *or innate force of matter, is a power of resisting, by which every body, as much as in it lies, continues in its present state, whether it be of rest, or of moving uniformly in a right line.*

This force is always proportional to the body whose force it is and differs nothing from the inactivity of the mass, but in our manner of conceiving it. A body, from the inert nature of matter, is not without difficulty put out of its state of rest or motion. Upon which account, this *vis insita* may, by a most significant name, be called inertia *(vis inertiae)* or force of inactivity. But a body only exerts this force when another force, impressed upon it, endeavors to change its condition; and the exercise of this force may be considered as both resistance and impulse; it is resistance so far as the body, for maintaining its present state, opposes the force impressed; it is impulse so far as the body, by not easily giving way to the impressed force of another, endeavors to change the state of that other. Resistance is usually ascribed to bodies at rest, and impulse to those in motion; but motion and rest, as commonly conceived, are only relatively distinguished; nor are those bodies always truly at rest, which commonly are taken to be so.

DEFINITION IV

An impressed force is an action exerted upon a body, in order to change its state, either of rest, or of uniform motion in a right line.

This force consists in the action only, and remains no longer in the body when the action is over. For a body maintains every new state it acquires, by its inertia only. But impressed forces are of different origins, as from percussion, from pressure, from centripetal force.

DEFINITION V

A centripetal force is that by which bodies are drawn or impelled, or any way tend, towards a point as to a center.

Of this sort is gravity, by which bodies tend to the center of the earth; magnetism, by which iron tends to the loadstone; and that force, whatever it is, by which the planets are continually drawn aside from the rectilinear motions, which otherwise they would pursue, and made to revolve in curvilinear orbits. A stone, whirled about in a sling, endeavors to recede from the hand that turns it; and by that endeavor, distends the sling, and that with so much the greater force, as it is revolved with the greater velocity, and as soon as it is let go, flies away. That force which opposes itself to this endeavor, and by which the sling continually draws back the stone towards the hand, and retains it in its orbit, because it is directed to the hand as the center of the orbit, I call the centripetal force. And the same thing is to be understood of all bodies, revolved in any orbits. They all endeavor to recede from the centers of their orbits; and were it not for the opposition of a contrary force which restrains them to, and detains them in their orbits, which I therefore call centripetal, would fly off in right lines, with a uniform motion.

A projectile, if it was not for the force of gravity, would not deviate towards the earth, but would go off from it in a right line, and that with a uniform motion, if the resistance of the air was taken away. It is by its gravity that it is drawn aside continually from its rectilinear course, and made to deviate towards the earth, more or less, according to the force of its gravity, and the velocity of its motion. The less its gravity is, or the quantity of its matter, or the greater the velocity with which it is projected, the less will it deviate from a rectilinear course, and the farther it will go. If a leaden ball, projected from the top of a mountain by the force of gunpowder, with a given velocity, and in a direction parallel to the horizon, is carried in a curved line to the distance of two miles before it falls to the ground; the same, if the resistance of the air were taken away, with a double or decuple velocity, would fly twice or ten times as far. And by increasing the velocity, we may at pleasure increase the distance to which it might be projected, and diminish the curvature of the line which it might describe, till at last it should fall at the distance of 10, 30, or 90 degrees, or even might go quite round the whole earth before it falls; or lastly, so that it might never fall to the earth, but go forwards into the celestial spaces, and proceed in its motion in infinitum.

And after the same manner that a projectile, by the force of gravity, may be made to revolve in an orbit, and go round the whole earth, the moon also, either by the force of gravity, if it is endued with gravity, or by any other force that impels it towards the earth, may be continually drawn aside towards the earth, out of the rectilinear way which by its innate force it would pursue; and would be made to revolve in the orbit which it now describes; nor could the moon without some such force be retained in its orbit. If this force was too small, it would not sufficiently turn the moon out of a rectilinear course; if it was too great, it would turn it too much, and draw down the moon from its orbit towards the earth.

In a scholium (explanatory discussion) following his opening definitions, Newton lays out assumptions concerning the absoluteness of time and space that stood for more than two centuries, until Einstein undermined them with his theories of relativity.

Hitherto I have laid down the definitions of such words as are less known and explained the sense in which I would have them to be understood in the following discourse. I do not define time, space, place, and motion, as being well known to all. Only I must observe that the common people conceive those quantities under no other notions but from the relation they bear to sensible objects. And thence arise certain prejudices, for the removing of which it will be convenient to distinguish them into absolute and relative, true and apparent, mathematical and common.

1. Absolute, true, and mathematical time, of itself and from its own nature, flows equably without relation to anything external, and by another name is called duration. Relative, apparent, and common time

is some sensible and external (whether accurate or unequable) measure of duration by the means of motion, which is commonly used instead of true time, such as an hour, a day, a month, a year.

2. Absolute space, in its own nature, without relation to anything external, remains always similar and immovable. Relative space is some movable dimension or measure of the absolute spaces, which our senses determine by its position to bodies and which is commonly taken for immovable space; such is the dimension of a subterraneous, an aerial, or celestial space, determined by its position in respect of the earth. Absolute and relative space are the same in figure and magnitude, but they do not remain always numerically the same. For if the earth, for instance, moves, a space of our air, which relatively and in respect of the earth remains always the same, will at one time be one part of the absolute space into which the air passes; at another time it will be another part of the same, and so, absolutely understood, it will be continually changed.

3. Place is a part of space which a body takes up and is, according to the space, either absolute or relative. I say a part of space, not the situation nor the external surface of the body. For the places of equal solids are always equal; but their surfaces, by reason of their dissimilar figures, are often unequal. . . .

4. Absolute motion is the translation of a body from one absolute place into another, and relative motion the translation from one relative place into another. Thus in a ship under sail, the relative place of a body is that part of the ship which the body possesses, or that part of the cavity which the body fills, and which therefore moves together with the ship; and relative rest is the continuance of the body in the same part of the ship, or of its cavity. But real, absolute rest is the continuance of the body in the same part of that immovable space in which the ship itself, its cavity, and all that it contains is moved. Wherefore, if the earth is really at rest, the body, which relatively rests in the ship, will really and absolutely move with the same velocity which the ship has on the earth. But if the earth also moves, the true and absolute motion of the body will arise partly from the true motion of the earth in immovable space, partly from the relative motion of the ship on the earth; and if the body moves also relatively in the ship, its true motion will arise partly from the true motion of the earth in immovable space, and partly from the relative motions as well of the ship on the earth as of the body in the ship; and from these relative motions will arise the relative motion of the body on the earth. . . .

Newton crucially concedes that there may be no accessible criteria for determining time and space absolutely, but asserts that a true philosophical treatment of these topics requires a careful distinction between the absolute and the relative.

Absolute time, in astronomy, is distinguished from relative by the equation or correction of the apparent time. For the natural days are truly unequal, though they are commonly considered as equal and used for the measure of time. Astronomers correct this inequality that they may measure the celestial motions by a more accurate time. It may be that there is no such thing as an equable motion whereby time may be accurately measured. All motions may be accelerated and retarded, but the flowing of absolute time is not liable to any change. . . .

As the order of the parts of time is immutable, so also is the order of the parts of space. . . . But because the parts of space cannot be seen or distinguished from one another by our senses, therefore in their stead we use sensible measures of them. For from the positions and distances of things from any body considered as immovable we define all places; and then, with respect to such places, we estimate all motions, considering bodies as transferred from some of those places into others. And so, instead of absolute places and motions, we use relative ones, and that without any inconvenience in common affairs. But in philosophical disquisitions we ought to abstract from our senses and consider things themselves, distinct from what are only sensible measures of them. For it may be that there is no body really at rest to which the places and motions of others may be referred.

And yet Newton does not entirely bend under the difficulty of determining absolute motion. The one criterion he considers reliable is that of centrifugal force. To take a modern (actual) example, an astronaut skilled in gymnastics conducts a perceptual experiment while on a space mission. Floating in a state of weightlessness, he tucks himself into a slow backwards somersault and, having done so, closes his eyes and imagines himself in a state of rest. When after a minute he opens his eyes, what he perceives is the space capsule in a slow forward rotation about his stationary body. According to Newton, however, anything other than a very slow rotation would indicate which body is rotating absolutely, the astronaut or the capsule; for the parts of a body truly in motion about an axis will tend away from that axis. (See Feynman's discussion, Chapter 59.) Newton's own experiment demonstrating this principle is terrestrial, but it does not take him long to move from a bucket of water hanging by a rope to the planets in their orbits.

The effects which distinguish absolute from relative motion are the forces of receding from the axis of circular motion. For there are no such forces in a circular motion purely relative; but in a true and absolute circular motion, they are greater or less according to the quantity of the motion.

If a vessel hung by a long cord is so often turned about that the cord is strongly twisted, then filled with water and held at rest together with the water, thereupon by the sudden action of another force it is whirled about the contrary way, and while the cord is untwisting itself the vessel continues for some time in this motion, the surface of the water will at first be plain, as before the vessel began to move. But after that the vessel, by gradually communicating its motion to the water, will make it begin sensibly to revolve and recede by little and little from the middle, and ascend to the sides of the vessel, forming itself into a concave figure (as I have experienced). And the swifter the motion becomes, the higher will the water rise, till at last, performing its revolutions in the same times with the vessel, it becomes relatively at rest in it. This ascent of the water shows its endeavor to recede from the axis of its motion; and the true and absolute circular motion of the water, which is here directly contrary to the relative, becomes known and may be measured by this endeavor. At first, when the relative motion of the water in the vessel was greatest, it produced no endeavor to recede from the axis; the water showed no tendency to the circumference, nor any ascent towards the sides of the vessel, but remained of a plain surface, and therefore its true circular motion had not yet begun. But afterwards, when the relative motion of the water had decreased, the ascent thereof towards the sides of the vessel proved its endeavor to recede from the axis; and this endeavor showed the real circular motion of the water continually increasing, till it had acquired its greatest quantity, when the water rested relatively in the vessel. And therefore this endeavor does not depend upon any translation of the water in respect of the ambient bodies, nor can true circular motion be defined by such translation.

There is only one real circular motion of any one revolving body, corresponding to only one power of endeavoring to recede from its axis of motion, as its proper and adequate effect. But relative motions in one and the same body are innumerable, according to the various relations it bears to external bodies, and, like other relations, are altogether destitute of any real effect, any otherwise than they may perhaps partake of that one only true motion. And therefore, in their system who suppose that our heavens, revolving below the sphere of the fixed stars, carry the planets along with them; the several parts of those heavens and the planets, which are indeed relatively at rest in their heavens, do yet really move. For they change their position one to another (which never happens to bodies truly at rest) and, being carried together with their heavens, partake of their motions and, as parts of revolving wholes, endeavor to recede from the axis of their motions.

In 1692, Newton corresponded with the young classicist and clergyman Richard Bentley, who was preparing his Boyle lectures, later published as A Confutation of Atheism. *Bentley was collecting evidence for design in the universe as part of the theistic apologetic indicated by his title, and Newton lent his support. The following extracts from a letter dated December 10, 1692 are notable both for Newton's acceptance of an argument from design—which reappears in more recent discussions of the "fine-tuning" of the universe—and for clarifications he makes regarding the behavior of matter, given gravity, across space either finite or infinite. (Clearly, the prospect of a "big crunch" is the main thing standing behind his preference for infinite space.) Newton even glances briefly in the direction of issues related to cosmic evolution and "lumpiness" in the universe.*

As to your first query, it seems to me that if the matter of our sun and planets and all the matter of the universe were evenly scattered throughout all the heavens, and every particle had an innate gravity towards all the rest, and the whole space throughout which this matter was scattered was but finite, the matter on the outside of this space would, by its gravity, tend toward all the matter on the inside and, by consequence, fall down into the middle of the whole space and there compose one great spherical mass. But if the matter was evenly disposed throughout an infinite space, it could never convene into one mass; but some of it would convene into one mass and some into another, so as to make an infinite number of great masses, scattered at great distances from one to another throughout all that infinite space. And thus might the sun and fixed stars be formed, supposing the matter were of a lucid nature. But how the matter should divide itself into two sorts, and that part of it which is fit to compose a shining body should fall down into one mass and make a sun and the rest, which is fit to compose an opaque body, should coalesce, not into one great body, like the shining matter, but into many little ones; or if the sun at first were an opaque body like the planets, or the planets lucid bodies like the sun, how he alone should be changed into a shining body whilst all they continue opaque, or all they be changed into opaque ones whilst he remains unchanged, I do not think explicable by mere natural causes, but am forced to ascribe it to the counsel and contrivance of a voluntary Agent. . . .

In the rising energy of the second part of his letter to Bentley, signalled by the hypotaxis of his style, Newton conveys his awe and his enthusiasm regarding the engineering feats of his geometer God.

To your second query, I answer that the motions which the planets now have could not spring from any natural cause alone but were impressed by an intelligent Agent. For since comets descend into the region of our planets and here move all manner of ways, going sometimes the same way with the planets, sometimes the contrary way, and sometimes crossways, in planes inclined to the plane of the ecliptic and at all kinds of angles, it is plain that there is no natural cause which could determine all the planets, both primary and secondary, to move the same way and in the same plane, without considerable variation; this must have been the effect of counsel. Nor is there any natural cause which could give the planets those just degrees of velocity, in proportion to their distances from the sun and other central bodies, which were requisite to make them move in such concentric orbs about those bodies. Had the planets been as swift as comets, in proportion to their distances from the sun (as they would have been had their motion been caused by their gravity, whereby the matter, at the first formation of the planets, might fall from the remotest regions towards the sun), they would not move in concentric orbs, but in such eccentric ones as the comets move in. Were all the planets as swift as Mercury or as slow as Saturn or his satellites, or were their several velocities otherwise much greater or less than they are, as they might have been had they arose from any other cause than their gravities, or had the distances from the centers about which they move been greater or less than they are, with the same velocities, or had the quantity of matter in the sun or in Saturn, Jupiter, and the earth, and by consequence their gravitating power, been greater or less than it is, the primary planets could not have revolved about the sun nor the secondary ones about Saturn, Jupiter, and the earth, in concentric circles, as they do, but would have moved in hyperbolas or parabolas or in ellipses very eccentric. To make this system, therefore, with all its motions, required a cause which understood and compared together the quantities of matter in the several bodies of the sun and planets and the gravitating powers resulting from thence, the several distances of the primary planets from the sun and of the secondary ones from Saturn, Jupiter, and the earth, and the velocities with which these planets could revolve about those quantities of matter in the central bodies; and to compare and adjust all these things together, in so great a variety of bodies, argues that cause to be, not blind and fortuitous, but very well skilled in mechanics and geometry.

SOURCES: Isaac Newton, *Mathematical Principles of Natural Philosophy and His System of the World*, translated by Andrew Mott, revised by Florian Cajori, Berkeley: U of California P, 1934; *Four Letters from Sir Isaac Newton to Doctor Bentley*, London, 1756.

35

Discernible Ends and Final Causes

Richard Bentley

The advent of Newtonian mechanics marked a sea change not only in physics itself but also in how physics was conceived in relation to religious issues such as belief in God and in divine providence. Late in 1692, Newton's correspondent Richard Bentley (1662–1742) delivered his Boyle lectures, A Confutation of Atheism from the Origin and Frame of the World. *In them he re-paints the "argument from design"—the traditional "proof" for the existence of God based on the order and fitness of the creation—on a Newtonian canvas. Facing the impediment both of ingrained Aristotelianism and of Cartesian rationalism, which deny the existence of empty space, Bentley finds himself defending a notion that for two millennia has been associated with the atomism of Democritus and Epicurus—not the sort of philosophical company he would normally mix in. Yet, for Newtonian gravity to work, there must be space between bodies.*

Since gravity is found proportional to the quantity of matter, there is a manifest necessity of admitting a vacuum, another principal doctrine of the Atomical philosophy. Because if there were everywhere an absolute plenitude and density without any empty pores and interstices between the particles of bodies, then all bodies of equal dimensions would contain an equal quantity of matter. . . .

From the point of view of modern cosmology, one of the most fascinating parts of Bentley's discussion is his attempt to grapple with the issue of the density and distribution of matter in space. That average density is today estimated at very roughly 10⁻³⁰ grams/cm²; and that distribution is now discussed by cosmologists under headings such as cosmic "smoothness" and "lumpiness" (see Chapter 76). Bentley's calculations deal with size rather than mass, but his question concerning how a smooth, thin distribution of matter might have been transformed into a complex and lumpy universe has not disappeared from the cosmological scene.

The sum of empty spaces within the concave of the firmament is 6860 million million million [6.86 x 10²¹] times bigger than all the matter contained in it.

Now from hence we are enabled to form a right conception and imagination of the supposed [original] chaos; and then we may proceed to determine the controversy with more certainty and satisfaction, whether a world like the present could possibly without a divine influence be formed in it or no. . . .

As to the state or condition of matter before the world was a-making, which is compendiously expressed by the word "chaos," [our adversaries] must suppose that either all the matter of our system was evenly or well-nigh evenly diffused through the region of the sun (this would represent a particular chaos), or all matter universally so spread through the whole mundane space (which would truly exhibit a general chaos), no part of the universe being rarer or denser than another. Which is agreeable to the ancient description of it, that the heavens and earth had one form, one texture and constitution, which could not be unless all the mundane matter were uniformly and evenly diffused. It is indifferent to our dispute whether they suppose it to have continued a long time or very little in the state of diffusion. For if there was but one single moment in all past eternity when matter was so diffused, we shall plainly and fully prove that it could never have convened afterwards into the present frame and order of things.

Even if Bentley does not in fact prove the impossibility of chaos bringing forth cosmos, he does skeptically put the question of how it might do so, satirizing the materialistic alternative to his own view that God himself designed and carried out the ordering of things. A specific target of Bentley's scorn is the Cartesian notion of vortices or "whirlpools" (French: tourbillions) of matter.

We have now represented the true scheme and condition of the chaos: how all the particles would be disunited, and what vast intervals of empty space would lie between each. To form a system, therefore, it is necessary that these squandered atoms should convene and unite into great and compact masses like the bodies of the earth and planets. Without such a coalition the diffused chaos must have continued and reigned to all eternity. But how could particles so widely dispersed combine into that closeness of texture?

Our adversaries can have only these two ways of accounting for it. Either by the common motion of matter proceeding from external impulse and conflict (without attraction) by which every body moves uniformly in a direct line according to the determination of the impelling force. For, they may say, the atoms of the chaos being variously moved according to this catholic law, must needs knock and interfere, by which means some that have convenient figures for mutual coherence might chance to stick together, and others might join to those, and so by degrees such huge masses might be formed as afterwards became suns and planets. Or there might arise some vertiginous motions or whirlpools in the matter of chaos, whereby the atoms might be thrust and crowded to the middle of those whirlpools, and there constipate one another into great solid globes, such as now appear in the world.

In part 2 of his sermon, Bentley summarizes the design argument and considers the ends for which creation was brought into being. In attempting to defend his position against possible charges of anthropocentrism, he speculates about rational life elsewhere in the universe.

The order and beauty of the systematical parts of the world, the discernible ends and final causes of them, the meliority above what was necessary to be, do evince by a reflex argument that it could not be produced by mechanism or chance, but by an intelligent and benign agent that by his excellent wisdom made the heavens.

But . . . we must offer one necessary caution: that we need not nor do not confine and determine the purposes of God in creating all mundane bodies merely to human ends and uses. Not that we believe it laborious and painful to Omnipotence to create a world out of nothing, or more laborious to create a great world than a small one, so as we might think it disagreeable to the majesty and tranquility of the Divine Nature to take so much pains for our sakes. Nor do we count it any absurdity that such a vast and immense universe should be made for the sole use of such mean and unworthy creatures as the children of men. For if we consider the dignity of an intelligent being and put that in the scales against brute inanimate matter, we may affirm without overvaluing human nature that the soul of one virtuous and religious man is of greater worth and excellency than the sun and his planets and all the stars in the world. . . . If all the heavenly bodies were thus serviceable to us, we should not be backward to assign their usefulness to mankind as the sole end of their creation. But we dare not undertake to show what advantage is brought to us by those innumerable stars in the galaxy and other parts of the firmament not discernible by naked eyes, and yet each many thousand times bigger than the whole body of the earth. If you say, they beget in us a great idea and veneration of the mighty Author and Governor of such stupendous bodies, and excite and elevate our minds to his adoration and praise, you say very truly and well. But would it not raise in us a higher apprehension of the infinite majesty and boundless beneficence of God to suppose that those remote and vast bodies were formed not merely upon our account to be peeped at through an optic glass, but for different ends and nobler purposes?

And yet who will deny but that there are great multitudes of lucid stars even beyond the reach of the best telescopes, and that every visible star may have opaque planets revolve about them, which we cannot discover? Now if they were not created for our sakes, it is certain and evident that they were not made for their own. For matter has no life nor perception, is not conscious of its own existence, nor capable of happiness, nor gives the sacrifice of praise and worship to the Author of its being. It remains, therefore, that all bodies were formed for the sake of intelligent minds. And as the earth was principally designed for the being and service and contemplation of men, why may not all other planets be created for the like uses, each for their own inhabitants which have life and understanding?

If any man will indulge himself in this speculation, . . . the Holy Scriptures do not forbid him to suppose as great a multitude of systems and as much inhabited as he pleases. . . . [We need not] be solicitous about the condition of those planetary people, nor raise frivolous disputes, how far they may participate in the miseries of Adam's fall, or in the benefits of Christ's incarnation. As if, because they are supposed to be rational they must needs be concluded to be men. For what is man? . . . God almighty by the inexhausted fecundity of his creative power may have made innumerable orders and classes of rational minds, some higher in natural perfections, others inferior to human souls. . . . God . . . may have joined immaterial souls, even of the same class and capacities in their separate state, to other kinds of bodies and in other laws of union . . . . So that we ought not upon any account to conclude that if there be rational inhabitants in the moon or Mars or any unknown planets of other systems, they must therefore have human nature, or be involved in the circumstances of our world. And thus much was necessary to be here inculcated (which will obviate and preclude the most considerable objections of our adversaries) that we do not determine the final causes and usefulness of the systematical parts of the world merely as they have respect to the exigencies or conveniences of human life.

SOURCE: Richard Bentley, *A Confutation of Atheism from the Origin and Frame of the World*, London, 1693.

36

The Planetarians, and This Small Speck of Dirt

Christiaan Huygens

Christiaan Huygens (1629–1695), an acquaintance of Descartes, Pascal, Leibniz, and (briefly) Newton, lived for many years in Paris and, though a Dutchman, became a founding member of the French Academy of Sciences. Huygens's contribution to descriptive astronomy is based largely on practical improvements he himself pioneered in the construction of telescopes. In 1659, Huygens published Systema Saturnium, *in which he described the shape of the rings of Saturn; he is also credited with founding the wave theory of light and discovering the usefulness of the pendulum as a regulator of clocks.*

Huygens's posthumous Cosmotheoros, *published in Latin in 1698 and translated into English the same year under the title* The Celestial Worlds Discovered, *is a retrospective and speculative work addressed familiarly to Huygens's brother Constantine.*

A man that is of Copernicus's opinion, that this earth of ours is a planet carried round and enlightened by the sun like the rest of them, cannot but sometimes have a fancy that it's not improbable that the rest of the planets have their dress and furniture, nay and their inhabitants too as well as this earth of ours: especially if he considers the later discoveries made since Copernicus's time of the attendants of Jupiter and Saturn, and the champaign and hilly countries in the moon, which are an argument of a relation and kin between our earth and them, as well as a proof of the truth of that system.

This has often been our talk, I remember, good brother, over a large telescope, when we have been viewing those bodies . . . . But we were always apt
to conclude that 'twas in vain to enquire after what nature had been pleased
to do there, seeing there was no likelihood of ever coming to an end of the
enquiry. Nor could I ever find that any philosophers, those bold heroes either
ancient or modern, ventured so far. At the very birth of astronomy, when the
earth was first asserted to be spherical and to be surrounded with air, even
then there were some men so bold as to affirm there were an innumerable
company of worlds in the stars. But later authors such as Cardinal Cusanus,
Brunus, Kepler (and if we may believe him, Tycho was of that opinion too)
have furnished the planets with inhabitants. . . . Some of them have coined
some pretty fairy stories of the men in the moon, just as probable as Lucian's
true history; among which I must count Kepler's, which he has diverted us
with in his Astronomical Dream.

But a while ago, thinking somewhat seriously of this matter . . . methought
the enquiry was not so impracticable, nor the way so stopped up with difficulties, but that there was very good room left for probable conjectures. As
they came into my head I clapped them down into commonplaces and shall
now try to digest them into some tolerable method for your better conception of them, and add somewhat of the sun and fixed stars, and the extent of
that universe of which our earth is but an inconsiderable point.

*In Huygens's musings we see the confluence of important themes both of ma-
ture Copernicanism and of what would come to be called Enlightenment
thought. The isotropic world of Galileo—in which there are mountains on
the moon just as there are on the earth, and in which earthshine brightens the
dark side of the moon even as moonshine brightens our night—is one that invites us to extrapolate from what we know on earth to other worlds. Huygens refers explicitly to such analogies as the basis of "our method in this
treatise, wherein, from the nature and circumstances of that planet which we
see before our eyes, we may guess at those that are farther distant from us."
The result, perhaps predictably, is cosmological speculation that is often
stunningly anthropocentric or geocentric, though of course no longer in any
narrowly geometrical sense. This central paradox of self-professing Copernicanism—the simultaneous reduction of the earth to a mere peripheral point
and its elevation as the measure of all things—is in Huygens complicated and
complemented by Leibnizian themes such as the principle of plenitude and
the (logical) plurality of worlds.*

*In offering what at first appears a standard declaration that astronomical
study does not conflict with scriptural teaching, Huygens lays the foundation*

for his further speculations concerning the existence of extraterrestrial intelligence that is human or quasi-human.

[The critics] have been answered so often that I am almost ashamed to repeat it: that it's evident God had no design to make a particular enumeration in the Holy Scriptures of all the works of his creation. When therefore it is plain that under the general name of *stars* and *earth* are comprehended all the heavenly bodies, even the little gentlemen round Jupiter and Saturn, why must all that multitude of beings which the almighty creator has been pleased to place upon them be excluded the privilege, and not suffered to have a share in the expression? And these men themselves can't but know in what sense it is that all things are said to be made for the use of man, not certainly for us to stare or peep through a telescope at, for that's little better than nonsense. Since then the greatest part of God's creation, that innumerable multitude of stars, is placed out of the reach of any man's eye, and many of them, it's likely, of the best glasses, so that they don't seem to belong to us; is it such an unreasonable opinion that there are some reasonable creatures who see and admire those glorious bodies at a nearer distance?

The more speculative Huygens's enquiry becomes, the more pronounced and careful is his defense of its utility and piety.

I must acknowledge still that what I here intend to treat of is not of that nature as to admit of a certain knowledge. I can't pretend to assert anything as positively true (for that would be madness) but only to advance a probable guess, the truth of which every one is at his own liberty to examine. If any one therefore shall gravely tell me, that I have spent my time idly in a vain and fruitless enquiry after what by my own acknowledgment I can never come to be sure of, the answer is that at this rate he would put down all natural philosophy as far as it concerns itself in searching into the nature of things. In such noble and sublime studies as these, 'tis a glory to arrive at probability, and the search itself rewards the pains. But there are many degrees of probable, some nearer truth than others, in the determining of which lies the chief exercise of our judgement.

But besides the nobleness and pleasure of the studies, may not we be so bold as to say, they are no small help to the advancement of wisdom and morality? So far are they from being of no use at all. For here we may mount from this dull earth and, viewing it from on high, consider whether Nature has laid out all her cost and finery upon this small speck of dirt. So, like travellers into other distant countries, we shall be better able to judge of what's done at home, know how to make a true estimate of, and set its own value upon everything. We shall be less apt to admire what this world calls great, shall nobly despise those trifles the generality of men set their affections on, when we know that there are a multitude of such earths inhabited and adorned as well as our own. And we shall worship and reverence that God, the maker of all these things; we shall admire and adore his providence and wonderful wisdom which is displayed and manifested all over the universe, to the confusion of those who would have the earth and all things formed by the shuffling concourse of atoms.

After a long discussion supporting the claim that the planets are provided with water, plants, animals, and other "furniture" such as the earth is fitted out with, Huygens addresses the issue of "spectators," and, surprisingly, appears to declare unanimity among those who have addressed it. Again the premises of the argument seem utterly torn between anthropocentrism and an impulse to reject it.

But still the main and most diverting point . . . is the placing some spectators in these new discoveries, to enjoy these creatures we have planted them with, and to admire their beauty and variety. And among all that have never so slightly meddled with these matters I don't find any that have scrupled to allow them their inhabitants: not men perhaps like ours, but some creatures or other endued with reason. For all this furniture and beauty the planets are stocked with seem to have been made in vain, without design or end, unless there were some in them that might at the same time enjoy the fruits and adore the wise Creator of them.

But this alone would be no prevailing argument with me to allow them such creatures. For what if we should say, that God made them for no other design but that he himself might see . . . and delight himself in the contemplation of them? . . . That which makes me of this opinion, that those worlds are not without such a creature endued with reason, is that otherwise our earth would have too much the advantage of them, in being the only part of the universe that could boast of such a creature so far above not only plants and trees but all animals whatsoever: a creature that has a divine somewhat within him, that knows, and understands, and remembers such an innumerable number of things; that deliberates, weighs and judges the truth; a creature upon whose account and for whose use whatsoever the earth brings forth seems to be provided. . . . If we should allow Jupiter a greater variety of other creatures, more trees, herbs and metals, all these would not advantage or dignify that planet so much as that one animal doth ours by the admirable productions of his penetrating wit.

Whereas previous ages may have posited that nonhuman rational creatures, angels in particular, dwell in or preside over the heavenly spheres—humankind, a little lower than the angels but lower nonetheless, being quarantined in the fallen realm, within the sublunary sphere—Huygens postulates extraterrestrials possessing even human vices. Moreover, armed with Leibnizian "optimism," including an instrumental view of evil as component of a greater good, he minimizes the seriousness of such moral evil within the scheme of the universe as a whole.

Nor let anyone say here that there's so much villainy and wickedness in this man that we have thus magnified, that it's a reasonable doubt whether he would not be so far from being the glory and ornament of the planet that enjoys his company that he would be rather its shame and disgrace. For first, the vices that most men are tainted with are no hindrance, but that those that follow the dictates of true reason and obey the rules of a rigid virtue are still a beauty and ornament to the place that has the happiness to harbor them. Besides, the vices of men themselves are of excellent use, and are not permitted and allowed in the world without wise design. . . . Virtues themselves . . . would be of no use if there were no dangers, no adversity, no afflictions for their exercise and trial.

If we should therefore imagine in the planets some such reasonable animal as man is, adorned with the same virtues and infected with the same vices, it would be so far from degrading or vilifying them that, while they want such a one, I must think them inferior to our earth.

Well, but allowing these planetarians some sort of reason, must it needs be the same with ours? Why, truly I think 'tis, and must be so, whether we consider it as applied to justice and morality, or exercised in the principles and foundations of science. For reason with us is that which gives us a true sense of justice and honesty, praise, kindness, and gratitude; 'tis that that teaches us to distinguish universally between good and bad, and renders us capable of knowledge and experience in it. And can there be anywhere a reason contrary to this, or can what we call just and generous in Jupiter or Mars be thought unjust villainy? This is not at all, I don't say probable, but possible.

Near the end of his little book, Huygens returns from speculation to retrospective reflection upon the early history of Copernicanism; and to a degree he criticizes, while trying to explain the motivations for, some of his predecessors' opinions. One may be tempted to adapt his concluding comment on Kepler as an appropriate appraisal of some aspects of Huygens's own work.

Before the invention of telescopes it seemed to contradict Copernicus's opinion to make the sun one of the fixed stars. For the stars of the first magnitude being esteemed to be about three minutes diameter, and Copernicus (observing that though the earth changed its place, they always kept the same distance from us) having ventured to say that the *magnus orbis* [the "great circle"] was but a point in respect of the sphere in which they were placed, it was a plain consequence that every one of them that appeared anything bright must be larger than the path or orbit of the earth, which is very absurd. This is the topping argument that Tycho Brahe set up against Copernicus. But when the telescopes shaved them of their fictitious rays, and showed them to us bare and naked (which they do best when the eye-glass is blacked with smoke) just like little shining points, then that difficulty vanished, and the stars might still be suns. Which is the more probable, because their light is certainly their own: for it's impossible that ever the sun should send or they reflect it at such a vast distance.

This is the opinion that commonly goes along with Copernicus's system. And the patrons of it do also with reason suppose that all these stars are not in the same sphere, as well because there's no argument for it, as that the sun, which is one of them, cannot be brought to this rule. But it's more likely they are scattered and dispersed all over the immense spaces of the heaven, and are as far distant perhaps from one another as the nearest of them are from the sun.

Here again too I know Kepler is of another opinion. . . . For though he agrees with us that the stars are diffused through all the vast profundity, yet he cannot allow that they have as large an empty space about them as our sun has. . . . But Kepler had a private design in making the sun thus superior to all the other stars, and planting it in the middle of the world, attended with planets, a favor that he did not desire to grant the rest. For his aim was by it to strengthen his Cosmographical Mystery, that the distances of the planets from the sun are in a certain proportion to the diameters of the spheres that are inscribed within, and circumscribed about Euclid's polyhedrical bodies. Which could never be so much as probable except there were but one chorus of planets moving round the sun, and so the sun were the only one of his kind.

But the whole Mystery is nothing but an idle dream taken from Pythagoras or Plato's philosophy . . . a mere fancy without any shadow of reason. I cannot but wonder how such things as these could fall from so ingenious a man, and so great an astronomer.

SOURCE: Christiaan Huygens, *The Celestial Worlds Discovered*, London, 1698.

Part Four

UNFURLING NEWTON'S UNIVERSE

37

A Signal of God

William Derham

William Derham (1657–1735) was an English clergyman who helped to popularize the new astronomy of his day and to declare its harmony with faith in the God of Christianity. His influential book Astro-Theology, *published in 1714 and reissued numerous times throughout the eighteenth century, not only contributes to the "argument from design" taken up again early in the next century by William Paley, but also provides evidence that even 170 years after* De Revolutionibus *not all doubts concerning the Copernican system had been quite swept away. Derham repeatedly appeals to the principle of economy in opposing the Ptolemaic system.*

[The Copernican system] is far more agreeable to nature, which never goes a roundabout way, but always acts by the most compendious, easy, and simple methods. And in the Copernican way, that is performed by one or a few easy revolutions, which in the other way is made the work of the whole heavens, and of many strange and unnatural orbs. Thus the diurnal motion is accounted for by one revolution of the earth which all the whole heavens are called for in the other way. So for the periodical motions of the planets, their stations, retrogradations, and direct motions, they are all accounted for by one easy, single motion round the sun, for which in the Ptolemaic way, they are forced to invent diverse strange, unnatural, interfering eccentrics and epicycles—a hypothesis so bungling and monstrous, as gave occasion to a certain king to say, if he had been of God's counsel when he made the heavens, he could have taught him how to have mended his work. . . .

The prodigious and inconceivable rapidity assigned by the Ptolemaics to the heavens is by the Copernican scheme taken off, and a far more easy and tolerable motion substituted in its room. For is it not a far more easy motion for the earth to revolve round its own axis in twenty-four hours than for so great a number of far more massy and far distant globes to revolve round the earth in the same space of time? If the maintainers of the Ptolemaic system do object against the motion of the earth, that it would make us dizzy and shatter our globe to pieces, what a precipitant, how terrible a rapidity, must that of the heavens be?

In the course of defending the basic principles of Copernicanism, Derham assisted in the popular propagation of Galileian-Newtonian physics. One of the most persistent arguments against any motion of the earth was based on "common sense": We do not perceive the earth to move. Derham answers with extensive quotation directly from Galileo—in which appears the classic example, beloved of both Copernicus and Einstein, of movement within an enclosed frame-of-reference.

The objections [to Copernicanism] from philosophy are too numerous to be distinctly answered, especially such as seem very frivolous, particularly those grounded on a supposition of the verity of the Aristotelian philosophy, as the immutability and incorruptibility of the heavens, etc. For answers to which I shall refer the reader to Galileo's *System of the World*. But for such objections as seem to have some reason in them, they are chiefly these: that if the earth be moved from west to east, a bullet shot westward would have a farther range than one shot eastward; or if shot north or south, it would miss the mark; or if perpendicularly upright, it would drop to the westward of the gun. That a weight dropped from the top of a tower would not fall down just at the bottom of the tower, as we see it doth. . . .

But not to enter into a detail of answers that might be given to the laws of motion, and the rules of mechanics and mathematics, I shall only make use of the most ingenious Galileo's plain experiment, which answereth all or most of the objections.

Shut yourself up with your friend in the great cabin of a ship, together with a parcel of gnats and flies, and other little winged creatures. Procure also a great tub of water, and put fishes therein. Hang also a bottle of water up, to empty itself drop by drop into another such bottle placed underneath with a narrow neck. whilst the ship lies still, diligently observe how those little winged creatures fly with the like swiftness towards every part of the cabin; how the fishes swim indifferently towards

all sides; and how the descending drops all fall into the bottle underneath. . . .

Having observed these particulars whilst the ship lies still, make the
ship to sail with whatever velocity you please, and so long as the motion
is uniform, not fluctuating this way and that way, you shall not perceive
there is any alteration in the aforesaid effects. Neither can you from
them conclude whether the ship moveth or standeth still. But in leaping
you shall reach as far on the floor as you did before; nor by reason of
the ship's motion shall you make a longer leap towards the poop than
the prow, notwithstanding that whilst you were up in the air, the floor
under your feet had run the contrary way to your leap. And if you cast
anything to your companion, you need use no more strength to make it
reach him, if he should be towards the prow, and you towards the poop,
than if you stood in the contrary position. The drops shall all fall into
the lower bottle, and not one towards the poop, although the ship shall
have run many feet whilst the drop was in the air. The fishes in the water
shall have no more trouble in swimming towards the forepart of the tub,
than towards the hinder part. . . . And lastly, the gnats and flies shall
continue their flight indifferently towards all parts and never be driven
together towards the side of the cabin next the prow, as if wearied with
following the swift motion of the ship. . . .

The cause of which correspondence of the effects is that the ship's motion is common to all things contained in it, and to the air also, I mean
when those things are shut up in the cabin. . . .

Thus Galileo by this one observation hath answered the most considerable
objections deduced from philosophy against the motion of the earth.

*Even more significant than Derham's defense of the Copernican system is his
recognition that already a more capacious "New System" has superseded it.*

Now I pass from the second system to the third, which is called the New System, which extends the universe to a far more immense compass than any of
the other systems do, even to an indefinite space, and replenishes it with a far
more grand retinue than ever was before ascribed unto it.

This new system is the same with the Copernican as to the system of the sun
and its planets. . . . But then whereas the Copernican hypothesis supposeth the
firmament of the fixed stars to be the bounds of the universe, and to be placed
at equal distance from its center the sun, the new system supposeth there are
many other systems of suns and planets besides that in which we have our res-
idence, namely, that every fixed star is a sun and encompassed with a system of
planets, both primary and secondary, as well as ours.

These several systems of the fixed stars, as they are at a great and sufficient distance from the sun and us, so they are imagined to be at as due and regular distances from one another. By which means it is that those multitudes of fixed stars appear to us of different magnitudes, the nearest to us large, those farther and farther away less and less.

Although he avoids dogmatism about the details of this new system, whose genealogy goes back to Cusanus and Bruno, Derham is attracted to it for two main reasons:

1. Because it is far the most magnificent of any, and worthy of an infinite Creator, whose power and wisdom, as they are without bounds and measure, so may in all probability exert themselves in the creation of many systems, as well as one. And as myriads of systems are more for the glory of God, and more demonstrate his attributes than one, so it is no less probable than possible there may be many besides this which we have the privilege of living in. . . .
2. We see it really so, as far as it is possible it can be discerned by us, at such immense distances as those systems of the fixed stars are from us. Our glasses are indeed too weak so to reach those systems as to give us any assurance of our seeing the planets themselves that encompass any of the fixed stars. We cannot say we see them actually moving round their respective suns or stars. But this we can discern, namely that the fixed stars have the nature of suns.

From other suns it is only a small step to other earths and all the speculation that naturally follows concerning other inhabitants of those earths. In such speculations Derham readily acknowledges the influence of Christiaan Huygens (see Chapter 36) but goes on to elaborate his own moral and theological applications for this new cosmology, which not only magnifies the glory of God as creator but also literally and figuratively puts human beings in their place. In making his case Derham, by identifying geocentrism with anthropocentrism, propagates a cliché that is still with us.

We have this farther to recommend those imaginations to us, that this account of the universe is far more magnificent, worthy of, and becoming the infinite Creator than any other of the narrower schemes. For here we have the works of the creation not confined to the more scanty limits of the orb, or arch of the fixed stars, or even the larger space of the primum mobile, which the ancients fancied were the utmost bounds of the universe, but they are extended to a far larger as well as more probable, even indefinite, space. . . .

Also in this prospect of the creation, as the earth is discarded from being the center of the universe, so neither do we make the uses and offices of all the glorious bodies of the universe to center therein, nay, in man alone, according to the old vulgar opinion that "all things were made for man." But in this our scheme we have a far more extensive, grand, and noble view of God's works: a far greater number of them, not those alone that former ages saw, but multitudes of others that the telescope hath discovered since, and all these far more orderly placed throughout the heavens, and at more due and agreeable distances, and made to serve to much more noble and proper ends. For here we have not one system of sun and planets alone, and one only habitable globe, but myriads of systems, and more of habitable worlds, and some even in our own solar system, as well as those of the fixed stars. And consequently if in the sun and its planets, although viewed only here upon the earth at a great distance, we find enough to entertain our eye, to captivate our understanding, to excite our admiration and praises of the infinite Creator and Contriver of them, what an augmentation of these glories shall we find in great multitudes of them!

Finally for our purposes, Derham gives expression to the "argument from design" by way of investing in the Newtonian mechanical universe, which runs like a clock.

Having shown that the giving motion to such immense, lifeless globes is the work of God, we shall find much greater demonstrations thereof if we consider that those motions are not at random, in inconvenient lines and orbs, but such as show wise design and counsel. I shall here specify but two examples. . . . One is that all the planets should (when their motions were impressed upon them) have their directions or tendencies given not in lines tending from the center to the circumference, or very obliquely, but perpendicularly to the radii. The other is that the motions and orbits of the planets should not interfere with one another but tend one and the same way, from west to east, and lie in planes but little inclined to one another, or when inclined, that it should be very beneficially so. . . . These and many other instances, and in a word, that every planet should have as many and various motions, and those as regularly and well-contrived and ordered, as the world and its inhabitants have occasion for: what could all this be but the work of a wise and kind as well as omnipotent Creator and orderer of the world's affairs?—a work which is as plain a signal of God, as that of a clock or other machine is of man.

SOURCE: William Derham, *Astro-Theology: Or, a Demonstration of the Being and Attributes of God, from a Survey of the Heavens*, 10th ed., London, 1767.

38

The Beautiful Pre-established Order

Gottfried Wilhelm Leibniz and Samuel Clarke

The broad contest in physical cosmology between Newtonianism and the mainly continental position represented by René Descartes and Gottfried Wilhelm Leibniz (1646–1716) is encapsulated by the famous correspondence between Leibniz and Samuel Clarke (1675–1729), a close associate of Newton's. In defending their particular physics, both Leibniz and Clarke see themselves as also defending the honor and attributes of God. Leibniz opens with the satirical view of Newton's mechanical universe as a watch—one that unfortunately needs regular maintenance. Clarke replies by challenging the assumptions concealed within this familiar analogy.

Leibniz:

Sir Isaac Newton and his followers have . . . a very odd opinion concerning the work of God. According to their doctrine God Almighty wants to wind up his watch from time to time, otherwise it would cease to move. He had not, it seems, sufficient foresight to make it a perpetual motion. Nay, the machine of God's making is so imperfect, according to these gentlemen, that he is obliged to clean it now and then by an extraordinary concourse, and even to mend it as a clockmaker mends his work, who must consequently be so much the more unskillful a workman as he is oftener obliged to mend his work and to set it right. According to my opinion, the same force and vigor remains always in the world and only passes from one part of matter to another, agreeably to the laws of nature and the beautiful pre-established order. And I hold that, when God works miracles, he does not do it in order to supply the wants of nature but those of grace. Whoever thinks otherwise must needs have a very mean notion of the wisdom and power of God.

Clarke:

... The reason why, among men, an artificer is justly esteemed so much the more skillful, as the machine of his composing will continue longer to move regularly without any farther interposition of the workman, is because the skill of all human artificers consists only in composing, adjusting or putting together certain movements, the principles of whose motion are altogether independent upon the artificer, such as are weights and springs and the like, whose forces are not made but only adjusted by the workman. But with regard to God, the case is quite different, because he not only composes or puts things together but is himself the author and continual preserver of their original forces or moving powers; and consequently it is not a diminution but the true glory of his workmanship that nothing is done without his continual government and inspection.

The notion of the world's being a great machine, going on without the interposition of God as a clock continues to go without the assistance of a clockmaker, is the notion of materialism and fate, and tends (under the pretence of making God a supra-mundane intelligence) to exclude providence and God's government in reality out of the world. And by the same reason that a philosopher can represent all things going on from the beginning of the creation without any government or interposition of providence, a skeptic will easily argue still farther backwards and suppose that things have from eternity gone on (as they now do) without any true creation or original author at all, but only what such arguers call all-wise and eternal nature. If a king had a kingdom wherein all things would continually go on without his government or interposition, or without his attending to and ordering what is done therein, it would be to him merely a nominal kingdom, nor would he in reality deserve at all the title of king or governor. And as those men who pretend that in an earthly government things may go on perfectly well without the king himself ordering or disposing of any thing may reasonably be suspected that they would like very well to set the king aside, so whosoever contends that the course of the world can go on without the continual direction of God, the Supreme Governor—his doctrine does in effect tend to exclude God out of the world.

Not all of Clarke and Leibniz's disagreements are theological. At the level of spatial and physical concepts one of their essential disagreements concerns the possibility of a vacuum. We may recall that Descartes' theory of vortices or "whirlpools" came about largely because he had to account for circular or elliptical planetary movements within what he thought were "liquid" heavens. Leibniz too refuses to admit a vacuum, and his reasoning employs what is sometimes called "the principle of plenitude," the rationalist assumption (partly traceable back to Neoplatonism) that a perfect God creates the best world possible, and that the best world is the one that is "fullest" and in which there are no gaps, and hence no vacuum. It is perhaps no accident that a discoverer of the infinitesimal calculus, one of whose assumptions is that number itself is infinitely divisible—and hence that there are no gaps in the series of real numbers—should develop a physical theory based on a denial of material indivisibility or discontinuity.

One need not be a partisan empiricist to note that the issue of the vacuum was in fact decisively settled not by metaphysical reflection but by physical experiments such as those already pioneered by Torricelli, von Guerricke, and Boyle.

Leibniz:

All those who maintain a vacuum are more influenced by imagination than by reason. When I was a young man, I also gave into the notion of a vacuum and atoms, but reason brought me into the right way. It was a pleasing imagination. Men carry on their inquiries no farther than those two things [i.e., vacuum and atoms]; they as it were nail down their thoughts to them; they fancy they have found out the first elements of things, a *non plus ultra*. We would have nature to go no farther, and to be finite, as our minds are; but this is being ignorant of the greatness and majesty of the author of things. The least corpuscle is actually subdivided *in infinitum*, and contains a world of other creatures which would be wanting in the universe if that corpuscle was an atom, that is, a body of one entire piece without subdivision.

In like manner, to admit a vacuum in nature is ascribing to God a very imperfect work; it is violating the grand principle of the necessity of a sufficient reason, which many have talked of without understanding its true meaning—as I have lately shown in proving by that principle that space is only an order of things, as time also is, and not at all an absolute being.

To omit many other arguments against a vacuum and atoms, I shall here mention those which I ground upon God's perfection, and upon the necessity of a sufficient reason. I lay it down as a principle that every perfection which God could impart to things without derogating from their other perfections has actually been imparted to them. Now let us fancy a space wholly empty. God could have placed some matter in it without derogating in any respect from all other things. Therefore he has actually placed some matter in that space; therefore there is no space wholly empty; therefore all is full. The same argument proves that there is no corpuscle but what is subdivided.

Another version of the "no gaps" theory as applied not to number or physics but to the whole "chain of being" is espoused by Alexander Pope in An Essay on Man *(1733), where he suggests that this continuity explains (with ironic double meaning) why there has to be such a "rank" as man.*

> Of systems possible, if 'tis confessed
> That Wisdom Infinite must form the best,
> Where all must full or not coherent be,
> And all that rises, rise in due degree,
> Then in the scale of reasoning life 'tis plain
> There must be, somewhere, such a rank as man.

Finally, as far as physical cosmology is concerned, in his correspondence with Clarke, Leibniz also expresses the greatest and most enduring objection to the Newtonian view of gravity, namely that such gravity involves action at a distance, and hence that its operation is "occult." This difficulty never was solved in Newtonian terms and was circumvented only with the advent of Einsteinian relativity, according to which all gravitational action is local action attributable to the grip of mass upon spacetime (see Chapter 60).

Leibniz:
I objected that an attraction, properly so called or in the scholastic sense, would be an operation at a distance, without any means of intervening. The author [Clarke] answers here that an attraction without any means of intervening would indeed be a contradiction. Very well! But then what does he mean, when he will have the sun to attract the globe of the earth through an empty space? Is it God himself that performs it? But this would be a miracle if ever there was any. This would surely exceed the powers of creatures.

Or, are perhaps some immaterial substances, or some spiritual rays, or some accident without a substance, or some kind of *species intentionalis*, or some other I know not what, the means by which this is pretended to be performed? Of which sort of things the author seems to have still a good stock in his head without explaining himself sufficiently.

That means of communication (says he) is invisible, intangible, not mechanical. He might as well have added inexplicable, unintelligible, precarious, groundless, and unexampled.

But it is regular (says the author); it is constant, and consequently natural. I answer, it cannot be regular without being reasonable, nor natural unless it can be explained by the natures of creatures.

If the means which causes an attraction properly so called be constant, and at the same time inexplicable by the powers of creatures, and yet be true, it must be a perpetual miracle; and if it is not miraculous, it is false. It is a chimerical thing, a scholastic occult quality.

The case would be the same as in a body going round without receding in the tangent, though nothing that can be explained hindered it from receding, which is an instance I have already alleged. And the author has not thought fit to answer it, because it shows too clearly the difference between what is truly natural on the one side and a chimerical occult quality of the schools on the other.

SOURCE: *A Collection of Papers, which passed between the late learned Mr. Leibnitz, and Dr. Clarke, in the years 1715 and 1716,* London, 1717.

39

An Event So Glorious to the Newtonian Doctrine of Gravity

Edmond Halley and "Astrophilus"

Edmond Halley (1656–1742) financed the publication of Newton's Principia *and went on to become one of the greatest astronomers ever, largely based on his careful application of Newtonian "doctrines" to the motions of celestial bodies. The most famous of these has now for over two hundred years borne his name: Halley's Comet.*

Halley, working with Newtonian mechanics and principles of the calculus, proposed in 1705 that comets were not one-off but instead recurring *phenomena whose return could be predicted. The geometry involved in this proposal is relatively simple. If one assumes that a comet comes as a one-time visitor streaking past the sun and makes a tight bend in its trajectory under the influence of the sun's gravitation, then one has in effect (under Newtonian rules) assumed that that curve is a parabola, with the sun as its focus. All Halley does is to adjust our conception of that trajectory, replacing parabola with ellipse, by describing the correspondence between the proportions of the pointy end of a parabola and those of the pointy end of an ellipse. To follow his elegant, compact, and historic proposal we require but three mildly technical definitions: (1) "latus rectum" is the line segment (or chord) drawn through the focus, and at right angles to the axis, of a parabola or ellipse; (2) "perihelion" is the point in the curve (in this case, orbit) closest to the focus (in this case, the sun); and (3) "aphelion" (pronounced "a-feely-un") is the point of an orbit farthest from the sun.*

Hitherto I have considered the orbits of comets as exactly parabolic, upon which supposition it would follow that comets, being impelled towards the sun by a centripetal force, would descend as from spaces infinitely distant and, by their so falling, acquire such a velocity as that they may again fly off into the remotest parts of the universe, moving upwards with a perpetual tendency so as never to return again to the sun. But since they appear frequently enough, and since some of them can be found to move with a hyperbolic motion, or a motion swifter than what a comet might acquire by its gravity to the sun, it is highly probable they rather move in very eccentric elliptical orbits, and make their returns after long periods of time; for so their number will be determinate, and perhaps not so very great. Besides, the space between the sun and the fixed stars is so immense that there is room enough for a comet to revolve, though the period of its revolution be vastly long.

Now the *latus rectum* of an ellipse is to the *latus rectum* of a parabola which has the same distance in its perihelion, as the distance in the aphelion in the ellipse is to the whole axis of the ellipse. And the velocities are in a subduplicate ratio of the same. Wherefore in very eccentric orbits the ratio comes very near to a ratio of equality; and the very small difference which happens on account of the greater velocity in the parabola is easily compensated in determining the situation of the orbit. The principal use therefore of this table of the elements of their motions [not shown here], and that which indeed induced me to construct it, is that whenever a new comet shall appear, we may be able to know, by comparing together the elements, whether it be any of those which has appeared before, and consequently to determine its period, and the axis of its orbit, and to foretell its return.

And indeed there are many things which make me believe that the comet which Apian observed in the year 1531 was the same with that which Kepler and Longomontanus more accurately described in the year 1607, and which I myself have seen return, and observed in the year 1682. All the elements agree, and nothing seems to contradict this my opinion, besides the inequality of the periodic revolutions. Which inequality is not so great neither, as that it may not be owing to physical causes. For the motion of Saturn is so disturbed by the rest of the planets, especially Jupiter, that the periodic time of that planet is uncertain for some whole days together. How much more, therefore, will a comet be subject to such like errors, which rises almost four times higher than Saturn, and whose velocity, though increased but a very little, would be sufficient to change its orbit from an elliptical to a parabolical one. And I am the more confirmed in my opinion of its being the same, for that in the year 1456, in the summertime, a comet was seen passing retrograde between the earth and the sun, much after the same manner; which, though nobody made observations upon it, yet from its period and the manner of its transit, I cannot think different from those I have just now mentioned. And since looking over the histories of comets I find, at an equal interval of time, a comet to have been seen about Easter in the year 1305, which is another double period of 151 years before the former.

Hence I think I may venture to foretell that it will return again in the year 1758. And, if it should then so return, we shall have no reason to doubt but the rest may return also. Therefore astronomers have a large field wherein to exercise themselves for many ages before they will be able to know the number of these many and great bodies revolving about the common center of the sun, and to reduce their motions to certain rules.

There is nothing quite so impressive in science as a theory that specifies and then satisfies the criteria for its own experimental confirmation. It is an added bonus, of course, when the person making the prediction can be claimed with pride by an entire nation, as we see in the following excerpt from The Gentleman's Magazine *for November 1759, a date just less than a year after the return of what the letter writer, who signs himself merely "Astrophilus," calls "the long expected comet." The writer also betrays a little of the attitude often encountered in popular science writing of whatever age: self-congratulation on account of what "we" know in contrast to the ignorance of previous ages.*

The return of a comet accurately foretold more than 50 years ago is the only phenomenon of its kind, and at the same time one of the most interesting to those who are desirous of forming the truest notions of the system of the world. It was a happiness reserved for the astronomers of this age to be favored with a more perfect and complete knowledge of these wandering bodies, which appear at certain times only, and after certain periods; of whose true motions, and even nature, the ancients were so entirely ignorant that the generality of them doubted even whether they were real bodies.

Dr. Halley, from a consideration of the great influence which the planet Jupiter must necessarily have upon the comet during its last return in 1682, according to the theory of gravity, had said that it was very probable that "its next return [would] not be until after the period of 76 years or more, about the end of the year 1758, or the beginning of 1759." Some astronomers, without examining or weighing Dr. Halley's reasons, and proceeding upon other principles, expected its return much sooner. But M. Clairaut, upon a farther prosecution of Dr. Halley's principles and calculations, thought he could with still greater precision foretell the time of its coming to its perihelion, which he fixed, after a series of very painful calculations, about the middle of April 1759, without pretending however to determine it within the space of a month, upon account of many small quantities which he was then obliged to neglect. . . .

The impatience of astronomers, and the desire of verifying the success of Dr. Halley's prediction, induced many to enquire in what part of the heavens the comet would first appear; but, as they were ignorant of the precise time of its return, they could only know the place of its appearance by making different suppositions concerning the time of its passing the perihelion. In consequence of this, M. Messier, an assistant to M. [J.-N.] Delisle, in his astronomical observations made at the Observatory de la Marine à l'Hotel de Clugny, had the happiness of first discovering this long-expected comet on Jan. 21, 1759, in the evening. Its very feeble light, equally extended round a luminous point, which was its nucleus, and its different situation in regard to the fixed stars, which he observed every day, soon convinced him that it was a comet. It remained then to know whether it was the very comet whose return had been predicted, and so long and impatiently expected. And of this we can only be assured after an exact determination of the velocity and direction of the apparent path of the new comet, which ought to be the same with that of the comet of 1682, supposing it to appear in the same place, and at the same season of the year, which the succeeding observations determined. For having compared this comet with the neighboring stars as often as the sky was clear, and having in some measure determined the position of those stars, M. Delisle was assured that its motion was retrograde, and that it had the same apparent direction and velocity which the comet of 1682 would have had if viewed at the same time and in the same place.

Astrophilus narrates the appearances of—and the various difficulties encountered in viewing—the comet in the northern hemisphere from January to March 1759, its descent into the southern hemisphere, and its reappearance low in the northern skies again in late April.

From the imperfect accounts which had been received in England that a comet had been observed by some of the French astronomers in the tail of Capricorn about the beginning of April, some of our most accurate observers here had reason to expect the appearance of this most remarkable comet above our horizon towards the end of that or the beginning of the following month. Accordingly on Monday, the 30th of April, between 8 and 9 in the evening, the comet was seen by many curious persons. . . . The comet, unfortunately for astronomers, now passed on in a direction almost perpendicular to the horizon, decreasing both in right ascension and declination, but through a tract of the heavens where there were few stars of the 1st, 2nd,

3rd, or 4th magnitude, but such as were at a considerable distance from it, and those which were near it as it moved along were so small that their position could not then be determined. But as the season is now advancing, when the places of those stars may be ascertained, we may soon hope to be favored with the elements of the comet, for the time of its last return, with some degree of precision, more particularly if the French astronomers were enabled to determine its situation soon after it had passed the perihelion; for without one or two good observations of the comet in that part of its orbit, we can never have its elements with any tolerable degree of exactness, as the comet had before the beginning of May run through the whole part of its orbit which coincides with the parabola, and continued to move, as long as it was visible, almost in a straight line. It might be seen with the naked eye till about the 4th or 5th of May, when the moon, which had then entered into its second quarter, shone with such brightness, and the light of the comet itself was so very faint, that they only who knew where to look for it could find it. Through the telescope nothing then could be seen but a nebulosity somewhat extended, in the middle of which the nucleus might easily be distinguished. . . .

Such is the history, however short and imperfect, of the observations of this comet, which have hitherto been made public, or which have come to my hands. How far its elements differ from those determined by Dr. Halley, from Flamsteed's observations during the time of its appearance in 1682 . . . must be determined by a more able hand, whenever a sufficient number of good observations taken in different parts of its orbit can be procured. However, from those already made it most certainly and demonstrably appears to be the same, the long expected comet, and therefore I cannot but congratulate my countrymen on an event so glorious to the Newtonian doctrine of gravity, and to the memory of that excellent philosopher Dr. Halley. And may it ever be remembered that the first instance of an event of this kind was foretold, and with accuracy too, by an Englishman.

SOURCE: David Gregory, *The Elements of Astronomy . . . to which is annexed Dr. Halley's synopsis of the astronomy of comets*, vol. 2, London, 1715; "Astrophilus," Untitled Letter, *The Gentleman's Magazine*, November, 1759.

40

A Voice from the Starry Heavens

Cotton Mather

Cotton Mather (1663–1728), one of the most prominent American writers of the early eighteenth century and a famous Puritan theologian, was also the first native-born American to become a member of the Royal Society. Although Puritan teaching often emphasized the importance of a Christian's separating from "the world" in a spiritual sense—that is to say, avoiding sin and the vanities of this life—Puritan regard for the doctrines of both salvation and creation underpinned a great interest in the created world as a product of God's wisdom. "The whole world is indeed a temple of God, built and fitted by that almighty Architect," says Mather in his introduction to The Christian Philosopher, *a work that meditates on nature and records Mather's wide reading in the sciences. To expound that architecture—and to extol its Architect—Mather begins with the light and with the structure of the universe at large.*

Even a pagan, Plutarch, will put the Christian philosopher in mind of this, that the world is no other than the temple of God, and that all the creatures are the glasses in which we may see the skill of him that is the Maker of all. And his brother Cicero has minded us, we know God from his works. 'Tis no wonder then that a Bernard should see this: "The true lover of God, wherever he turns, receives a familiar remembrance of his Creator." The famous hermit's book—of those three leaves: the heaven, the water, and the earth—well studied, how nobly would it fill the chambers of the soul with the most precious and pleasant riches? Clement of Alexandria calls the world "a scripture of those three leaves." And the creatures therein speaking to us have been justly called true orators by those who have best understood them.

In keeping with the traditional "two books" approach (God's words and God's works) and with the suggestion in the book of Genesis that God made the heavenly lights "for signs," Mather devotes two of his first essays to the stars.

Let us proceed and, conforming to the end of our erect stature, behold the heavens, and lift up our eyes unto the stars.

The learned Huygens has a suspicion that every star may be a sun to other worlds in their several vortices. Consider then the vast extent of our solar vortex, and into what astonishments must we find the grandeur and glory of the Creator to grow upon us! Especially if it should be so (as he thinks) that all these worlds have their inhabitants, whose praises are offered up unto our God! . . .

The Jews have a fancy among them, that when the Almighty first bespangled the heavens with stars, he left a spot near the North Pole unfinished and unfurnished, so that if any other should set up for a God, there might be this trial made of his pretensions: Go, fill up if you can that part of the heavens which is yet left imperfect. But without any such suppositions we may see enough in the heavens to proclaim this unto us: Lift up your eyes on high, and behold: Who has created these things? None but an infinitely glorious God could be the Creator of them!

The telescope, invented the beginning of the last century and improved now to the dimensions even of eighty feet, whereby objects of a mighty distance are brought much nearer to us, is an instrument wherewith our good God has in a singular manner favored and enriched us: a messenger that has brought unto us, from very distant regions, most wonderful discoveries.

My God, I cannot look upon our glasses without uttering thy praises. By them I see thy goodness to the children of men!

By this enlightener of our world it is particularly discovered:

That all the planets at least, excepting the sun, are dense and dark bodies, and that what light these opaque bodies have is borrowed from the sun.

That every one of the planets, excepting the sun, do change their faces like the moon. Venus and Mercury appear sometimes like a half-moon, and sometimes quite round, according as they are more or less opposite to the sun. Mars has his times of appearing in a curvi-lined figure. Jupiter has four little stars that continually move about him, and in doing so cast a shadow upon him. Saturn has a ring encompassing of him.

That each of these planets have spots in their superficies, like those of the moon.

That not only each of these planets, but the sun also, besides whatever other motion they may have, do move themselves upon their own centers, some of them with a motion of revolution, others by that of libration.

It was a good remark made by one of the ancients, What are the heavens and all the beauties of nature but a kind of mirror in which the whole work of the Master is reflected?

The pagan Tully, contemplating the admirable order and the constancy of the heavenly bodies and their motions, adds upon it: Whosoever thinks this is not governed by mind and understanding, is himself to be accounted void of all mind and understanding.

Mather's chapters on stars also show us an early eighteenth-century intellectual looking for ways to express what is measurable but almost inconceivable. His (or Huygens's) estimate of the distance of the sun from the earth is quite close to the presently accepted figure, but his estimate of the distance to the nearest fixed star is under by a factor of ten.

According to Mr. Huygens, the distance of the sun from us is 12,000 diameters of the earth. A diameter of the earth is 7,846 miles. The distance of the nearest fixed stars from us, compared with that of the sun, is as 27,664 to 1. So then the distance of the nearest fixed stars is at least 2,404,520,928,000 miles, which is so great that if a cannon-ball (going all the way with the same velocity it has when it parts from the mouth of the gun) would scarce arrive there in 700,000 years. Great God, what is thy immensity!

Mather ends his first chapter on the stars with an unreferenced quotation— perhaps his own prayer and reflection—that blends praise of science, speculation on other worlds, and honor to the Creator.

"Glorious God, I give thanks unto thee for the benefits and improvements of the sciences granted by thee unto these our latter ages. The glasses, which our God has given us the discretion to invent and apply for the most noble purposes, are favors of heaven most thankfully to be acknowledged.

"The world has much longer enjoyed the scriptures, which are glasses that bring the best of heavens much nearer to us. But though the object-glasses are here, the eye-glasses are wanting. My God, bestow thou that faith upon me, which, using the perspective [i.e., telescope] of thy Word, may discover the heavenly world, and acquaint me with what is in that world, which I hope I am going to.

"I hear a great voice from the starry heavens, 'Ascribe ye greatness to our God.' Great God, what a variety of worlds hast thou created! How astonishing are the dimensions of them! How stupendous are the displays of thy greatness, and of thy glory, in the creatures, with which thou hast replenished those worlds! Who can tell what angelical inhabitants may there see and sing the praises of the Lord! Who can tell what uses those marvelous globes may be designed for! Of these unknown worlds I know thus much, 'tis our great God that has made them all."

The realization (assisted by the telescope) that the universe beyond the circle of the moon is subject, like this earth, to change and decay, may also be given a theological application. Although Mather did not have use of the word "entropy," he recognized that the running down of the sun must lead to questions about its origin. A similar backwards extrapolation in present-day Big Bang cosmology raises but leaves unanswered the radical question of origin.

The quantity of light and heat in the sun is daily decreasing. It is perpetually emitting millions of rays, which do not return to it. Bodies attract them and suffocate them and imprison them, and they go no more back into their fountain.

Mr. Bernoulli, from the flashes of the light in the vacuity of a tube accommodated with mercury, whereby a dark room is enlightened, renders it likely that our atmosphere and all the bodies of our globe are saturated at all times with rays of light which never do return unto their fountain.

'Tis true, this decrease of the sun is very inconsiderable. It shows that the particles of light are extremely small, since the sun for so many ages has been constantly emitting oceans of rays without any sensible diminution. However, 'tis from hence evident that the sun had a beginning. It could not have been from eternity. Eternity must have wasted it. It had long e'er now been reduced unto less than the light of a candle.

Glorious God, thou art the Father of Lights, the Maker of the sun!

SOURCE: Cotton Mather, *The Christian Philosopher: A Collection of the Best Discoveries in Nature, with Religious Improvements*, London, 1721.

41

This Most Surprising Zone of Light

Thomas Wright of Durham

Thomas Wright of Durham (1711–1786) is a curious figure in the history of cosmology. He is best known for a theory of the Milky Way, which he only fleetingly adumbrated but which Immanuel Kant credited him with (see Chapter 42).

In his epistolary An Original Theory or New Hypothesis of the Universe *(1750), Wright displays an attraction to poetry, quoting numerous excerpts from Ovid to Pope. But his imagination is perhaps most notably draftsmanlike and visual. He had made numerous large-scale drawings or "schemes" up to nine feet long and six feet wide illustrating various celestial configurations such as eclipses and the course of comets. What the graphic and the poetic most importantly have in common, for Wright, is a deep sense of analogy, simile, or, one might say, synecdoche. The scale drawing, like a poetic figure,* models *the larger reality.*

Just as an earlier generation had revelled in the analogy between the microcosm "man" and the macrocosm—the variously defined larger world—so Wright works out possible correspondences between the now thoroughly Copernican solar system and the larger universe that contains it. The longest poetic quotation in the first letter of An Original Theory *is from the eighth book of* Paradise Lost, *where Milton's angel Raphael tells Adam:*

> other suns perhaps,
> With their attendant moons, thou wilt descry,
> Communicating male and female light . . .
> (for a fuller quotation, see Chapter 32)

What is significant for Wright is how "sun" becomes a generic noun and so permits an extrapolation from our sun with its habitable planet or planets, to other suns with theirs. This process of reasoning is itself an extrapolation of the Copernican process: As the earth is de-centered and becomes a planet, so the sun is de-centered and becomes a star.

Against this backdrop, Wright proposes to model the Milky Way:

This luminous circle has often engrossed my thoughts, and of late has taken up all my idle hours; and I am now in great hopes I have not only at last found out the real cause of it, but also by the same hypothesis, which solves this appearance, shall be able to demonstrate a much more rational theory of the creation than hitherto has been anywhere advanced, and at the same time give you an entire new idea of the universe, or infinite system of things. This most surprising zone of light, which has employed successively for many ages past the wisest heads amongst the ancients to no other purpose than barely to describe it, we find to be a perfect circle, and nearly bisecting the celestial sphere, but very irregular in breadth and brightness, and in many places divided into double streams. . . . Eratosthenes supposed it Juno's milk, spilt whilst giving suck to Hercules . . . but I think it is plain Ovid judged them to be stars . . . :

> A way there is in heaven's expanded plain
> Which when the skies are clear is seen below,
> And mortals by the name of *milky* know,
> The ground-work is of stars . . .
>   Ovid's *Met.* lib.i.

But Democritus long ago believed them to be an infinite number of small stars; and such of late years they have been discovered to be, first by Galileo, next by Kepler, and now confirmed by all modern astronomers who have ever had an opportunity of seeing them through a good telescope.

Given that the Milky Way consists of stars, Wright immediately puts the observation together with the "other suns" hypothesis and makes the following calculation:

Now admitting the breadth of the *Via Lactea* to be at a mean but nine degrees, and supposing only twelve hundred stars in every square degree, there will be nearly in the whole orbicular area 3,888,000 stars, and all these in a very minute portion of the great expanse of heaven. What a vast idea of endless beings must this produce and generate in our minds! And when we consider them all as flaming suns, progenitors, and *Primum Mobiles* of a still much greater number of peopled worlds, what less than an infinity can circumscribe them, less than an eternity comprehend them, or less than Omnipotence produce and support them, and where can our wonder cease?

After various digressions, Wright in his seventh letter tackles his task directly:

I shall now endeavor to ... solve the phenomena of the *Via Lactea*; and in order thereto, I want nothing to be granted but what may easily be allowed, namely that the Milky Way is formed of an infinite number of small stars.

Let us imagine a vast infinite gulf or medium every way extended like a plane, and inclosed between two surfaces, nearly even on both sides, but of such a depth or thickness as to occupy a space equal to the double radius, or diameter, of the visible creation, that is to take in one of the smallest stars each way, from the middle station, perpendicular to the plane's direction, and as near as possible according to our idea of their true distance.

But to bring this image a little lower, and as near as possible level to every capacity ... let us suppose the whole frame of nature in the form of an artificial horizon of a globe, I don't mean to affirm that it really is so in fact, but only state the question thus, to help your imagination to conceive more aptly what I would explain.

This "supposition" is very important, but note that it is quite deliberately intended as imaginative scaffolding, not edifice. It is a model, and merely a model.

[Plate XXI—see the next page] will then represent a just section of it. Now in this space let us imagine all the stars scattered promiscuously, but at such an adjusted distance from one another, as to fill up the whole medium with a kind of regular irregularity of objects. And next let us consider what the consequence would be to an eye situated near the center point, or anywhere about the middle Plane, as at the point A. Is it not, think you, very evident, that the stars would there appear promiscuously dispersed on each side, and more and more inclining to disorder as the observer would advance his station towards either surface, and nearer to B or C, but in the direction of the general plane towards H or D, by the continual approximation of the visual rays, crowding together as at H, betwixt the limits D and G, they must infallibly terminate in the utmost confusion. If your optics fails you before you arrive at these external regions, only imagine how infinitely greater the number of stars would be in those remote parts, arising thus from their continual crowding behind one another, as all other objects do towards the horizon

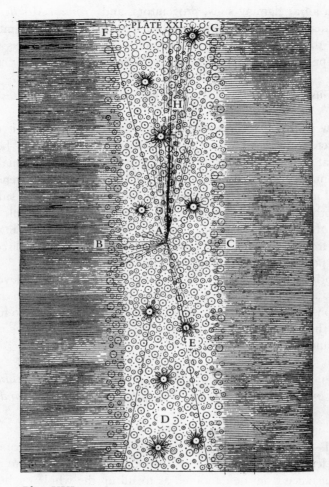

Plate XXI

point of their perspective, which ends but with infinity. Thus, all their rays at last so near uniting, must meeting in the eye appear, as almost in contact, and form a perfect zone of light. This I take to be the real case, and the true nature of our Milky Way, and all the irregularity we observe in it at the earth I judge to be entirely owing to our sun's position in this great firmament, and may easily be solved by his eccentricity, and the diversity of motion that may naturally be conceived amongst the stars themselves, which may here and there in different parts of the heavens occasion a cloudy knot of stars, as perhaps at E.

But now to apply this hypothesis to our present purpose . . . our reason must now have recourse to the analogy of things. It being once agreed that the stars are in motion, which . . . is not far from an undeniable truth, we must next consider in what manner they move. First then, to suppose them to move in right lines you know is contrary to all the laws and principles we at present know of; and since there are but two ways that they can possibly move in any natural order, that is either in right lines or in curves, this being one, it must of course be the other, i.e. in an orbit. And consequently, were we able to view them from their middle position . . . we might expect to find them separately moving in all manner of directions round a general center. . . .

It only now remains to show how a number of stars so disposed in a circular manner round any given center may solve the phenomena before us. There are but two ways possible . . . but which of the two will meet your approbation I shall not venture to determine, only here inclosed I intend to send you both.

The first is in the manner I have above described, i.e., all moving the same way, and not much deviating from the same plane, as the planets in their heliocentric motion do round the solar body. . . . The second method of solving this phenomena is by a spherical order of the stars all moving with different direction round one common center, as the planets and comets together do round the sun, but in a kind of shell, or concave orb.

Wright expatiates on various versions of these two general explanations. What they all have in common, however, is the analogy with the solar system (in that stars revolve around a common center more or less as planets do around the sun), and a general congruence with the notion that stars occur in some sort of layers, so that what we see when we look at the Milky Way is a result of our looking as it were horizontally into the layer of stars rather than vertically through the layer and out the other side. Individual features of our solar system continue to provide Wright with explanatory analogies, most notably here the rings of Saturn:

If all the stars we see moved in one vast ring, like those of Saturn, round any central body, or point, the general phenomena of our stars would be solved by it. . . . Not only the phenomena of the Milky Way may be thus accounted for, but also all the cloudy spots and irregular distribution of them; and I cannot help being of opinion, that could we view Saturn through a telescope capable of it, we should find his rings no other than an infinite number of lesser planets, inferior to those we call his satellites.

Wright's impressive educated guess about Saturn's rings and the logical (and analogical) power he displays should not cause us to minimize the mercurial and in some ways medieval elements of his thought. At the end of his eighth letter, having calculated that there may be "within the whole celestial area 60,000,000 planetary worlds like ours," Wright offers the following medita-

tion, which can hardly but remind us of Geoffrey Chaucer's Troilus, *who finally learns to laugh at earthly fortune only when he is transported after death to the eighth sphere of the heavens:*

In this great celestial creation, the catastrophe of a world such as ours, or even the total dissolution of a system of worlds, may possibly be no more to the great Author of nature than the most common accident in life with us, and in all probability such final and general doom-days may be as frequent there as even birth-days or mortality with us upon the earth.

This idea has something so cheerful in it that I own I can never look upon the stars without wondering why the whole world does not become astronomers; and that men endowed with sense and reason should neglect a science . . . so capable of enlarging the understanding . . . [and] dignifying our natures with something analogous to the knowledge we attribute to angels; from whence we ought to despise all the vicissitudes of adverse fortune, which make so many narrow-minded mortals miserable.

This medieval flavor, finally, is accented by the fact that Wright followed An Original Theory *with a series of retractions in his* Second or Singular Thoughts, *discovered and published for the first time in the 1960s. Here he abandons the hypotheses of* An Original Theory *in favor of an utterly contrary and breathtakingly idiosyncratic view of the heavens.*

The visible heavens or starry firmament might prove to be no other than a solid orb . . . and the fixed stars no more than perpetual lumination or vast eruptions and of refulgent or inflammable matter promiscuously distributed as celestial volcanoes all round the starry regions emitting an etherial and intense fire of various magnitudes. . . . This idea, which naturally solves from like visible causes and effects all the several phenomena of the celestial regions without excepting any one of them, I am very much inclined to conclude will be found in the end to give the truest construction of the visible creation in all its movements, modes, and consequences.

SOURCES: Thomas Wright of Durham, *An Original Theory or New Hypothesis of the Universe*, London, 1750; *Second or Singular Thoughts Upon the Theory of the Universe*, ed. M. A. Hoskin, London: Dawsons of Pall Mall, 1968.

42

How Fortunate Is This Globe!

Immanuel Kant

Immanuel Kant (1724–1804), an icon in the history of philosophy yet still an enigma to many, made his main contribution to cosmology in an early work, published anonymously in 1755, his Universal Natural History and Theory of the Heavens. *The subtitle of this work announces from the outset Kant's desire to appear scientifically informed and up-to-date:* An Essay concerning the Constitution and Mechanical Origins of the entire World-Edifice based on Newtonian Principles.

Kant, a theist, begins his discussion of the universe by defending, as so many have felt they must, the piety of the undertaking. The defense is required in Kant's case especially because the materialism of his cosmology may appear to eliminate the need for Providence.

The defender of religion is concerned that those harmonies which can be explained from a natural tendency of matter might prove to be independent of divine Providence. He confesses unambiguously that if one can discover natural causes for the entire order of the universe and in turn derive these causes from the most general and essential properties of matter, then it is unnecessary to have recourse to a higher government. The advocate of naturalism settles his account merely by not disputing this assumption.

This assumption has been boldly repeated since Kant's time in explicit comments, for example, by Pierre Laplace (see Chapter 45), who had no need of "that hypothesis" (that is, of a God), and by Stephen Hawking, whose "no boundary" universe produces the question "What place, then, for a creator?" (see Chapter 74). Nevertheless, Kant's cosmogony is a picture painted on a Lucretian canvas within a nominally theistic frame, and its brushstrokes reveal aesthetic motivation amid the colors of rational method.

I accept that the matter of the whole world was at the beginning in a state of general dispersion, and take that to have been a complete chaos. I see this substance forming itself in accordance with the established laws of attraction and modifying its movement by repulsion. I enjoy the pleasure, without any recourse to arbitrary hypotheses, of witnessing a well-ordered whole produced under the regulation of the established laws of motion. Moreover, this whole looks so like that system of the world which we have before our eyes that I cannot but identify the one with the other. Initially this unexpected evolution of the order of nature overall strikes me as dubious, for it establishes such a complexly woven reality upon so poor and simple a foundation. At last I learn, however, from the view already indicated, that such natural evolution is not a thing unheard of in nature. Rather, her own inherent striving necessarily produces such a result, and this process is the most splendid evidence of her dependency upon that primordial Being which indeed encompasses within itself both the source of these beings and their fundamental laws of operation. This insight redoubles my confidence in the sketch of the system which I have drawn. My assurance grows with every step I take, and my faintheartedness vanishes altogether.

"But," someone will say, "to defend your system is to defend the opinions of Epicurus, which sound so much like yours." Nor will I utterly deny all agreement with him. Many have become atheists based on the appearance of reasons which, had they been pondered more carefully, might have convinced them most powerfully regarding the reality of the supreme Being. The implications which a perverse understanding draws from irreproachable principles are often highly reproachable; and so were the conclusions of Epicurus, despite the fact that his picture of the world befitted the brilliance of a great mind.

I will therefore not dispute the claim that the theory of Lucretius, or his predecessors, Epicurus, Leucippus, and Democritus, much resembles mine. I too, like them, see the primordial condition of nature as a universal dispersion of the primary stuff of all heavenly bodies, or of the atoms, as they call them.

When Kant moves into the section of the Universal Natural History *entitled "Of the Systematic Constitution among the Fixed Stars," he turns his attention to the problem of the Milky Way. To Thomas Wright, whose* An Original Theory *he knew only from a lengthy German book review, Kant gives somewhat ambiguous credit for stimulating his thinking on this topic.*

It was reserved for an Englishman, Mr. Wright of Durham, to take a happy step towards a notion which he himself seems not to have put to any real purpose and whose useful application he has not adequately observed. He did not regard the fixed stars as a scattered swarm, random and disordered; rather he found a systematic constitution in the cosmic whole and a universal relation of these stars to the fundamental spatial configuration which they occupy. We want to improve upon the idea he has proposed and to apply it in such a way that it may be fruitful in producing significant implications whose full confirmation only the future will provide.

Inaccurately assuming that Wright proposed a disk pattern for the Milky Way, Kant pursues this conception with more enthusiasm than consistency.

Now, the better to fathom the structure of the universal connection that holds sway in the universe, we want to try to discover why the positions of the fixed stars have come to be interrelated in the configuration of a common plane.

The extent of the sun's attractive force is not limited to the narrow locale of the planetary system. By all appearances this extends into infinity. The comets, which rise very far beyond the orbit of Saturn, are compelled by the attraction of the sun to turn back once more and to move in orbits. Perhaps the characteristic of a force apparently embodied within the nature of matter may more properly be treated as unlimited—and is indeed so treated by those who accept Newton's principles. Yet we ask one merely to grant that the attraction of the sun reaches approximately to the nearest fixed star, and that the influence of the fixed stars, as of so many suns, operates within a similar compass, with the result that the whole host of them are striving to approach each other on account of this mutual attraction. Thus every world-system is constituted by a mutual drawing together, which persists unceasingly and utterly unhindered, and whereby sooner or later each implodes into a single clump, unless this cataclysm is prevented, as it is with the spheres of our planetary system, by the action of centrifugal forces. These forces, by deflecting the heavenly spheres from falling in a straight line, produce, in combination with those forces of attraction, the eternal orbital revolutions whereby the edifice of the creation is secured from destruction and fitted out for endless duration.

It is worth noting that the orbits of Kant's own discussion often mediate between the impulses of a rational systematizer and those of someone almost mystically drawn to concepts such as infinity and the eternal. The result is not always a perfectly coherent thought-system. For example, his materialis-

*tic cosmogony apparently presupposes process, including ongoing process.
And yet, as in the preceding paragraph, the planetary orbits are spoken of as
"eternal." Moreover, in the following paragraph, he flirts with the possibility of a "universal center" in a discussion that simultaneously postulates an
infinite universe.*

Thus all the suns of the firmament have orbital motions either around a universal center or around many centers. But here one may derive an analogy
from what we observe in the revolutions of our solar system: the same cause
that imparted to the planets the centrifugal force whereby they pursue their
orbits has also arranged these in such a way that they are all mutually related
within one plane. In the same manner, the cause, whatever it may be, that has
imparted the power of revolving to the suns of the upper world (these being
like so many stellar planets within higher systems of worlds) has likewise as
much as possible brought about an alignment of their orbits within one plane
and has striven to limit deviations from it.

This way of presenting things lets us view the system of the fixed stars on
the model of the planetary system, assuming this were infinitely enlarged. For
if instead of the six planets with their ten attendant moons we postulate so
many thousands of these, and also the twenty-eight or thirty observed comets
multiplied by the hundreds and thousands, and if we think of these bodies as
self-luminous, then the eye of the observer, viewing them from earth, would
see before it precisely that appearance which the fixed stars of the Milky Way
do produce. For the "planets," on account of their proximity to the common
configuration within which they are related—we on earth also being located
in that same plane—would appear as a densely illuminated belt of innumerable stars aligned like the greatest of great circles. This streak of light would
be seen everywhere copiously set about with stars, although according to this
hypothesis they are wandering stars and not fixed in one place. . . .

The form of the starry heavens is therefore due to no other cause than the
same systematic constitution on a large scale that governs our planetary system on a small scale—all the suns constituting a system whose universal relative plane is the Milky Way. Those suns which are least closely related to this
plane will be seen at the side of it; but on that account they are less congre-
gated, much more scattered and fewer in number. They are so to speak the
comets among the suns.

This new theory, however, ascribes to the suns an advancing movement,
and yet everyone considers them unmoving and fixed in their places from the
very beginning. Their designation accordingly as fixed stars seems confirmed
and unshaken by centuries of observation. If it had any basis, this difficulty
would annihilate my proposed theory. But all the evidence suggests that this

lack of movement is something merely apparent. Either it is only an extraordinary slowness caused by the great distance from the common center around which the stars revolve, or it is mere imperceptibility occasioned by the distance of that place from which the observation is being made.

Having by means of analogy accounted for the shape and coherence of what we would call a galaxy, Kant pushes the analogy one step further in relation to the nebulae. This is Kant's famous postulate concerning extragalactic "island universes."

Because these nebulous stars must undoubtedly be removed at least as far from us as the other fixed stars, not only would their magnitude be astonishing—since it would have to exceed that of the largest stars many thousand times—but also it would be most peculiar if, despite their extraordinary size, such self-luminous bodies and suns were to give off the dullest and feeblest light.

It is far more natural and understandable to regard them as being not such enormous single stars but rather systems of many stars, whose distance presents them within such a narrow space that their light, which from each one individually is imperceptible, strikes us on account of their immense multitude as a pale glimmer. Their analogy with the stellar system in which we are located; their shape, which is just what it ought to be according to our theory; the feebleness of their light, which demands an assumption of infinite distance—all these correspond perfectly with the assertion that these elliptical figures are just such world-systems and, so to speak, Milky Ways, whose structure we have just unfolded. And if conjectures in which analogy and observation perfectly agree in mutual support have the same value as formal proofs, then the certainty of these systems must be considered as proven.

So the attention of the astronomical observer has reasons enough for being applied to this project. The fixed stars, as we know, are all related within a common configuration and thereby form a coordinated whole, which is a world of worlds. We see that at immeasurable distances there are more such star-systems, and that the creation in the whole infinite extent of its greatness is everywhere systematic and interrelated.

The assertion of the interconnectedness of all physical reality is one that happily comports with both Newtonian and Kantian principles. The almost apocalyptic conclusion of Kant's cosmological treatise, however, gives us a keen taste—in connection with a concentrated sampling of Enlightenment and Romantic flavors—of a subject-oriented universe more profoundly anthropocentric than anything Ptolemy ever dreamed of.

CONCLUSION

... Is it likely that the immortal soul in the entire everlastingness of her future duration, which the grave itself merely transforms but does not interrupt, shall remain forever bound to this point of universal space, to our earth? Shall she never come to participate in a closer contemplation of the other wonders of the creation? Who knows but that she may be intended some day to experience close up those distant spheres of the universal edifice and their excellent furnishings, which from afar have so excited her curiosity. Perhaps other spheres of the circling universe are evolving, therefore, so that, when the full time of our allotted stay here has expired, new dwelling places beneath other skies may await us. Who knows but that those satellites are revolving about Jupiter in order that one day they may shine upon us?

It is fitting and just to entertain ourselves with such speculations, except that no one will base his hope for the future upon such uncertain pictures of the imagination. The immortal spirit, when vanity has exacted her toll upon human nature, with a sudden leap will soar heavenwards, rising above everything that is transitory. And, within a new relationship to all of creation, whose source is a closer bond with the highest Being, it will continue its existence. ...

Indeed, if one has filled one's spirit with notions such as these and the others I have presented, then the sight of a starry sky on a clear night affords a quality of pleasure which none but a noble soul can savor. In the universal stillness of nature and in the tranquility of the senses, the immortal spirit's hidden power of knowing speaks a language with no name and imparts secret thoughts which can be surely tasted but not revealed. If among the rational creatures of this planet there be lowly beings who are oblivious to all the charms whereby so great an object can captivate them, and who instead have bound themselves over to the service of vanity, then how unfortunate is this globe to have managed the nurture of such pitiful creatures! But how fortunate is this globe, on the other hand, to whose inhabitants a way has been opened, under the most propitious conditions, whereby they may achieve a happiness and grandeur infinitely more precious than even the most extraordinary benefit that nature can bestow upon any world in the whole universe!

Thirty-two years later, when he wrote the second preface to his Critique of Pure Reason *in 1787, Kant continued to speak in prophetic tones, presenting his revolution in epistemology and metaphysics as one of Copernican proportions—and perhaps it was. Yet on the face of it, given the subject-orientation which is Kant's legacy, the specific analogy between his work and Copernicus's rejection of geocentrism may appear ironic or merely baffling.*

Until now people have assumed that all our knowledge must orient itself to objects. . . . [But] now we would like to try to improve our progress in the projects of metaphysics by assuming that objects must be oriented to our knowledge. . . . We thus proceed exactly along the main line of thought of Copernicus. When he couldn't make any progress explaining the movements of the heavens by assuming that the whole starry host revolved around the observer, he pinned his hopes for success on a reversal of this assumption, thus making the observer revolve and letting the stars rest in peace. Here we can attempt the same sort of procedure in metaphysics as regards the intuition of objects.

SOURCES: Immanuel Kant, *Allgemeine Naturgeschichte und Theorie des Himmels*, Königsberg and Leipzig, 1755 (translation, except for the "Conclusion," adapted from *Kant's Cosmogony*, translated by W. Hastie, Glasgow, 1900); excerpt of *Critique of Pure Reason*, translated from *Kritik der reinen Vernunft*, ed. Erich Adikes, Berlin, 1889.

43

To Become Adequately Copernican

Johann Heinrich Lambert

Johann Heinrich Lambert (1728–1777) is best known for his contributions to mathematics (he was the first to devise a proof for the irrationality of Pi) and to photometry, where his name has come to denote a unit of brightness (1/π candle per cm² = 1 lambert). When it comes to cosmology, Lambert is often lumped together with Kant because both men account for the Milky Way in a similar manner and because both describe the universe as comprising systems within systems within systems. However, Kant's cosmos is evolving and infinite, whereas Lambert's is cyclic and finite—though still larger than anyone can readily conceive.

The following selection from letter 19 of Lambert's Cosmological Letters *is notable for its attempt to conceptualize Copernicanism as a kind of language that needs to expand to higher levels of expression as subsequent fixed points are in turn displaced and replaced with still further fixed points in a very long (though not infinite) series. Lambert's discussion of waves upon waves (in which, as he remarks, nature takes great pleasure) almost reminds us of discussions in our own time of fractals: for example, the individual leaf of a fern reveals a feathered pattern that is repeated on a larger scale in the frond of which the leaf itself is a part. In the same way, what Lambert calls a cycloid—such as the moon's pattern of revolution around the earth—is repeated exponentially in a higher-order cycloid, namely, in the earth's revolution around the sun. Lambert extrapolates this recursive pattern on and on into the cosmos at large, perhaps to the thousandth degree, until a true fixed point (in principle) is arrived at.*

Now indeed for once we have an opportunity to become adequately Copernican—or else we shall never be such, nor ever have been. For this is my dilemma since you raise the issue of cycloids of the thousandth degree, universal bodies of the thousandth rank, a thousand sub-categories of hypotheses, and as many languages as astronomy will demand of us and as will have to be invented, so it seems, for every investigation and controversy, so that we know in advance which language to speak, and in which tone.

How the world is turned upside down! At first, the earth was at rest. Then it started to perambulate while the sun sat down to rest. Now the sun too is on the march and the body of the fourth degree rests. But this body also will get disturbed, and rest will then pass on to that of the fifth degree—and so on until order itself descends upon the last degree, where rest prevails not hypothetically but in fact.

It pleases me no end how we can choose to put any given heavenly body at rest or set it in motion. In much the same way we stipulate a tone from which members of a choir can find their notes. If the earth is at rest, we can stick with spherical astronomy and make all celestial phenomena conform accordingly. In this way we translate any astronomical vocabulary into the common tongue. But when the sun is at rest, the translation concerns our planets, satellites, and comets, and here the language of ellipses and hyperbolas will suffice. Should the process continue so as to include our neighboring suns and fixed stars, then new ellipses and hyperbolas will appear and our language must expand to encompass cycloids of the first degree—and subsequently of the second, third, and so on.

However, let me return to my first question: Now are we really Copernican, or how are we progressing? If I am conceiving the business correctly, this is how it looks: Ptolemy stayed with the popular common idiom. Copernicus started out, taking the first step and introducing us to the alphabet of astronomy's language. But he did not know that this language was merely provisional and would lead us only by many stages to the true language. Tycho Brahe mistook the matter and confused a vowel with a consonant. This mistake had the effect of impeding pronunciation and preventing fluency. Kepler and Newton utterly removed this confusion and found means for raising the still crude language to its proper level of refinement, as well as for composing their epic poems in keeping with the mythology of their era. Yet times change, and our nearest posterity will restrict the currency of this language to poems alone, or else use it as a shorthand for expressions whose precise meaning is unimportant. Where meaning is important, however, they will speak in an even more refined idiom.

Let me develop this idea one stage further. Insofar as Copernicus took the *first* step, thousands of steps remain to be taken by us and our posterity, and we shall not be perfectly Copernican for a long time to come. Yet insofar as we know that the last of those steps will end at the body that directs the entire creation about itself, I do believe we are thinking in a manner that is Copernican enough. Or do you want to go still further than this? I simply cannot see where else you would go from here. Or is it that we never ever could be truly Copernican? But to be truly Copernican in a literal sense would require us to believe firmly that the sun remains forever in its place and moves only about its own axis. I grant that this is an error that we should avoid, since the sun no more than the fixed stars is really fixed in one place. The scenery of the heavens is undergoing a transformation, and perhaps a time will come when the constellation of Orion will come to resemble the Great Bear.

I think I shall keep speaking the Copernican language, for it is the easiest one to translate into common speech. Moreover, as you remark, we can relate the next language to it in a way that allows us to reduce the anomalies of the cycloids to ellipses just as we do in the case of the moon. I trust that astronomers on the moon have achieved greater facility in these various languages than we have, even though they had an extra step to take on account of their inhabiting a satellite. According to their common speech, no doubt, the moon does not move. And precisely because the earth appears to them always in the same spot in their sky, it must seem to them not to move either, or hardly to move, or to move only in the sense that they see it turning about its own axis. It would be more natural for them than for us to graduate to another language and to postulate the revolution of the moon about the earth. Yet before long, both moon and earth would have to be seen as revolving around the sun. And given this twofold analogy, they would also have good reason for doubting that the sun was stationary, and in the end would refuse to grant rest to anything else either. If somewhere beyond Saturn there be planets or comets whose satellites have satellites of their own, their resident astronomers would have three steps to take before they could get the sun moving as well. But these three steps would come that much more naturally to them, since their proofs would almost necessarily entice them farther out into space.

Nonetheless, although I am sticking with the Copernican idiom, I keep trying to achieve some mastery of the subsequent languages so that I may imagine more or less how they ought to be translated. With nothing could I more fittingly compare the various degrees of cycloids than with waves in water. When water is moved by whatever cause, its surface becomes uneven. There arises a series of waves, one after the other, and between each two a trough that becomes filled in when the waves settle. Looked at horizontally, this sequential rising and falling of the surface of the water describes a kind of serpentine curve, which to me portrays a picture of the cycloids. If the waves are small, then the rise and fall is quite simple, like a cycloid of the first degree. As they become noticeably bigger, however, each large wave comprises many smaller ones. And although the smaller curves certainly remain, they follow the rise and fall of the larger wave up and down.

It would seem that nature takes great pleasure in such series of waves and oscillations, for we discover them in almost every motion. In the same manner a ship travels through a series of oscillating wave-forms upon the sea. The larger the ship is, the larger the billows required to make it rise and fall. A small skiff moves with each wave, whether large or small; and in the time it takes the warship to rise and fall only once, the skiff endures several minor risings and fallings that together make up the one large one. And this is how all the spheres appear to pursue their courses through cosmic space. All finally revolve about the universal midpoint, but the larger each body is, the less yielding it is and the simpler are its oscillations. Naturally, things there are quieter and more uniform than they are on the surface of the ocean. The universal wind which sets all things in motion is more reliable and in itself steadier than is the east wind that blows between the tropics here on earth, and more fitted to the preservation of the bodies that it propels.

The pattern of waves within waves or cycloids within cycloids brings to mind, of course, the mechanism of a clock. The "order" beyond that of a given planetary year is the so-called Platonic year, when all the planets together return to the same relative position, like the second hand, the minute hand, and the hour hand of a clock coming back together at midnight.

Time and space, through the movement of each body, are bound up together within the structure of the cosmos. Thus, apparent order ought to be the simplest order; and the larger the expanse of time we wish to make sense of, the more complex will be the order we have to comprehend. How systematically and precisely does the daily revolution of the heavens mete out for us the days and hours and all their parts, to whatever purpose we may devote them. No clockwork is more precise than the daily course of each fixed star! Do we need to measure times of greater duration—to calibrate the orbits of comets and planets, which reveal most readily and visibly their deviation from the primary order? Then we must take into account the next cosmic drive wheel and yet still remain true to that exemplary account which Copernicus sketched out for us. However, if periods of many centuries present themselves, then also at this level new deviations will appear, which we, like Copernicus himself, must harmonize into a new system. We must then progress to the third drive wheel, whose revolution is measured in Platonic years. Thus by means of merely apparent regularities we possess a precise clockwork that tells the hours, days, millennia, and every other time period as well, which step by step expand a millionfold upon those which precede them.

SOURCE: Translated (with kind advice from Peter Loeffler) from Johann Heinrich Lambert, *Cosmologische Briefe über die Einrichtung des Weltbaues*, Augsburg, 1761.

44

Laboratories of the Universe

William Herschel

William Herschel (1738–1822) was probably the greatest astronomer of the eighteenth century. He was born Friedrich Wilhelm Herschel, of Hanover, but fled to England at the age of nineteen after the French occupied his native city in 1757. In England he pursued a career as a musician. Music practice (teaching, playing, and composing) led to music theory, which led to optics, which led to astronomy. Aided by his sister, Caroline Herschel, who joined him in 1772 and remained his partner in astronomy for the rest of his life, Herschel built a series of larger and larger telescopes, not superseded in size or quality until almost the middle of the next century. With these tools, and with a mind remarkable for its combination of boldness and punctiliousness, Herschel pioneered the realm of stellar astronomy.

The event that allowed Herschel to become a full-time astronomer occurred in 1781 when, now almost forty-three, he discovered the planet Uranus, which he prudently suggested be named Georgium Sidus, *the Georgian star, after King George III. Soon thereafter he was named King's astronomer and granted funding for the unimpeded pursuit of his study of the heavens.*

Herschel's writings reveal a quite conscious attempt to mediate between observation and theory, the former being predominant in its bulk, but the latter standing out for its keenness of imagination, as in the following paper delivered to the Royal Society in 1785.

The subject of the Construction of the Heavens . . . is of so extensive and important a nature that we cannot exert too much attention in our endeavors to throw all possible light upon it. I shall therefore now attempt to pursue the delineations of which a faint outline was begun in my former paper.

By continuing to observe the heavens with my last-constructed and since that time much-improved instrument, I am now enabled to bring more confirmation to several parts that were before but weakly supported, and also to offer a few still further extended hints, such as they present themselves to my present view. But first let me mention that, if we would hope to make any progress in an investigation of this delicate nature, we ought to avoid two opposite extremes of which I can hardly say which is the most dangerous. If we indulge a fanciful imagination and build worlds of our own, we must not wonder at our going wide from the path of truth and nature; but these will vanish like the Cartesian vortices, that soon gave way when better theories were offered. On the other hand, if we add observation to observation, without attempting to draw not only certain conclusions, but also conjectural views from them, we offend against the very end for which only observations ought to be made. I will endeavor to keep a proper medium; but if I should deviate from that, I could wish not to fall into the latter error.

Herschel now begins to develop an account of the Milky Way and to unfold his seminal theory of development in the heavens—we may call it stellar evolution—based on his observation of nebulae. The theory derives from (1) his combination of meticulous counting and mapping of stars ("gaging"), (2) his application of Newtonian gravitational theory to astronomy, (3) his recognition concerning the relationship between distances of space and stretches of time, and (4) his imaginative manipulation of point of view.

That the Milky Way is a most extensive stratum of stars of various sizes admits no longer of the least doubt; and that our sun is actually one of the heavenly bodies belonging to it is as evident. I have now viewed and gaged this shining zone in almost every direction and find it composed of stars whose number, by the account of these gages, constantly increases and decreases in proportion to its apparent brightness to the naked eye. But in order to develop the ideas of the universe that have been suggested by my late observations, it will be best to take the subject from a point of view at a considerable distance both of space and time.

THEORETICAL VIEW

Let us then suppose numberless stars of various sizes, scattered over an indefinite portion of space in such a manner as to be almost equally distributed throughout the whole. The laws of attraction, which no doubt extend to the remotest regions of the fixed stars, will operate in such a manner as most probably to produce the following remarkable effects.

FORMATION OF NEBULAE

Form I. In the first place, since we have supposed the stars to be of various sizes, it will frequently happen that a star, being considerably larger than its neighboring ones, will attract them more than they will be attracted by others that are immediately around them; by which means they will be, in time, as it were, condensed about a center; or, in other words, form themselves into a cluster of stars of almost a globular figure, more or less regularly so, according to the size and original distance of the surrounding stars. The perturbations of these mutual attractions must undoubtedly be very intricate, as we may easily comprehend by considering what Sir Isaac Newton says in the first book of his *Principia*, in the 38th and following problems. . . .

Form II. The next case, which will also happen almost as frequently as the former, is where a few stars, though not superior in size to the rest, may chance to be rather nearer each other than the surrounding ones; for here also will be formed a prevailing attraction in the combined center of gravity of them all, which will occasion the neighboring stars to draw together; not indeed so as to form a regular or globular figure, but however in such a manner as to be condensed towards the common center of gravity of the whole irregular cluster. And this construction admits of the utmost variety of shapes, according to the number and situation of the stars which first gave rise to the condensation of the rest.

Herschel presents two further forms of nebulae that are mainly combinations of the first two, and concludes with a fifth.

In the last place, as a natural consequence of the former cases, there will be formed great cavities or vacancies by the retreat of the stars towards the various centers which attract them; so that upon the whole there is evidently a field of the greatest variety for the mutual and combined attractions of the heavenly bodies to exert themselves in. I shall therefore, without extending myself farther upon this subject, proceed to a few considerations that will naturally occur to every one who may view this subject in the light I have here done.

OBJECTIONS CONSIDERED

At first sight, then, it will seem as if a system such as it has been displayed in the foregoing paragraphs would evidently tend to a general destruction, by

the shock of one star's falling upon another. It would here be sufficient answer to say that, if observation should prove this really to be the system of
the universe, there is no doubt but that the great Author of it has amply provided for the preservation of the whole, though it should not appear to us in
what manner this is effected. But I shall moreover point out several circumstances that do manifestly tend to a general preservation; as, in the first
place, the indefinite extent of the sidereal heavens, which must produce a
balance that will effectually secure all the great parts of the whole from approaching to each other. There remains then only to see how the particular
stars belonging to separate clusters will be preserved from rushing on to their
centers of attraction. And here I must observe that, though I have before, by
way of rendering the case more simple, considered the stars as being originally at rest, I intended not to exclude projectile forces; and the admission of
them will prove such a barrier against the seeming destructive power of attraction as to secure from it all the stars belonging to a cluster, if not for ever,
at least for millions of ages. Besides, we ought perhaps to look upon such
clusters, and the destruction of now and then a star, in some thousands of
ages, as perhaps the very means by which the whole is preserved and renewed. These clusters may be the laboratories of the universe, if I may so express myself, wherein the most salutary remedies for the decay of the whole
are prepared.

From this theoretical view of the heavens, which has been taken, as we observed, from a point not less distant in time than in space, we will now retreat to our own retired station in one of the planets attending a star in its
great combination with numberless others. And in order to investigate what
will be the appearances from this contracted situation, let us begin with the
naked eye. The stars of the first magnitude being in all probability the nearest
will furnish us with a step to begin our scale. Setting off, therefore, with the
distance of Sirius or Arcturus, for instance, as unity, we will at present suppose that those of the second magnitude are at double, and those of the third
at treble the distance, and so forth. . . . Taking it then for granted that a star
of the seventh magnitude is about seven times as far as one of the first, it follows that an observer who is enclosed in a globular cluster of stars, and not
far from the center, will never be able with the naked eye to see to the end of
it. For, since, according to the above estimations, he can only extend his view
to about seven times the distance of Sirius, it cannot be expected that his eyes
should reach the borders of a cluster which has perhaps not less than fifty
stars in depth everywhere around him. The whole universe, therefore, to him
will be comprised in a set of constellations, richly ornamented with scattered
stars of all sizes. Or if the united brightness of a neighboring cluster of stars
should, in a remarkable clear night, reach his sight, it will put on the appear-

ance of a small, faint, whitish, nebulous cloud, not to be perceived without the greatest attention.

To pass by other situations, let him be placed in a much extended stratum, or branching cluster of millions of stars. . . . Here also the heavens will not only be richly scattered over with brilliant constellations, but a shining zone or Milky Way will be perceived to surround the whole sphere of the heavens, owing to the combined light of those stars which are too small, that is too remote, to be seen. Our observer's sight will be so confined that he will imagine this single collection of stars, of which he does not even perceive the thousandth part, to be the whole contents of the heavens.

Allowing him now the use of a common telescope, he begins to suspect that all the milkiness of the bright path which surrounds the sphere may be owing to stars. He perceives a few clusters of them in various parts of the heavens, and finds also that there are a kind of nebulous patches. But still his views are not extended so far as to reach to the end of the stratum in which he is situated, so that he looks upon these patches as belonging to that system which to him seems to comprehend every celestial object.

He now increases his power of vision and, applying himself to a close observation, finds that the Milky Way is indeed no other than a collection of very small stars. He perceives that those objects which had been called nebulae are evidently nothing but clusters of stars. He finds their number increase upon him, and when he resolves one nebula into stars he discovers ten new ones which he cannot resolve. He then forms the idea of immense strata of fixed stars, of clusters of stars and of nebulae; till, going on with such interesting observations, he now perceives that all these appearances must naturally arise from the confined situation in which we are placed. *Confined* it may justly be called, though in no less a space than what before appeared to be the whole region of the fixed stars, but which now has assumed the shape of a crookedly branching nebula, not indeed one of the least, but perhaps very far from being the most considerable of those numberless clusters that enter into the construction of the heavens.

Herschel goes on in his paper to employ an analogy to which he would return throughout his career: the analogy between the growth and aging of earthly biological organisms, and the growth and aging of stars and star systems. "The nebula we inhabit," he says, "might be said to be one that has fewer marks of profound antiquity upon it than the rest." Indeed, he ascribes to our "sidereal stratum" "a certain air of youth and vigor." Moreover, just as Wright, earlier in the century, echoing Milton, had extrapolated from the sun to "suns," so Herschel, like Kant, speaks of other "milky-ways." The establishment of suns and milky-ways as broad classes of things thus under-

girds a study of them akin to the study of species on earth. Herschel returns
to this strategy in a paper given to the Royal Society in 1789.

I need not repeat that by my analysis it appears that the heavens consist of regions where suns are gathered into separate systems, and that the catalogues
I have given comprehend a list of such systems. But may we not hope that
our knowledge will not stop short at the bare enumeration of phenomena capable of giving us so much instruction? Why should we be less inquisitive
than the natural philosopher, who sometimes, even from an inconsiderable
number of specimens of a plant or an animal is enabled to present us with the
history of its rise, progress, and decay? Let us then compare together and
class some of these numerous sidereal groups, that we may trace the operations of natural causes as far as we can perceive their agency.

The paper concludes by restating the same biological parallel juxtaposed
with another persistent analogy: that of space and time, distance and dura-
tion, whose relationship implies that objects of the same class viewed at vary-
ing distances present us with a chronological cross-section whereby we may
view the life and death of a certain kind of specimen, even of galaxies.

This method of viewing the heavens seems to throw them into a new kind of
light. They now are seen to resemble a luxuriant garden which contains the
greatest variety of productions in different flourishing beds; and one advantage we may at least reap from it is that we can, as it were, extend the range
of our experience to an immense duration. For, to continue the simile I have
borrowed from the vegetable kingdom, is it not almost the same thing
whether we live successively to witness the germination, blooming, foliage,
fecundity, fading, withering, and corruption of a plant, or whether a vast
number of specimens, selected from every stage through which the plant
passes in the course of its existence, be brought at once to our view?

If Herschel was a great pioneer of astronomical process, he was one also who
recognized process across the course of his own long career. In the following
excerpt from a paper delivered to the Royal Society in 1811, he describes
how he came to modify his theory of nebulae:

A knowledge of the construction of the heavens has always been the ultimate
object of my observations, and having been many years engaged in applying
my forty, twenty, and large ten feet telescopes, on account of their great
space-penetrating power, to review the most interesting objects discovered in
my sweeps. . . . I find that by arranging these objects in a certain successive

regular order, they may be viewed in a new light, and, if I am not mistaken, an examination of them will lead to consequences which cannot be indifferent to an enquiring mind.

If it should be remarked that in this new arrangement I am not entirely consistent with what I have already in former papers said on the nature of some objects that have come under my observation, I must freely confess that by continuing my sweeps of the heavens my opinion of the arrangement of the stars and their magnitudes, and of some other particulars, has undergone a gradual change. And indeed when the novelty of the subject is considered, we cannot be surprised that many things formerly taken for granted should, on examination, prove to be different from what they were generally, but incautiously, supposed to be.

For instance, an equal scattering of the stars may be admitted in certain calculations; but when we examine the Milky Way, or the closely compressed clusters of stars of which my catalogues have recorded so many instances, this supposed equality of scattering must be given up. We may also have surmised nebulae to be no other than clusters of stars disguised by their very great distance, but a longer experience and better acquaintance with the nature of nebulae will not allow a general admission of such a principle, although undoubtedly a cluster of stars may assume a nebulous appearance when it is too remote for us to discern the stars of which it is composed.

Impressed with an idea that nebulae properly speaking were clusters of stars, I used to call the nebulosity of which some were composed, when it was of a certain appearance, *resolvable*. But when I perceived that additional light, so far from resolving these nebulae into stars, seemed to prove that their nebulosity was not different from what I had called milky, this conception was set aside as erroneous. In consequence of this, such nebulae as afterwards were suspected to consist of stars, or in which a few might be seen, were called *easily resolvable*. But even this expression must be received with caution, because an object may not only contain stars, but also nebulosity not composed of them.

Herschel came, then, to see various nebular phenomena as markers of the diachronic development of stars and star systems—a kind of time-lapse photography. Thus, by tacit recourse to the traditional parallel of macrocosm and microcosm, he defends his having catalogued such a great number of nebulae:

This consideration will be a sufficient apology for the great number of assortments into which I have thrown the objects under consideration. And it will be found that those contained in one article are so closely allied to those in the next that there is perhaps not so much difference between them, if I may use the comparison, as there would be in an annual description of the human figure, were it given from the birth of a child till he comes to be a man in his prime.

SOURCE: *Philosophical Transactions of the Royal Society*, vol. 75 (1785); vol. 79 (1789); [no vol.] (1811), London.

45

As Certain as the Planetary Orbits

Pierre Simon Laplace

Pierre Simon Laplace (1749–1827) for a time taught mathematics to prospective artillery officers at l'École Militaire in Paris, where one of his students was the young Napoleon. Later, after the French Revolution, Laplace became one of the founding professors of l'École Polytechnique and was a member of the group that invented the metric system. His greatest mathematical contributions were in the study of probability—both theoretical and practical—and in calculus, which he applied to Newtonian mechanics. Laplace's success in using this physics to explain the world, especially in the realm of solar astronomy, led him to conclude that it would prove universally explanatory. This "radical mechanism" thus transformed itself in the thought of Laplace into an assumption, one most famously expressed in his answer to Napoleon's enquiry concerning the place of God in his system: "Sire, I have no need of that hypothesis."

The swift journey from assumption to assertion is compactly expressed early in Laplace's Philosophical Essay on Probabilities *(first published 1812):*

All events, even those which on account of their insignificance do not seem to follow the great laws of nature, are a result of it just as necessarily as the revolutions of the sun. In ignorance of the ties which unite such events to the entire system of the universe, they have been made to depend upon final causes or upon hazard, according as they occur and are repeated with regularity, or appear without regard to order. But these imaginary causes have gradually receded with the widening bounds of knowledge and disappear entirely be-

fore sound philosophy, which sees in them only the expression of our ignorance of the true causes.

Present events are connected with preceding ones by a tie based upon the
evident principle that a thing cannot occur without a cause which produces
it. This axiom, known by the name of *the principle of sufficient reason*, extends even to actions which are considered indifferent. The freest will is unable without a determinative motive to give them birth. If we assume two
positions with exactly similar circumstances and find that the will is active in
the one and inactive in the other, we say that its choice is an effect without a
cause. It is then, says Leibniz, the blind chance of the Epicureans. The contrary opinion is an illusion of the mind, which, losing sight of the evasive reasons of the choice of the will in indifferent things, believes that choice is
determined of itself and without motives.

We ought then to regard the present state of the universe as the effect of its
anterior state and as the cause of the one which is to follow. Given for one instant an intelligence which could comprehend all the forces by which nature
is animated and the respective situation of the beings who compose it—an intelligence sufficiently vast to submit these data to analysis—it would embrace
in the same formula the movements of the greatest bodies of the universe and
those of the lightest atom. For it, nothing would be uncertain and the future,
as the past, would be present to its eyes.

The human mind offers, in the perfection which it has been able to give to
astronomy, a feeble idea of this intelligence. Its discoveries in mechanics and
geometry, added to that of universal gravity, have enabled it to comprehend
in the same analytical expressions the past and future states of the system of
the world. Applying the same method to some other objects of its knowledge, it has succeeded in referring to general laws observed phenomena and
in foreseeing those which given circumstances ought to produce. All these efforts in the search for truth tend to lead it back continually to the vast intelligence which we have just mentioned, but from which it will always remain
infinitely removed. This tendency, peculiar to the human race, is that which
renders it superior to animals; and their progress in this respect distinguishes
nations and ages and constitutes their true glory.

Let us recall that formerly, and at no remote epoch, an unusual rain or an
extreme drought, a comet having in train a very long tail, the eclipses, the aurora borealis, and in general all the usual phenomena were regarded as so
many signs of celestial wrath. Heaven was invoked in order to avert their
baneful influence. No one prayed to have the planets and the sun arrested in
their courses: observation had soon made apparent the futility of such
prayers. But as these phenomena, occurring and disappearing at long intervals, seemed to oppose the order of nature, it was supposed that Heaven, irri-

tated by the crimes of the earth, had created them to announce its vengeance. Thus the long tail of the comet of 1456 spread terror through Europe, already thrown into consternation by the rapid successes of the Turks, who had just overthrown the Lower Empire. This star after four revolutions has excited among us a very different interest. The knowledge of the laws of the system of the world acquired in the interval had dissipated the fears begotten by the ignorance of the true relationship of man to the universe. And Halley, having recognized the identity of this comet with those of the years 1531, 1607, and 1682, announced its next return for the end of the year 1758 or the beginning of the year 1759. The learned world awaited with impatience this return which was to confirm one of the greatest discoveries that have been made in the sciences, and fulfil the prediction of Seneca when he said, in speaking of the revolutions of those stars which fall from an enormous height: "The day will come when, by study pursued through several ages, the things now concealed will appear with evidence; and posterity will be astonished that truths so clear had escaped us." Clairaut then undertook to submit to analysis the perturbations which the comet had experienced by the action of the two great planets, Jupiter and Saturn. After immense calculations he fixed its next passage at the perihelion toward the beginning of April, 1759, which was actually verified by observation. The regularity which astronomy shows us in the movements of the comets doubtless exists in all phenomena.

The curve described by a simple molecule of air or vapor is regulated in a manner just as certain as the planetary orbits. The only difference between them is that which comes from our ignorance.

Laplace's comprehensive determinism may appear philosophically antithetical to the Romanticism often associated with late eighteenth- and early nineteenth-century Europe. If Laplace's system has no room for the "hypothesis" of God, then surely it has no room for the human spirit either. His "mechanism" seems to have very little indeed to do with any motion or spirit (in Wordsworth's words) "that impels / All thinking things, all objects of all thought, / And rolls through all things." Yet it is striking how often Laplace uses terms such as genius, glory, and imagination. In his System of the World *(1798) Laplace surveys the history of astronomy and remarks, as a modern writer like Robert Osserman does in* Poetry of the Universe, *on the way in which discoveries are often arrived at as a result of long efforts to make the data fit one's theory.*

Without the speculations of the Greeks on the curves formed from the section of a cone by a plane, these beautiful laws [of planetary motion] might have been still unknown. The ellipse being one of these curves, its oblong figure gave rise, in the mind of Kepler, to the idea of supposing the planet Mars, whose orbit he had discovered to be oval, to move on it, and soon, by means of the numerous properties which the ancient geometricians had found in the conic sections, he became convinced of the truth of this hypothesis. The history of the sciences offers us many examples of these applications of pure geometry, and of its advantages. For everything is connected in the immense chain of truths, and often a single observation has been sufficient to show the connection between a proposition apparently the most sterile, and the phenomena of nature which are only mathematical results of general laws.

The perception of this truth probably gave birth to the mysterious analogies of the Pythagoreans. They had seduced Kepler, and he owed to them one of his most beautiful discoveries. Persuaded that the mean distances of the planets from the Sun ought to be regulated conformably to these analogies, he compared them a long time both with the regular geometrical solids and with the intervals of tones. At length, after seventeen years of meditations and calculation, conceiving the idea of comparing the powers of the numbers which expressed them, he found that the squares of the times of the planetary revolutions are to each other as the cubes of the major axes of their orbits, a most important law.

Nevertheless, while granting a role to speculation and imagination, Laplace puts great emphasis on patient induction—on what came to be called the scientific method.

We might be astonished that Kepler should not have applied the general laws of elliptic motion to comets. But, misled by an ardent imagination, he lost the clue of the analogy, which should have conducted him to this great discovery. The comets, according to him, being only meteors, engendered in the ether, he neglected to study their motions, and thus stopped in the middle of the career which was open to him, abandoning to his successors a part of the glory which he might yet have acquired. In his time the world had just begun to get a glimpse of the proper method of proceeding in the search for truth, at which genius only arrived by instinct, frequently connecting errors with its discoveries. Instead of passing slowly by a succession of inductions, from insulated phenomena to others more extended, and from these to the general laws of nature, it was more easy and more agreeable to subject all the phenomena to the relations of convenience and harmony, which the imagination could create and modify at pleasure.

Thus Kepler explained the disposition of the solar system by the laws of musical harmony. We behold him even in his latest works amusing himself with these chimerical speculations, even so far as to regard them as the "life and soul" of astronomy.

In spite of his strictures upon the imagination, Laplace's own enthusiasm for the story of astronomical discovery is evident in his writing. For him, artistic and literary sublimity are not so much replaced as surpassed by science. Certainly the categories of beauty and sublimity remain, as in his account of the achievements of Huygens and preeminently of Newton.

The labors of Huygens followed soon after those of Kepler and Galileo. Very few men have deserved so well of the sciences, by the importance and sublimity of their researches. The application of the pendulum to clocks is one of the most beautiful acquisitions which astronomy and geography have made. . . .

Huygens made numerous discoveries in geometry and mechanics; and if this extraordinary genius had conceived the idea of combining his theorems on centrifugal forces with his beautiful investigation on involutes, and with the laws of Kepler, he would have preceded Newton in his theory of curvilinear motion, and in that of universal gravitation.

To Newton's Principia *Laplace apparently wishes to accord literary immortality—"a preeminence over all other productions of human intellect"—that outwears the inevitable fact that any contribution to science will be superseded by further discoveries and refinements. Laplace's rhetoric, even as he describes the achievements of cold science, rises to the apocalyptic.*

The case is not the same with the sciences as with literature. Literature has limits which a man of genius may reach when he employs a language brought to perfection; he is read with the same interest in all ages; and time only adds to his reputation by the vain efforts of those who try to imitate him.

The sciences, on the contrary, without bounds like nature herself, increase infinitely by the labors of successive generations the most perfect work, and, by raising them to a height from which they can never again descend, give birth to new discoveries which produce in their turn new works which efface the former from which they originated. Others will present in a point of view more general and more simple the theories described in the *Principia*, and all the truths which it has displayed; but it will remain as an eternal monument of the profundity of that genius which has revealed to us the greatest law of the universe.

The theme of the monumental reappears in the concluding paragraphs of The System of the World. *In a maneuver that has become a cliché in elementary history of science but whose ironies are still not widely perceived, Laplace conflates geocentrism with anthropocentrism, using the occasion proudly to cast scorn on the pride of earlier humankind and to declare the sublimity of astronomy as a human achievement. It is as if the psalmic excla-*

mation "*the heavens declare the glory of God*" were retranslated to read: "*Astronomy declares the glory of man.*"

Contemplated as one grand whole, astronomy is the most beautiful monument of the human mind, the noblest record of its intelligence. Seduced by the illusions of the senses and of self-love, man considered himself for a long time as the center of the motion of the celestial bodies, and his pride was justly punished by the vain terrors they inspired. The labor of many ages has at length withdrawn the veil which covered the system. Man appears upon a small planet, almost imperceptible in the vast extent of the solar system, itself only an insensible point in the immensity of space. The sublime results to which this discovery has led may console him for the limited place assigned him in the universe. Let us carefully preserve and even augment the number of these sublime discoveries, which form the delight of thinking beings.

SOURCES: Adapted from Pierre Simon Laplace, *A Philosophical Essay on Probabilities*, trans. F. W. Truscott and F. L. Emory, New York: Dover Publications, 1951; and from *The System of the World*, trans. J. Pond, London, 1809.

46

The Intelligence of the Watch-Maker

William Paley

William Paley (1743–1805) was an English theologian whose name it is still compulsory to mention in any discussion (or refutation) of divine or "teleological" cosmic design. Barrow and Tipler call his Natural Theology *"synonymous with the gospel according to anthropocentric design" (*The Anthropic Cosmological Principle, *p. 76). And while almost everyone says his arguments have been discredited, mainly by Darwin, even Darwinians seem to recognize something deeply significant in Paley's elegant grappling with what looks like teleology. Some value, if also deficiency, is implied in Richard Dawkins's claim that, "When it comes to complexity and beauty of design, Paley hardly even began to state his case" (*The Blind Watchmaker, *p. 2). Although questions of biology lead beyond the bounds of this anthology, both biological and cosmological design raise the issue of the world's "fine tuning"—and Paley likewise engages them both by means of his famous metaphor of the watch.*

In crossing a heath, suppose I pitched my foot against a *stone*, and were asked how the stone came to be there. I might possibly answer that, for anything I knew to the contrary, it had lain there for ever; nor would it perhaps be very easy to show the absurdity of this answer. But suppose I had found a *watch* upon the ground, and it should be inquired how the watch happened to be in that place. I should hardly think of the answer which I had before given—that, for anything I knew, the watch might have always been there. Yet why should not this answer serve for the watch as well as for the stone? Why is it not as admissible in the second case as in the first? For this reason, and for no other, [namely] that, when we come to inspect the watch, we perceive (what we could not discover in the stone) that its several parts are framed and put together for a purpose, e.g., that they are so formed and adjusted as to produce motion, and that motion so regulated as to point out the hour of the day; that, if the different parts had been differently shaped from what they are, of a different size from what they are, or placed after any other manner, or in any other order, than that in which they are placed, either no motion at all would have been carried on in the machine, or none which would have answered the use that is now served by it. . . . This mechanism being observed . . . , the inference we think is inevitable, that the watch must have had a maker: that there must have existed, at some time, and at some place or other, an artificer who formed it for the purpose which we find it actually to answer, who comprehended its construction, and designed its use.

Nor would it, I apprehend, weaken the conclusion, that we had never seen a watch made; that we had never known an artist capable of making one; that we were altogether incapable of executing such a piece of workmanship ourselves. . . .

Neither, secondly, would it invalidate our conclusion, that the watch sometimes went wrong, or that it seldom went exactly right. The purpose of the machinery, the design, and the designer, might be evident, and in the case supposed would be evident, in whatever way we accounted for the irregularity of the movement, or whether we could account for it or not. . . .

Nor, thirdly, would it bring any uncertainty into the argument, if there were a few parts of the watch, concerning which we could not discover, or had not yet discovered, in what manner they conduced to the general effect; or even some parts concerning which we could not ascertain whether they conduced to that effect in any manner whatever. . . .

Nor, fourthly, would any man in his senses think the existence of the watch, with its various machinery, accounted for, by being told that it was one out of possible combinations of material forms; that whatever he had found in the place where he found the watch, must have contained some internal configuration or other; and that this configuration might be the structure now exhibited, [namely] of the works of a watch, as well as a different structure.

Nor, fifthly, would it yield his inquiry more satisfaction to be answered, that there existed in things a principle of order, which had disposed the parts of the watch into their present form and situation. He never knew a watch made by the principle of order; nor can he even form to himself an idea of what is meant by a principle of order, distinct from the intelligence of the watch-maker.

In spite of Paley's reputation for pushing interesting arguments too far, in the second half of his Natural Theology *he sounds a note of modesty when he turns his discussion to astronomy.*

My opinion of astronomy has always been that it is *not* the best medium through which to prove the agency of an intelligent Creator; but that, this being proved, it shows, beyond all other sciences, the magnificence of his operations. The mind which is once convinced, it raises to sublimer views of the Deity than any other subject affords; but it is not so well adapted, as some other subjects are, to the purpose of argument.

One point in Paley's discussion of astronomy that has drawn recent attention is his explicit recognition of change in the universe and of one's ability in principle, given that change, to extrapolate both forwards and backwards in time. Put simply, he recognizes the impossibility of accounting for the cosmos as he might naively have accounted for the stone on the heath: "For anything I knew to the contrary, it had lain there for ever." On the contrary, the sun itself, like planets and watches, must ultimately be seen as a contingent and transitory thing.

If these masses [the planets], partaking of the nature and substance of the sun's body, have in process of time lost their heat, that body itself, in time likewise, no matter in how much longer time, must lose its heat also, and therefore be incapable of an eternal duration in the state in which we see it, either for the time to come, or the time past.

SOURCE: William Paley, *Natural Theology: or, evidences of the existence and attributes of the Deity, collected from the appearances of nature*, London, 1802.

47

Must We Then Reject the Infinitude of the Stars?

H. W. M. Olbers

The name of H. W. M. Olbers (1758–1840) is associated with one of the most famous riddles of Newtonian cosmology: Why is the sky dark at night? For if we accept that space is infinite and that matter—the stars and galaxies—are uniformly or homogeneously distributed throughout all of space, then an answer to that question is indeed problematical. This is "Olbers' Paradox," so named not because Olbers (director of the Bremen observatory) was the first to formulate it, but because he wrote a concise, rigorous statement of the problem, to which he also proposed a solution.

As the perfection of our instruments increases, we behold more and more stars of ever smaller magnitudes. And reason must admit, however difficult it may still be for our imaginations to grasp, that Herschel has glimpsed through his giant telescopes heavenly objects that are 1500 or indeed several thousand times more distant from us than are Sirius or Arcturus.

But has the penetrating gaze of the immortal Herschel thereby approached, or even begun to approach, the limits of the universe? Who can believe it? Is space not infinite? Can we even conceive of its having limits? And is it conceivable that the almighty Creator should have left this infinite space empty? . . .

It is highly likely, then, that not only that part of space which the powerfully equipped human eye has observed, or can observe, but also the whole infinitude of space is occupied by suns and their retinue of planets and comets. I say "highly likely," for our limited reason is incapable of certainty.

There may be other locations in space occupied by creations other than suns, planets, comets, and luminous material, creations of which we perhaps have no conception whatsoever.

Of course Halley tried to adduce a proof for the infinite quantity of suns. "If their quantity were not infinite," he says, "then the space they do occupy would have a center of gravity towards which every object in the universe would fall with increasing acceleration, and so collapse. Only because the universal frame is infinite can everything remain in equilibrium and continue as it is." But Halley considered only gravitation and not momentum. Our planetary system would surely not collapse into the sun, even if there were no other stars nearby, or if it were totally isolated in space. And that the stars too are affected by momentum, their own movement appears to demonstrate. This alone is enough to establish the invalidity of Halley's proof, against which many other objections could likewise be raised.

Even if Halley's proof does not hold, however, it is still highly likely that the ordered cosmos which we observe to the limits of our powers of sight carries on through all the endless reaches of space, and we have only to enquire whether there are other reasons that undermine this assumption.

And indeed one very important objection does arise: if there really are suns throughout all of infinite space, whether at roughly equal distances from each other or distributed into galactic systems, then they themselves are infinite in number, and it follows that the entire heavens would have to be as bright as the sun. For every line I can conceive of as radiating from my sight will necessarily strike a star, and so also must every point in the sky radiate starlight, and consequently sunlight.

It hardly need be pointed out how contrary this is to experience.

Olbers' chain of reasoning is relatively easy to follow—even for the non-mathematician! He asks us to imagine a large spherical shell of a given radius (=1) about the sun, this shell as it were containing the stars of the first apparent magnitude. Then imagine a second shell with twice the radius of the first, also containing the appropriate number of stars for its size, and then still other shells, on and on, with radii of even multiples (=2, 3, 4, etc.) of the first. One need grasp only two pieces of mathematical information to see how this scenario unfolds:

1. *The surface area of a sphere is $4\pi r^2$ (four times pi times the radius squared)—so that when you increase a sphere's radius (r) twofold, you increase its area by 2^2 (i.e., fourfold); when you increase the radius threefold, you increase the area by by 3^2 (i.e., ninefold); and so on.*

2. *When you increase the distance of a light source twofold, you de-crease its apparent luminosity fourfold, and so on (that is, by an in-verse square proportion).*

Thus the increase in area according to (1) and the decrease in apparent lumi-nosity according to (2) balance each other exactly, so that the stars in each of the successive shells increase geometrically in number proportionate to the sequence 1, 4, 9, 16 and so on; while the apparent luminosity of individ-ual stars in the same successive shells decreases proportionate to the se-quence 1, 1/4, 1/9, 1/16 and so on. Olbers concedes that only a small amount of light may be shed upon us by each shell of stars. Nevertheless, be-cause the stars in each of the successive shells send us the same amount of light as the stars in the first shell do, and because the number of shells is in-finite (given an infinite homogeneous universe), therefore "not only does the entire vault of heaven become covered with stars, but they must also line up behind each other without end. . . . Clearly the same conclusion also obtains if the stars are distributed not uniformly in space but in individual systems separated by vast spaces." Olbers continues:

How fortunate for us that Nature has ordained things otherwise! How fortu-nate for us that not every point on the vault of heaven is beaming sunlight upon the earth! To say nothing of the unendurable brightness and the heat beyond all compare that would prevail, I draw attention merely to the most imperfect astronomy that we earth-dwellers would be left with. We should know nothing of the starry heavens. Only with great pains could we locate our own sun, and we should barely discern the moon and the planets as darker disks against the sun-bright background of the heavens. That is to say, the planets, themselves beamed down on by the utter brightness of the sky, would in proportion to their greater or lesser albedo [reflecting power] ap-pear more darkly than the rest of the sky.

But must we then reject the infinitude of the star systems because the entire sky does not appear to us as bright as the sun? Must we therefore confine these star systems to a small location within infinite space? By no means. That deduction from the infinitude of the stars rested on the assumption that space is utterly transparent—that light, consisting of parallel beams, remains wholly undiminished from whatever distance. This absolute transparency of space is not only wholly unproven, but also quite improbable.

Olbers goes on to argue that it is certain that space is not absolutely transparent, and that "it requires only the minutest degree of nontransparency to annihilate utterly that deduction from the infinitude of the stars, so totally contrary to actual experience." However, Olbers' solution is really no solution at all. For as John Herschel later asserted, whatever matter were responsible for the lack of utter transparency in space, it too, given enough time—and in Newton's universe there is enough time—would heat up to the point where it would glow white hot and so compensate for the radiation it was supposed to block out.

In an Einsteinian universe, by contrast, space is boundless but not infinite, and Olbers' paradox simply does not arise. It remains interesting, however, not only as a neat thought experiment, but also as an example of how virtually unthinkable it once was to relinquish such a central tenet of Newtonian doctrine—the infinitude of space.

SOURCE: Translated from H. W. M. Olbers, "Ueber die Durchsichtigkeit des Weltraums," *Astronomisches Jahrbuch*, vol. 51, ed. Johann Elert Bode, Berlin, 1826.

48

The Great Principle That Governs the Universe

Mary Fairfax Somerville

Mary Fairfax Somerville (1780–1872) was a mathematician, scientist, and translator who in a most serious and significant sense popularized the New-tonian universe, in particular as expounded by Laplace. The story is told of Laplace, over dinner in Paris, remarking to Somerville that "only two women have ever read the Mécanique Céleste; *both are Scotch women: Mrs. Greig and yourself." (During her first marriage, Somerville was known as Mrs. Greig.) In fact Somerville not only had read it but also translated it into English—and "into common language"—under the title* The Mechanism of the Heavens. *The success of her second book,* The Connection of the Physi-cal Sciences, *excerpted here, led to her election to the Royal Astronomical Society, of which she and Caroline Herschel became the first women mem-bers, in 1835. In her own day her work was also endorsed by the likes of John Herschel and Alexander von Humboldt.*

Like Humboldt (see following chapter), Somerville pursues a vision of the interconnectedness of the sciences and of the universe itself. Moreover, while introducing this vision, she communicates a sense of Romantic sublimity along with her Laplacian rigor.

Astronomy affords the most extensive example of the connection of the physical sciences. In it are combined the sciences of number and quantity, of rest and motion. In it we perceive the operation of a force which is mixed up with every thing that exists in the heavens or on earth; which pervades every atom, rules the motions of animate and inanimate beings, and is as sensible in the descent of a rain drop as in the falls of Niagara; in the weight of the air, as in the periods of the moon. Gravitation not only binds satellites to their planet, and planets to the sun, but it connects sun with sun throughout the wide extent of creation, and is the cause of the disturbances, as well as of the order, of nature: since every tremor it excites in any one planet is immediately transmitted to the farthest limits of the system, in oscillations, which correspond in their periods with the cause producing them, like sympathetic notes in music, or vibrations from the deep tones of an organ.

The heavens afford the most sublime subject of study which can be derived from science. The magnitude and splendor of the objects, the inconceivable rapidity with which they move, and the enormous distances between them impress the mind with some notion of the energy that maintains them in their motions, with a durability to which we can see no limit. Equally conspicuous is the goodness of the great First Cause, in having endowed man with faculties by which he can not only appreciate the magnificence of His works, but trace, with precision, the operation of His laws, use the globe he inhabits as a base wherewith to measure the magnitude and distance of the sun and planets, and make the diameter of the earth's orbit the first step of a scale by which he may ascend to the starry firmament. Such pursuits, while they ennoble the mind, at the same time inculcate humility, by showing that there is a barrier which no energy, mental or physical, can ever enable us to pass: that, however profoundly we may penetrate the depths of space, there still remain innumerable systems compared with which, those apparently so vast must dwindle into insignificance, or even become invisible; and that not only man, but the globe he inhabits—nay, the whole system of which it forms so small a part—might be annihilated, and its extinction be unperceived in the immensity of creation.

Nowhere is Somerville's pursuit of coherence clearer than when she writes about gravitation itself, whose full name in the Newtonian system is of course universal *gravitation. It is as if, in the mid-nineteenth century, one still tasted how awesome was the claim, then already more than two centuries old, that (in Galileo's words) earth and its physics are part of "the dance of the stars." Such was her confidence in Newtonian gravitation—"the great principle that governs the universe"—that she dared to assert: "every motion in the solar system has been so completely explained, that the laws of any astronomical phenomena that may hereafter occur, are already determined."*

From our perspective, Somerville's discussion of "ether" may seem merely antiquated; but it is worth noticing her concern with things that are still the subject of lively discussion, such as the question of whether the forces of physics can at some level be unified, or the recognition of how the mechanics of the universe actually presuppose its immense size and the almost incredible diffuseness of matter within it.

The known quantity of matter bears a very small proportion to the immensity of space. Large as the bodies are, the distances which separate them are immeasurably greater; but as design is manifest in every part of creation, it is probable that if the various systems in the universe had been nearer to one another, their mutual disturbances would have been inconsistent with the harmony and stability of the whole. It is clear that space is not pervaded by atmospheric air, since its resistance would, long ere this, have destroyed the velocity of the planets; neither can we affirm it to be a void, since it seems to be replete with ether, and traversed in all directions by light, heat, gravitation, and possibly by influences whereof we can form no idea.

Whatever the laws may be that obtain in the more distant regions of creation, we are assured that one alone regulates the motions, not only of our own system, but also of the binary systems of the fixed stars; and as general laws form the ultimate object of philosophical research, we cannot conclude these remarks without considering the nature of gravitation—that extraordinary power, whose effects we have been endeavoring to trace through some of their mazes. . . .

The curves in which the celestial bodies move by the force of gravitation are only lines of the second order. The attraction of spheroids, according to any other law of force than that of gravitation, would be much more complicated; and as it is easy to prove that matter might have been moved according to an infinite variety of laws, it may be concluded that gravitation must have been selected by Divine Wisdom out of an infinity of others, as being the most simple, and that which gives the greatest stability to the celestial motions.

It is a singular result of the simplicity of the laws of nature, which admit only of the observation and comparison of ratios, that the gravitation and theory of the motions of the celestial bodies are independent of their absolute magnitudes and distances. Consequently, if all the bodies of the solar system, their mutual distances, and their velocities, were to diminish proportionally, they would describe curves in all respects similar to those in which they now move; and the system might be successively reduced to the smallest sensible dimensions, and still exhibit the same appearances. We learn by experience that a very different law of attraction prevails when the particles of matter are placed within inappreciable distances from each other, as in chemical and capillary attraction, the attraction of cohesion, and molecular repulsion, yet it has been shown that in all probability not only these, but even gravitation itself, is only a particular case of the still more general principle of electric action.

The action of the gravitating force is not impeded by the intervention even of the densest substances. If the attraction of the sun for the center of the earth, and of the hemisphere diametrically opposite to him, were diminished by a difficulty in penetrating the interposed matter, the tides would be more obviously affected. Its attraction is the same also, whatever the substances of the celestial bodies may be; for if the action of the sun upon the earth differed by a millionth part from his action upon the moon, the difference would occasion a periodical variation in the moon's parallax, whose maximum would be the 1/15 of a second, and also a variation in her longitude amounting to several seconds, a supposition proved to be impossible by the agreement of theory with observation. Thus all matter is pervious to gravitation and is equally attracted by it.

Gravitation is a feeble force, vastly inferior to electric action, chemical affinity, and cohesion, yet as far as human knowledge extends, the intensity of gravitation has never varied within the limits of the solar system; nor does even analogy lead us to expect that it should. On the contrary, there is every reason to be assured that the great laws of the universe are immutable, like their Author. Not only the sun and planets, but the minutest particles, in all the varieties of their attractions and repulsions—nay, even the imponderable matter of the electric, galvanic, or magnetic fluid—are all obedient to permanent laws, though we may not be able in every case to resolve their phenomena into general principles. Nor can we suppose the structure of the globe alone to be exempt from the universal fiat, though ages may pass before the changes it has undergone, or that are now in progress, can be referred to existing causes with the same certainty with which the motions of the planets, and all their periodic and secular [i.e., temporal] variations, are referable to the law of gravitation. The traces of extreme antiquity perpetually occurring to the geologist give that information, as to the origin of things, in vain looked for in the other parts of the universe. They date the beginning of time with regard to our system, since there is ground to believe that the formation of the earth was contemporaneous with that of the rest of the planets; but they show that creation is the work of Him with whom "a thousand years are as one day, and one day as a thousand years."

SOURCE: Mary Somerville, *On the Connexion of the Physical Sciences*, 5th edition, London, 1840.

49

The Unfailing Connection and Course of Events

Alexander von Humboldt

Alexander von Humboldt (1769–1859) was a pioneer in the development of the sciences of geography and climatology. His work demonstrating the connections between life forms and their physical surroundings—most notably in the accounts of his journeys of investigation in Central and South America between 1799 and 1804—is also recognized as a founding contribution to what is now called the science of ecology.

Humboldt's educational, cultural, and scientific endowments ideally fitted him to become one of the greatest popularizers of science in his century, and the five volumes of his crowning work, entitled Cosmos *(1845–1862), is founded on his view that things must be seen in their complementarity and connectedness. The aim of Humboldt's account is to comprehend the whole in both its coherence and its beauty.*

When the mind of man attempts to subject to itself the world of physical phenomena, when in meditative contemplation of existing things he strives to penetrate the rich fullness of the life of nature and the free or restricted operations of natural forces, he feels himself raised to a height from whence, as he glances round the far horizon, details disappear, and groups or masses are alone beheld, in which the outlines of individual objects are rendered indistinct as by an effect of aerial perspective. This illustration is purposely selected in order to indicate the point of view from whence we design to consider the material universe and to present it as the object of contemplation in both its divisions, celestial and terrestrial. . . . We strive by classification and due subordination of phenomena, by penetration into the play of obscure forces, and by an animated representation in which the visible spectacle may be reflected back as in a faithful mirror—to conceive and to describe the whole creation (Greek, *to pan*) in a manner befitting the dignity of the word Cosmos in its sense of *universe, order of the material world,* and beauty or *ornament* of that universal order.

Humboldt's easy use of analogies such as that of the "aerial view" and his emphasis on the beauty as well as the order which "cosmos" connotes are characteristic of the aesthetic and subjective dimension of his cosmology. His long overview of poetic descriptions of nature and of the "history of the physical contemplation of the universe" reveal his awareness that he is not merely laying out "the facts" of the universe in some pseudoscientific way but is himself participating in, as well as observing, the history of consciousness. Moreover, as he traces subjective and literary dimensions of cosmology Humboldt does not himself remain aloof: He writes as a scientist who is also an engaged and at times enthusiastic literary critic, as for example when expositing an Old Testament Psalm:

It is the characteristic of Hebrew poetry in reference to nature that, as a reflex of monothesim, it always embraces the whole world in its unity, comprehending the life of the terrestrial globe as well as the shining regions of space. . . .

A single Psalm [104] may be said to present a picture of the entire Cosmos: "The Lord covereth himself with light as with a garment, he hath stretched out the heavens like a canopy. He laid the foundations of the round earth that it should not be removed for ever. The waters springing in the mountains descend to the valleys, unto the places which the Lord hath appointed for them, that they may never pass the bounds which he has set them, but may give drink to every beast of the field. Beside them the birds of the air sing among the branches. The trees of the Lord are full of sap, the cedars of Lebanon which he hath planted, wherein the birds make their nests, and the fir trees wherein the stork builds her house." The great and wide sea is also described, "wherein are living things innumerable; there move the ships, and there is that leviathan which thou hast made to sport therein." The fruits of the field, the objects of the labor of man, are also introduced, the corn, the cheerful vine, and the olive garden. The heavenly bodies complete this picture of nature: "The Lord appointed the moon for seasons, and the sun knoweth the term of his course. He bringeth darkness, and it is night, wherein the wild beasts roam. The young lions roar after their prey, and seek their meat from God. The sun ariseth and they get them away together, and

lay them down in their dens." And then "man goeth forth unto his work and
to his labor until the evening."

We are astonished to see, within the compass of a poem of such small dimension, the universe, the heavens and the earth, thus drawn with a few
grand strokes. The moving life of the elements is here placed in opposition to
the quiet laborious life of man, from the rising of the sun, to the evening
when his daily work is done. This contrast, the generality in the conception
of the mutual influence of phenomena, the glance reverting to the omnipresent invisible Power, which can renew the face of the earth, or cause the
creature to return again to the dust, give to the whole a character of solem-
nity and sublimity rather than of warmth and softness.

*Humboldt's emphasis on the connectedness of things appears also in his view
of the Copernican revolution, in which the spirits of geographical and cosmographical discovery mirror each other. The account is likewise notable for
its clear if unobtrusive adherence to that Romantic theory whereby the eye is
no passive observer of phenomena but imparts color and form to whatever it
beholds. So too, accordingly, the telescope, which Humboldt describes in organic, even phallic or perhaps imperialistic terms, "penetrates" space and
calls forth "a new world of ideas."*

The epoch of the most extensive discoveries upon the surface of our planet
was immediately succeeded by man's first taking possession of a considerable
part of the celestial spaces by the telescope. The application of a newly
formed organ, of an instrument of space-penetrating power, called forth a
new world of ideas. Now began a brilliant age of astronomy and mathematics. . . .

By the discovery of telescopic vision there was lent to the eye—the organ
of the sensuous contemplation of the visible universe—a power of which we
are yet far from having reached the limit, but of which the first feeble com-
mencement (magnifying hardly as much as thirty-two times in linear dimension) sufficed to penetrate into cosmical depths before unknown. The exact
knowledge of many heavenly bodies belonging to our solar system, the unchanging laws according to which they revolve in their orbits, and the perfected insight into the true structure of the universe, are the characteristics of
the epoch which we here attempt to describe. The results which this age produced have defined the leading outlines of the picture of nature or sketch of
the Cosmos, and have added an intelligent recognition of the contents of the
celestial spaces—at least in the well-understood arrangement of one planetary group—to the earlier explored contents of terrestrial space.

Reading Copernicus's de Revolutionibus as a Romantic literary critic and not only as a historian of science, Humboldt rejects the notion that Copernicus intended his system merely as a model, to save the appearances:

The free and powerful language employed by Copernicus, the evident out-pouring of deep internal conviction, sufficiently refutes the assertion that the system which bears his immortal name was proposed as an hypothesis convenient to calculating astronomers, but which might very well be without foundation. "By no other arrangement," he exclaims with inspired enthusiasm, "have I been able to discover so admirable a symmetry of the universe, so harmonious a combination of orbits, than by placing the light of the world, the sun, as on a kingly throne, in the midst of the beautiful temple of nature, guiding from thence the entire family of circum-revolving planets." Even the idea of universal gravitation or attraction towards the center of the world, the sun, inferred from the force of gravity in spherical bodies, appears to have floated before the mind of this great man, as is shown by a remarkable passage in the ninth chapter of the first book of the *Revolutions* [see the end of Chapter 22 for Digges's translation].

Likewise in his account of the role of Kepler, Humboldt points to the role of enthusiasm and imagination in combination with the relatively more passive exercise of observation.

In [my] remarks on the influence exerted by the direct visible contemplation of particular heavenly bodies I have named Kepler more particularly for the sake of recalling how, in this great, richly gifted, and extraordinary man the love for imaginative combinations was united with a remarkable talent for observation, a grave and severe method of induction, a courageous and almost unexampled perseverance in calculation, and a depth of mathematical thought which . . . exercised a happy influence on Fermat, and through him on the invention of the infinitesimal calculus. The possessor of such a mind was pre-eminently suited by the richness and mobility of his ideas, and even by the boldness of the cosmological speculations which he hazarded, to promote and animate the movement which carried the seventeenth century uninterruptedly forward towards the attainment of its exalted object, the enlarged contemplation of the universe.

Humboldt's summary of his narrative adumbrates a parallel between Kepler's age and his own in their advancement of geography, cosmography, and physics. Yet this parallel, Humboldt suggests, flattering as it may be to the nineteenth century, in fact leads to the somewhat humbling recognition that the advances of any age are but steps in an ongoing journey.

The series of external events which suddenly enlarged the intellectual horizon, stimulating men to the research of physical laws and animating them to the endeavor to rise to the ultimate apprehension of the Universe as a Whole, closed, according to my view, with those geographical discoveries—the greatest ever achieved—which placed the nations of the Old Continent in possession of an entire terrestrial hemisphere till then concealed. From thenceforward . . . the human intellect produces great results, no longer from the incitement of external events but through the operation of its own internal power, and this simultaneously in all directions. Nevertheless, amongst the instruments which men formed for themselves, constituting as it were new organs augmenting their powers of sensuous perception, there was one which acted like a great and sudden event. By the space-penetrating power of the telescope, a considerable portion of the heavens was explored as it were at once, the number of known celestial bodies was increased, and their form and orbits began to be determined. Mankind now first entered on the possession of the "celestial space" of the Cosmos. . . .

Every age dreams that it has approached near to the culminating point of the knowledge and comprehension of nature. I doubt whether upon serious reflection such a belief will really appear to enhance the enjoyment of the present. A more animating conviction . . . is that the possessions yet achieved are but a very inconsiderable portion of those which, in the advance of activity and of general cultivation, mankind in their freedom will attain in succeeding ages. In the unfailing connection and course of events every successful investigation becomes a step to the attainment of something beyond.

SOURCE: Adapted from Alexander von Humboldt, *Cosmos: Sketch of a Physical Description of the Universe*, vols. 1–2, translated under the supervision of Edward Sabine, London, 1847, 1848.

50

The Primordial Particle

Edgar Allan Poe

Edgar Allan Poe (1809–1849) is probably best known for his inimitable combination of the realistic and the fantastic, and for his sense of plot. These qualities, had he lived a century later, might have placed him more conspicuously among the cosmologists. Yet, in his own day, he produced one notably bold and imaginative discussion of cosmic themes: Eureka, *subtitled* A Prose Poem, *which Poe dedicated to Alexander von Humboldt. Some of the concepts Poe there unfolds, however cryptically and poetically, appear strangely akin to those that undergird cosmology today: naked singularity, the Big Bang, a universe boundless but not infinite.*

Hitherto, the universe of stars has always been considered as coincident with the universe proper. . . . It has been always either directly or indirectly assumed—at least since the dawn of intelligible astronomy—that, were it possible for us to attain any given point in space, we should still find, on all sides of us, an interminable succession of stars. This was the untenable idea of Pascal when making perhaps the most successful attempt ever made at periphrasing [sic] the conception for which we struggle in the word "universe." "It is a sphere," he says, "of which the center is everywhere, the circumference, nowhere." But although this intended definition is, in fact, no definition of the universe of stars, we may accept it, with some mental reservation, as a definition . . . of the universe proper—that is to say, of the universe of space. This latter, then, let us regard as "a sphere of which the center is everywhere, the circumference nowhere." In fact, while we find it impossible to fancy an end to space, we have no difficulty in picturing to ourselves any one of an infinity of beginnings.

As our starting point, then, let us adopt the Godhead. Of this Godhead, in itself, he alone is not imbecile—he alone is not impious who propounds----nothing. . . . "We know absolutely nothing of the nature or essence of God: in order to comprehend what he is, we should have to be God ourselves."

By him, however—now, at least, the Incomprehensible—by him—assuming him as spirit—that is to say, as *not matter*—a distinction which, for all intelligible purposes, will stand well instead of a definition—by him, then, existing as spirit, let us content ourselves, tonight, with supposing to have been created, or made out of nothing, by dint of his volition—at some point of space which we will take as a center—at some period into which we do not pretend to inquire, but at all events immensely remote—by him, then again, let us suppose to have been created----what? This is a vitally momentous epoch in our considerations. What is it that we are justified—that alone we are justified in supposing to have been, primarily and solely, created?

We have attained a point where only intuition can aid us—but now let me recur to the idea which I have already suggested as that alone which we can properly entertain of intuition. It is but the conviction arising from those inductions or deductions of which the processes are so shadowy as to escape our consciousness, elude our reason, or defy our capacity of expression. With this understanding, I now assert—that an intuition altogether irresistible, although inexpressible, forces me to the conclusion that what God originally created—that that matter which, by dint of his volition, he first made from his spirit, or from nihility, could have been nothing but matter in its utmost conceivable state of----what?—of *simplicity*? . . .

Let us now endeavor to conceive what matter must be, when, or if, in its absolute extreme of simplicity. Here the reason flies at once to imparticularity—to a particle—to one particle—a particle of one kind—of one character—of one nature—of one size—of one form—a particle, therefore, "without form and void"—a particle positively a particle at all points—a particle absolutely unique, individual, undivided, and not indivisible only because he who created it, by dint of his will, can by an infinitely less energetic exercise of the same will, as a matter of course, divide it.

Oneness, then, is all that I predicate of the originally created matter; but I propose to show that this oneness is a principle abundantly sufficient to account for the constitution, the existing phenomena and the plainly inevitable annihilation of at least the material universe.

The willing into being the primordial particle, has completed the act, or more properly the conception, of creation. We now proceed to the ultimate purpose for which we are to suppose the particle created—that is to say, the ultimate purpose so far as our considerations yet enable us to see it—the constitution of the universe from it, the particle. . . .

The assumption of absolute unity in the primordial particle includes that of infinite divisibility. Let us conceive the particle, then, to be only not totally exhausted by diffusion into space. From the one particle, as a center, let us suppose to be irradiated spherically—in all directions—to immeasurable but still to definite distances in the previously vacant space—a certain inexpressibly great yet limited number of unimaginably yet not infinitely minute atoms.

Thus, Poe begins to move from the notion of a primordial particle to its expansion outwards, and to questions concerning the resulting shape, diffusion, and consistency of the material universe.

A very slight inspection of the heavens assures us that the stars have a certain general uniformity, equability, or equidistance, of distribution through that region of space in which, collectively, and in a roughly globular form, they are situated—this species of very general, rather than absolute, equability, being in full keeping with my deduction of inequidistance, within certain limits, among the originally diffused atoms, as a corollary from the evident design of infinite complexity of relation out of irrelation. I started, it will be remembered, with the idea of a generally uniform but particularly *un*uniform distribution of atoms—an idea, I repeat, which an inspection of the stars, as they exist, confirms.

But even in the merely general equability of distribution, as regards the atoms, there appears a difficulty which, no doubt, has already suggested itself to those among my readers who have borne in mind that I suppose this equability of distribution effected through radiation from a center. The very first glance at the hitherto unseparated and seemingly inseparable idea of agglomeration about a center, with dispersion as we recede from it—the idea, in a word, of *in*equability of distribution in respect to the matter irradiated.

Now . . . by just such difficulties as the one now in question—such roughness—such peculiarities—such protuberances above the plane of the ordinary—that reason feels her way, if at all, in her search for the True. By the difficulty—the "peculiarity"—now presented, I leap at once to the secret—a secret which I might never have attained but for the peculiarity and the inferences which, in its mere character of peculiarity, it affords me.

The process of thought at this point may be thus roughly sketched: I say to myself—"Unity, as I have explained it, is a truth—I feel it. Diffusion is a truth—I see it. Irradiation, by which alone these two truths are reconciled, is the consequent truth—I perceive it. Equability of diffusion, first deduced a priori and then corroborated by the inspection of phenomena, is also a truth—I fully admit it. So far all is clear around me: there are no clouds behind which the secret—the great secret of the gravitating modus operandi—can possibly lie hidden; but this secret lies hereabouts, most assuredly; and were there but a cloud in view, I should be driven to suspicion of that cloud." And now, just as I say this, there actually comes a cloud into view. This cloud is the seeming impossibility of reconciling my truth, irradiation, with my truth, equability of diffusion. I say now: "Behind this seeming impossibility is to be found what I desire." I do not say "real impossibility"; for invincible faith in my truths assures me that it is a mere difficulty after all—but I go on to say, with unflinching confidence, that, when this difficulty shall be solved, we shall find, wrapped up in the process of solution, the key to the secret at which we aim. Moreover—I feel that we shall discover but one possible solution of the difficulty; this for the reason that, were there two, one would be supererogatory—would be fruitless—would be empty—would contain no key—since no duplicate key can be needed to any secret of Nature.

And now, let us see: Our usual notions of irradiation—in fact all our distinct notions of it—are caught merely from the process as we see it exemplified in light. Here there is a continuous outpouring of ray-streams, and with a force which we have at least no right to suppose varies at all. Now, in any such irradiation as this—continuous and of unvarying force—the regions nearer the center must inevitably be always more crowded with the irradiated matter than the regions more remote. But I have assumed no such irradiation as this. I assumed no continuous irradiation; and for the simple reason that such an assumption would have involved, first, the necessity of entertaining a conception which I have shown no man can entertain, and which . . . all observation of the firmament refutes—the conception of the absolute infinity of the universe of stars—and would have involved, secondly, the impossibility of understanding a reaction—that is, gravitation—as existing now—since, while an act is continued, no reaction, of course, can take place. My assumption, then, or rather my inevitable deduction from just premises—was that of a *determinate* irradiation—one finally *dis*continued.

Let me now describe the sole possible mode in which it is conceivable that matter could have been diffused through space, so as to fulfil the conditions at once of irradiation and of generally equable distribution.

For convenience of illustration, let us imagine, in the first place, a hollow sphere of glass, or of anything else, occupying the space throughout which the universal matter is to be thus equally diffused, by means of irradiation, from the absolute, irrelative, unconditional particle, placed in the center of the sphere.

Now, a certain exertion of the diffusive power (presumed to be the Divine Volition)—in other words, a certain force—whose measure is the quantity of matter—that is to say, the number of atoms—emitted; emits, by irradiation, this certain number of atoms; forcing them in all directions outwardly from the center—their proximity to each other diminishing as they proceed—until, finally, they are distributed, loosely, over the interior surface of the sphere.

When these atoms have attained this position, or while proceeding to attain it, a second and inferior exercise of the same force—or a second and inferior force of the same character—emits, in the same manner—that is to say, by irradiation as before—a second stratum of atoms which proceeds to deposit itself upon the first; the number of atoms, in this case as in the former, being of course the measure of the force which emitted them, in other words the force being precisely adapted to the purpose it effects—the force and the number of atoms sent out by the force, being directly proportional.

When this second stratum has reached its destined position—or while approaching it—a third still inferior exertion of the force, or a third inferior force of a similar character—the number of atoms emitted being in all cases the measure of the force—proceeds to deposit a third stratum upon the second—and so on, until these concentric strata, growing gradually less and less, come down at length to the central point; and the diffusive matter, simultaneously with the diffusive force, is exhausted.

We have now the sphere filled, through means of irradiation, with atoms equably diffused. The two necessary conditions—those of irradiation and of equable diffusion—are satisfied; and by the sole process in which the possibility of their simultaneous satisfaction is conceivable.

SOURCE: *Eureka: A Prose Poem*, in *The Complete Works of Edgar Allan Poe*, vol. 16, ed. James A. Harrison, New York, 1902.

51

The Shadow! The Shadow!

Maria Mitchell

*Maria Mitchell (1818–1889) is famous as the first American female as-
tronomer and the first professor of astronomy at the first American women's
college, Vassar, where the chair of astronomy still bears her name. In 1847,
at the age of twenty-nine, she was also the first to discover a comet using a
telescope. For this accomplishment, she received from the king of Denmark a
gold medal that since 1831 had been on offer to the first discoverer of a
"telescopic comet." In 1848 she became the first (and for another fifty years
the only) woman elected to the American Academy of Arts and Sciences.
Mitchell's diary, from which the following two excerpts are taken, shows her
as possessing a mind at once incisively scientific and powerfully poetic. Here
she looks back on her 1847 discovery and reflects on circumstances both
sublime and ridiculous that surrounded her observation.*

October, [1854]. I have just gone over my comet computations again, and
it is humiliating to perceive how very little more I know than I did seven
years ago when I first did this kind of work. To be sure, I have only once in
the time computed a parabolic orbit; but it seems to me that I know more
in general. I think I am a little better thinker, that I take things less upon
trust, but at the same time I trust myself much less. The world of learning is
so broad, and the human soul is so limited in power! We reach forth and
strain every nerve, but we seize only a bit of the curtain that hides the infinite from us.

Will it really unroll to us at some future time? . . .

Dec. 5, 1854. . . . The comet looked in upon us on the 29th. It made a twilight call, looking sunny and bright, as if it had just warmed itself in the
equinoctial rays. A boy on the street called my attention to it, but I found on
hurrying home that father had already seen it, and had ranged it behind
buildings so as to get a rough position.

It was piping cold, but we went to work in good earnest that night, and the
next night on which we could see it, which was not until April.

I was dreadfully busy, and a host of little annoyances crowded upon me. I
had a good star near it in the field of my comet-seeker, but *what* star?

On that rested everything, and I could not be sure even from the catalogue,
for the comet and the star were so much in the twilight that I could get no
good neighboring stars. . . .

Then came a waxing moon, and we waxed weary in trying to trace the
fainter and fainter comet in the mists of twilight and the glare of moonlight.

Next I broke a screw of my instrument, and found that no screw of that
description could be bought in the town.

I started off to find a man who could make one, and engaged him to do so
the next day. The next day was Fast Day; all the world fasted, at least from
labor.

However, the screw was made, and it fitted nicely. The clouds cleared, and
we were likely to have a good night. I put up my instrument, but scarcely had
the screwdriver touched the new screw than out it flew from its socket, rolled
along the floor of the "walk," dropped quietly through a crack into the gut-
ter of the house-roof. I heard it click, and felt very much like using language
unbecoming a woman's mouth.

I put my eye down to the crack, but could not see it. There was but one
thing to be done—the floor-boards must come up. I got a hatchet, but could
do nothing. I called father; he brought a crowbar and pried up the board,
then crawled under it and found the screw. I took good care not to lose it a
second time.

The instrument was fairly mounted when the clouds mounted to keep it
company, and the comet and I again parted.

In all observations, the blowing out of a light by a gust of wind is a very
common and very annoying accident; but I once met with a much worse one,
for I dropped a chronometer, and it rolled out of its box onto the ground. We
picked it up in a great panic, but it had not even altered its rate, as we found
by later observations.

The glaring eyes of the cat, who nightly visited me, were at one time very
annoying, and a man who climbed up a fence and spoke to me, in the stillness of the small hours, fairly shook not only my equanimity, but the pencil

which I held in my hand. He was quite innocent of any intention to do me harm, but he gave me a great fright.

The spiders and bugs which swarm in my observing-houses I have rather an attachment for, but they must not crawl over my recording-paper. Rats are my abhorrence, and I learned with pleasure that some poison had been placed under the transit-house.

One gets attached (if I may use the term) to certain midnight apparitions. The Aurora Borealis is always a pleasant companion; a meteor seems to come like a messenger from departed spirits; and the blossoming of trees in the moonlight becomes a sight looked for with pleasure.

Aside from the study of astronomy, there is the same enjoyment in a night upon the housetop, with the stars, as in the midst of other grand scenery; there is the same subdued quiet and grateful seriousness; a calm to the troubled spirit, and a hope to the desponding.

More than twenty years later, Mitchell described her observations of the "Denver" eclipse of the sun (1878), and again one glimpses from the account elements of the event's practical, social, and poetic as well as scientific character.

When we look at any other object than the sun, we stimulate our vision. A good observer will remain in the dark for a short time before he makes a delicate observation on a faint star, and will then throw a cap over his head to keep out strong lights.

When we look at the sun, we at once try to deaden its light. We protect our eyes by dark glasses—the less sunlight we can get the better. We calculate exactly at what point the moon will touch the sun, and we watch that point only. The exact second by the chronometer when the figure of the moon touches that of the sun is always noted. It is not only valuable for the determination of longitude, but it is a check on our knowledge of the moon's motions. Therefore, we try for the impossible.

One of our party, a young lady from California, was placed at the chronometer. She was to count aloud the seconds, to which the three others were to listen. Two others, one a young woman from Missouri, who brought with her a fine telescope, and another from Ohio, besides myself, stood at the three telescopes. A fourth, from Illinois, was stationed to watch general effects, and one special artist, pencil in hand, to sketch views.

Absolute silence was imposed upon the whole party a few minutes before each phenomenon.

Of course we began full a minute too soon, and the constrained position was irksome enough, for even time is relative, and the minute of suspense is longer than the hour of satisfaction.

The moon, so white in the sky, becomes densely black when it is closely ranging with the sun, and it shows itself as a black notch on the burning disc when the eclipse begins.

Each observer made her record in silence, and then we turned and faced one another, with record in hand—we differed more than a second; it was a large difference.

Between first contact and totality there was more than an hour, and we had little to do but look at the beautiful scenery and watch the slow motion of a few clouds, on a height which was cloud-land to dwellers by the sea.

Our photographer begged us to keep our positions while he made a picture of us. The only value to the picture is the record that it preserves of the parallelism of the three telescopes. You would say it was stiff and unnatural, did you not know that it was the ordering of Nature herself—they all point to the center of the solar system.

As totality approached, all again took their positions. The corona, which is the "glory" seen around the sun, was visible at least thirteen minutes before totality; each of the party took a look at this, and then all was silent, only the count, on and on, of the young woman at the chronometer. When totality came, even that ceased.

How still it was!

As the last rays of sunlight disappeared, the corona burst out all around the sun, so intensely bright near the sun that the eye could scarcely bear it; extending less dazzlingly bright around the sun for the space of about half the sun's diameter, and in some directions sending off streamers for millions of miles.

It was now quick work. Each observer at the telescopes gave a furtive glance at the un-sunlike sun, moved the dark eye-piece from the instrument, replaced it by a more powerful white glass, and prepared to see all that could be seen in two minutes forty seconds. They must note the shape of the corona, its color, its seeming substance, and they must look all around the sun for the "interior planet."

There was certainly not the beauty of the eclipse of 1869. Then immense radiations shot out in all directions, and threw themselves over half the sky. In 1869, the rosy prominences were so many, so brilliant, so fantastic, so weirdly changing, that the eye must follow them; now, scarcely a protuberance of color, only a roseate light around the sun as the totality ended. But if streamers and prominences were absent, the corona itself was a great glory.

Our special artist, who made the sketch for my party, could not bear the light.

When the two minutes forty seconds were over, each observer left her instrument, turned in silence from the sun, and wrote down brief notes. Happily, some one broke through all rules of order, and shouted out, "The shadow! the shadow!" And looking toward the south-east we saw the black band of shadow moving from us, a hundred and sixty miles over the plain, and toward Indian Territory. It was not the flitting of the closer shadow over the hill and dale: it was a picture which the sun threw at our feet of the dignified march of the moon in its orbit.

And now we looked around. What a strange orange light there was in the north-east! What a spectral hue to the whole landscape! Was it really the same old earth, and not another planet?

SOURCE: *Maria Mitchell: Life, Letters, and Journals*, ed. Phebe Mitchell Kendall, Boston, 1896.

52

Unraveled Starlight

William Huggins

*Near the middle of the nineteenth century, a new means of scientific investi-
gation emerged which changed the face of astronomy. Spectral analysis or
"spectroscopy" ushered in what William Huggins (1824–1910), appropriat-
ing Kepler's term, called "The New Astronomy."*

*To get a sense of how unexpected the discoveries of the New Astronomy
were, we may consider the declarations of Auguste Comte (1798–1857).
Comte, the founder of Positivism, insists in his* Course of Positive Philosophy
*(1830–42) on a sharp distinction between planetary and stellar ("sidereal")
astronomy.*

It is easy to describe clearly the character of astronomical science, for in our
time it has been thoroughly set free from every theological and metaphysical
influence. When we look at the simple facts of the case, it is evident that
though three of our senses take cognizance of distant objects, only one of the
three perceives the stars. The blind could know nothing of them; and we who
see, after all our preparation, know nothing of stars hidden by distance, except by induction. Of all objects, the planets are those which appear to us under the least varied aspect. We see how we may determine their forms, their
distances, their bulk, and their motions, but we can never know anything of
their chemical or mineralogical structure, and much less that of organized beings living on their surface. We may obtain positive knowledge of their geometrical and mechanical phenomena. But all physical, chemical,
physiological, and social researches, for which our powers fit us on our own
earth, are out of the question in regard to the planets. Whatever knowledge is
obtainable by means of the sense of sight, this we may hope to attain with re-

gard to the stars, whether we at present see the method or not. But whatever knowledge requires the aid of other senses we must at once exclude from our expectations, despite any appearances to the contrary.

Having excluded things such as chemistry and mineralogy from astronomy, Comte offers the following definition with an attendant restriction:

We may therefore define astronomy as the science by which we discover the laws of the geometrical and mechanical phenomena presented by the heavenly bodies.

It is desirable to add a limitation which is important, though not of primary necessity. The part of the science which we command from what we may call the solar point of view is distinct, and evidently capable of being made complete and satisfactory; while that which is regarded from the universal point of view is in its infancy to us now, and must ever be illimitable to our successors of the remotest generations. Men will never compass in their conceptions the whole of the stars. The difference is very striking now to us who find a perfect knowledge of the solar system at our command, while we have not obtained the first and most simple element in sidereal astronomy—the determination of the stellar intervals. Whatever may be the ultimate progress of our knowledge in certain portions of the larger field, it will leave us always at an immeasurable distance from understanding the universe. . . .

We must carefully separate the idea of the solar system from that of the universe, and be always assured that our only true interest is in the former. Within this boundary alone is astronomy the supreme and positive science that we have determined it to be. In fact, the innumerable stars that are scattered through space are of no other scientific interest other than to provide positions which may be called fixed, with which we may compare the interior movements of our system.

As logical and understandable as Comte's caution may have appeared, hindsight allows us to wish he had turned his skepticism upon his own pronouncements that so magisterially ruled certain researches "out of the question." For the new science of spectroscopy did indeed permit astronomers to extend the bounds of knowledge beyond what Comte assumed were the limits of vision.

Building on studies of light initiated by Newton in the seventeenth century and Josef Fraunhofer early in the nineteenth, John Draper in the United States and Gustav Robert Kirchhof in Germany pioneered the recognition that information concerning objects' physical state, temperature, and chemical composition can be inferred from the spectrum of light which they emit or absorb. (For example, metal as it is heated emits red light, then white, like the filament in a light bulb. Ozone, as is well known, absorbs ultraviolet light.) The spectrum emitted by each element and compound serves moreover as a unique visual fingerprint, often including distinct patterns of "dark lines" and "bright lines," which mark absorption and emission respectively. By this means the element helium (named after helios, *the sun) was identified as a new element after its spectral fingerprint was observed during a solar eclipse in 1868. (For more on the uses of spectroscopy, see Chapter 61.)*

In an article published in 1897, Huggins narrates the development of astronomical spectroscopy and some of its cardinal achievements to date. After citing a skeptical passage from Comte as his foil, Huggins traces the powers of light as a vehicle for information.

We could never know for certain, it seemed, whether the matter and the forces with which we are familiar are peculiar to the earth, or are common with it to the midnight sky,

> All sowed with glistering stars more thick than grass,
> Whereof each other doth in brightness pass.

For how could we extend the methods of the laboratory to bodies at distances so great that even the imagination fails to realize them?

The only communication from them which reaches us across the gulf of space is the light which tells us of their existence. Fortunately this light is not so simple in its nature as it seems to be to the unaided eye. In reality it is very complex; like a cable of many strands, it is made up of light rays of many kinds. Let this light-cable pass from air obliquely through a piece of glass, and its separate strand-rays all go astray, each turning its own way, and then go on apart. Make the glass into the shape of a wedge or prism, and the rays are twice widely scattered. . . . Within this unraveled starlight exists a strange cryptography. Some of the rays may be blotted out, others may be enhanced in brilliancy. These differences, countless in variety, form a code of signals, in which is conveyed to us, when once we have made out the cipher in which it is written, information of the chemical nature of the celestial gases by which the different light rays have been blotted out, or by which they have been enhanced. In the hands of the astronomer a prism has now become more potent in revealing the unknown than even was said to be "Agrippa's magic glass."

It was the discovery of this code of signals, and of its interpretation, which made possible the rise of the new astronomy.

Huggins describes in moving terms how in the late 1850s this new cryptography changed the course of his own researches.

I soon became a little dissatisfied with the routine character of ordinary astronomical work, and in a vague way sought about in my mind for the possibility of research upon the heavens in a new direction or by new methods. It was just at this time, when a vague longing after newer methods of observation for attacking many of the problems of the heavenly bodies filled my mind, that the news reached me of Kirchhof's great discovery of the true nature and the chemical constitution of the sun from his interpretation of the Fraunhofer lines.

This news was to me like the coming upon a spring of water in a dry and thirsty land. Here at last presented itself the very order of work for which in an indefinite way I was looking—namely, to extend his novel methods of research upon the sun to the other heavenly bodies. A feeling as of inspiration seized me: I felt as if I had it now in my power to lift a veil which had never before been lifted; as if a key had been put into my hands which would unlock a door which had been regarded as forever closed to man—the veil and door behind which lay the unknown mystery of the true nature of the heavenly bodies.

Of course an important part of the story of any new science is the technical struggle to apply new theories and methods with precision and consistency.

It is scarcely possible at the present day, when all these points are as familiar as household words, for any astronomer to realize the large amount of time and labor which had to be devoted to the successful construction of the first star spectroscope. Especially was it difficult to provide for the satisfactory introduction of the light for the comparison spectrum. We soon found to our dismay how easily the comparison lines might become instrumentally shifted, and so be no longer strictly fiducial. As a test we used the solar lines as reflected to us from the moon—a test of more than sufficient delicacy with the resolving power at our command.

Then it was that an astronomical observatory began for the first time to take on the appearance of a laboratory. Primary batteries, giving forth noxious gases, were arranged outside one of the windows. A large induction coil stood mounted on a stand on wheels so as to follow the positions of the eye-end of the telescope, together with a battery of several Leyden jars. Shelves with Bunsen burners, vacuum tubes, and bottles of chemicals, especially of specimens of pure metals, lined its walls.

The observatory became a meeting place where terrestrial chemistry was brought into direct touch with celestial chemistry. The characteristic light-rays from earthly hydrogen shone side by side with the corresponding radiations from starry hydrogen, or else fell upon the dark lines due to the absorption of the hydrogen in Sirius or in Vega. Iron from our mines was line-matched, light for dark, with stellar iron from opposite parts of the celestial sphere. Sodium, which upon the earth is always present with us, was found to be widely diffused through the celestial spaces.

This time was indeed one of strained expectation and of scientific exaltation for the astronomer, almost without parallel. For nearly every observation revealed a new fact, and almost every night's work was red-lettered by some discovery.

One of the discoveries of Kirchhof in the late 1850s was that, whereas glowing solids and liquids produce continuous spectra, a glowing gas exhibits a discontinuous "bright line" spectrum. This was the key that allowed Huggins to solve one of the greatest riddles of his age concerning the nebulae: Are they all ultimately resolvable into stars, or are some of them indeed gaseous?

Working alone, I was fortunate in the early autumn of . . . 1864 to begin some observations in a region hitherto unexplored; and which, to this day, remain in my memory with the profound awe which I felt on looking for the first time at that which no eye of man had seen, and which even the scientific imagination could not foreshow.

The attempt seemed almost hopeless. For not only are the nebulae very faintly luminous—as Marius put it, "like a rushlight shining through a horn"—but their feeble shining cannot be increased in brightness, as can be that of the stars, neither to the eye nor in the spectroscope, by any optic tube however great.

Shortly after making the observations of which I am about to speak, I dined at Greenwich, Otto Struve being also a guest, when, on telling of my recent work on the nebulae, Sir George Airy said, "It seems to me a case of 'Eyes and No Eyes.'" Such work indeed it was, as we shall see, on certain of the nebulae.

The nature of these mysterious bodies was still an unread riddle. Towards the end of the last century the elder Herschel, from his observations at Slough, came very near suggesting what is doubtless the true nature, and place in the Cosmos, of the nebulae. I will let him speak in his own words:

A shining fluid of a nature unknown to us.

What a field of novelty is here opened to our conceptions! . . . We may now explain that very extensive nebulosity, expanded over more than sixty degrees of the heavens, about the constellation of Orion; a luminous matter accounting much better for it than clustering stars at a distance. . . .

If this matter is self luminous, it seems more fit to produce a star by its condensation, than to depend on the star for its existence.

This view of the nebulae as parts of a fiery mist out of which the heavens had been slowly fashioned began, a little before the middle of the present century, at least in many minds, to give way before the revelations of the giant telescopes which had come into use, and especially of the telescope, six feet in diameter, constructed by the late Earl of Rosse at a cost of not less than £12,000.

Nebula after nebula yielded, being resolved apparently into innumerable stars, as the optical power was increased; and so the opinion began to gain ground that all nebulae may be capable of resolution into stars. According to this view, nebulae would have to be regarded, not as early stages of an evolutional progress, but rather as stellar galaxies already formed, external to our system—cosmical "sandheaps" too remote to be separated into their component stars. Lord Rosse himself was careful to point out that it would be unsafe from his observations to conclude that all nebulosity is but the glare of stars too remote to be resolved by our instruments. In 1858 Herbert Spencer showed clearly that, notwithstanding the Parsonstown revelations, the evidence from the observation of nebulae up to that time was really in favor of their being early stages of an evolutional progression.

On the evening of the 29th of August, 1864, I directed the telescope for the first time to a planetary nebula in Draco. The reader may now be able to picture to himself to some extent the feeling of excited suspense, mingled with a degree of awe, with which, after a few moments of hesitation, I put my eye to the spectroscope. Was I not about to look into a secret place of creation?

I looked into the spectroscope. No spectrum such as I expected! A single bright line only! At first I suspected some displacement of the prism, and that I was looking at a reflection of the illuminated slit from one of its faces. This thought was scarcely more than momentary; then the true interpretation flashed upon me. The light of the nebula was monochromatic, and so, unlike any other light I had as yet subjected to prismatic examination, could not be extended out to form a complete spectrum. After passing through the two prisms it remained concentrated into a single bright line, having a width corresponding to the width of the slit, and occupying in the instrument a position at that part of the spectrum to which its light belongs in refrangibility. A little closer looking showed two other bright lines on the side towards the blue, all the three lines being separated by intervals relatively dark.

The riddle of the nebulae was solved. The answer, which had come to us in the light itself, read: Not an aggregation of stars, but a luminous gas. Stars after the order of our own sun, and of the brighter stars, would give a different spectrum; the light of this nebula had clearly been emitted by a luminous gas.

With an excess of caution, at the moment I did not venture to go further than to point out that we had here to do with bodies of an order quite different from that of the stars. Further observations soon convinced me that, though the short span of human life is far too minute relatively to cosmical events for us to expect to see in succession any distinct steps in so august a process, the probability is indeed overwhelming in favor of an evolution in the past, and still going on, of the heavenly hosts. A time surely existed when the matter now condensed into the sun and planets filled the whole space occupied by the solar system, in the condition of gas, which then appeared as a glowing nebula, after the order, it may be, of some now existing in the heavens. There remained no room for doubt that the nebulae, which our telescopes reveal to us, are the early stages of long processions of cosmical events, which correspond broadly to those required by the nebular hypothesis in one or other of its forms.

Not, indeed, that the philosophical astronomer would venture to dogmatize in matters of detail, or profess to be able to tell you pat off by heart exactly how everything has taken place in the universe, with the flippant tongue of a Lady Constance after reading "The Revelations of Chaos":

"It shows you exactly how a star is formed; nothing could be so pretty. A cluster of vapor—the cream of the Milky Way; a sort of celestial cheese churned into light."

It is necessary to bear distinctly in mind that the old view which made the matter of the nebulae to consist of an original fiery mist—in the words of the poet [Milton]:

> . . . a tumultuous cloud
> Instinct with fire and nitre—

could no longer hold its place after Helmholtz had shown, in 1854, that such an originally fiery condition of the nebulous stuff was quite unnecessary, since in the mutual gravitation of widely separated matter we have a store of potential energy sufficient to generate the high temperature of the sun and stars.

Spectral analysis proved equally important in the solution of the problem of "radial motion," or motion in the line of sight—a solution which in the twentieth century played a decisive role in the debate concerning the expansion of the universe. Huggins, a great teacher as well as astronomer, explains the problem and its solution.

From the beginning of our work upon the spectra of the stars, I saw in vision the application of the new knowledge to the creation of a great method of as-

tronomical observation which could not fail in future to have a powerful influence on the progress of astronomy, indeed in some respects greater than
the more direct one of the investigation of the chemical nature and the relative physical conditions of the stars.

It was the opprobrium of the older astronomy . . . that only that part of
the motions of the stars which is across the line of sight could be seen and directly measured. The direct observation of the other component in the line of
sight, since it caused no change of place and, from the great distance of the
stars, no appreciable change of size or of brightness within an observer's lifetime, seemed to lie hopelessly quite outside the limits of a man's powers. Still,
it was only too clear that, so long as we were unable to ascertain directly
those components of the stars' motions which lie in the line of sight, the
speed and direction of the solar motion in space, and many of the great problems of the constitution of the heavens, must remain more or less imperfectly
known.

Now as the color of a given kind of light, and the exact position it would
take up in a spectrum, depends directly upon the length of the waves, or, to
put it differently, upon the number of waves which would pass into the eye in
a second of time, it seemed more than probable that motion between the
source of the light and the observer must change the apparent length of the
waves to him, and the number reaching his eye in a second. To a swimmer
striking out from the shore each wave is shorter, and the number he goes
through in a given time is greater than would be the case if he had stood still
in the water. Such a change of wave-length would transform any given kind
of light, so that it would take a new place in the spectrum, and from the
amount of this change to a higher or to a lower place, we could determine
the velocity per second of the relative motion between the star and the
earth. . . .

Now . . . the idea of a change of color in light from motion between the
source of light and the observer was announced for the first time by Doppler
in 1841. Later, various experiments were made in connection with this view
by Ballot, Sestini, Klinkerfues, Clerk Maxwell, and Fizeau. But no attempts
had been made, nor were indeed possible, to discover by this principle the
motions of the heavenly bodies in the line of sight. For, to learn whether any
change in the light had taken place from motion in the line of sight, it was
clearly necessary to know the original wave-length of the light before it left
the star.

As soon as our observations had shown that certain earthly substances
were present in the stars, the original wave-lengths of their lines became
known, and any small want of coincidence of the stellar lines with the same

lines produced upon the earth might safely be interpreted as revealing the velocity of approach or of recession between the star and the earth.

These considerations were present to my mind from the first, and helped me to bear up under many toilsome disappointments. . . . It was not until 1866 that I found time to construct a spectroscope of greater power for this research. It would be scarcely possible, even with greater space, to convey to the reader any true conception of the difficulties which presented themselves in this work, from various instrumental causes, and the extreme care and caution which were needful to distinguish spurious instrumental shifts of a line from the true shift due to the star's motion.

At last, in 1868, I felt able to announce in a paper printed in the *Transactions of the Royal Society* for that year, the foundation of this new method of research, which, transcending the wildest dreams of an earlier time, enabled the astronomer to measure off directly in terrestrial units the invisible motions in the line of sight of the heavenly bodies.

To pure astronomers the method came before its time, since they were unfamiliar with spectrum analysis, which lay completely outside the routine work of an observatory. It would be easy to mention the names of men well known, to whom I was "as a very lovely song of one that hath a pleasant voice." They heard my words, but for a time were very slow to avail themselves of this new power of research.

SOURCES: Selection from Comte adapted from *The Positive Philosophy of Auguste Comte*, trans. Harriet Martineau, 3rd ed., London, 1893; William Huggins, "The New Astronomy: A Personal Retrospect," *The Nineteenth Century: A Monthly Review*, ed. James Knowles, 41 (1897).

53

Astronomy Still Young

Agnes Mary Clerke

Agnes Mary Clerke was born in County Cork, Ireland, in 1842 and became one of the keenest minds and most eloquent voices among astronomers and historians of astronomy of her age. (She died in 1907.) From her vantage point near the end of the Victorian era she conveys deep appreciation of progress that has been made in the study of the heavens without the "chronological snobbery" that even today often accompanies expositions of "how far we've come." Moreover, in her awareness of the relationship between scientific progress and social and economic conditions, Clerke seems positively ahead of her time. The "popular" in the title of one of her major works, Popular History of Astronomy During the Nineteenth Century, *does double duty: it indicates of course that the book is intended for a non-specialist audience, but it also hints at Clerke's contention that progress in science depends in large measure on public participation. For her, William Herschel was a bellwether in an age when technical professionalism threatened astronomy with loss of what its future depended upon: popularity.*

The rise of Herschel was the one conspicuous anomaly in the astronomical history of the eighteenth century. It proved decisive of the course of events in the nineteenth. It was unexplained by anything that had gone before; yet all that came after hinged upon it. It gave a new direction to effort; it lent a fresh impulse to thought. It opened a channel for the widespread public interest which was gathering towards astronomical subjects to flow in.

Much of this interest was due to the occurrence of events calculated to arrest the attention and excite the wonder of the uninitiated. The predicted return of Halley's comet in 1759 verified, after an unprecedented fashion, the computations of astronomers. It deprived such bodies forever of their portentous character; it ranked them as denizens of the solar system. Again, the transits of Venus in 1761 and 1769 were the first occurrences of the kind since the awakening of science to their consequence. Imposing preparations, journeys to remote and hardly accessible regions, official expeditions, international communications, all for the purpose of observing them to the best advantage, brought their high significance vividly to the public consciousness; a result aided by the facile pen of Lalande, in rendering intelligible the means by which these elaborate arrangements were to issue in an accurate knowledge of the sun's distance. Lastly, Herschel's discovery of Uranus, March 13, 1781, had the surprising effect of utter novelty. Since the human race had become acquainted with the company of the planets, no addition had been made to their number. The event thus broke with immemorial traditions, and seemed to show astronomy as still young, and full of unlooked-for possibilities.

Further popularity accrued to the science from the sequel of a career so strikingly opened. Herschel's huge telescopes, his detection by their means of two Saturnian and as many Uranian moons, his piercing scrutiny of the sun, picturesque theory of its constitution, and sagacious indication of the route pursued by it through space; his discovery of stellar revolving systems, his bold soundings of the universe, his grandiose ideas, and the elevated yet simple language in which they were conveyed—formed a combination powerfully effective to those least susceptible of new impressions. . . .

This great accession of popularity gave the impulse to the extraordinarily rapid progress of astronomy in the nineteenth century. Official patronage combined with individual zeal sufficed for the elder branches of science. A few well-endowed institutions could accumulate the materials needed by a few isolated thinkers for the construction of theories of wonderful beauty and elaboration, yet precluded, by their abstract nature, from winning general applause. But the new physical astronomy depends for its prosperity upon the favor of the multitude whom its striking results are well fitted to attract. It is, in a special manner, the science of amateurs. It welcomes the most unpretending cooperation. There is no one "with a true eye and a faithful hand" but can do good work in watching the heavens. And not infrequently prizes of discovery which the most perfect appliances failed to grasp have fallen to the share of ignorant or ill-provided assiduity.

Observers, accordingly, have multiplied; observatories have been founded in all parts of the world; associations have been constituted for mutual help and counsel. . . . Modern facilities of communication have helped to impress more deeply upon modern astronomy its associative character. The electric telegraph gives a certain ubiquity which is invaluable to an observer of the skies. With the help of a wire, a battery, and a code of signals, he sees whatever is visible from any portion of our globe, depending, however, upon other eyes than his own, and so entering as a unit into a widespread organization of intelligence. The press, again, has been a potent agent of cooperation. . . .

Clerke cites the United States in particular as an example of how popular favor can spur scientific endeavor.

Public favor brings in its train material resources. It is represented by individual enterprise, and finds expression in an ample liberality. The first regular observatory in the southern hemisphere was founded at Paramatta by Sir Thomas Makdougall Brisbane in 1821. The Royal Observatory at the Cape of Good Hope was completed in 1829. Similar establishments were set to work by the East India Company at Madras, Bombay, and St. Helena, during the first third of the nineteenth century. The organization of astronomy in the United States of America was due to a strong wave of popular enthusiasm. In 1825 John Quincy Adams vainly urged upon Congress the foundation of a National Observatory; but in 1843 the lectures of Ormsby MacKnight Mitchel on celestial phenomena stirred an impressionable audience to the pitch of providing him with the means of erecting at Cincinnati the first astronomical establishment worthy the name in that great country. On the 1st of January 1882 no less than one hundred and forty-four were active within its boundaries.

The apparition of the great comet of 1843 gave an additional fillip to the movement. To the excitement caused by it the Cambridge Observatory—called the "American Pulkowa"—directly owed its origin; and the example was not ineffective elsewhere. Corporations, universities, municipalities, vied with each other in the creation of similar institutions; private subscriptions poured in; emissaries were sent to Europe to purchase instruments and procure instruction in their use. In a few years the young Republic was, in point of astronomical efficiency, at least on a level with countries where the science had been fostered since the dawn of civilization.

Clerke concludes her introduction to the popular history of astronomy in the nineteenth century with a passage that seems accurate, if perhaps no longer complete, more than a century later. Astronomical contemplation continues to lead us beyond our solar system and yet also, in an apparently contrary direction, into the infinitesimal structure of matter.

A vast widening of the scope of astronomy has accompanied, and in part occasioned, the great extension of its area of cultivation which our age has witnessed. In the last century its purview was a comparatively narrow one. Problems lying beyond the range of the solar system were almost unheeded, because they seemed inscrutable. Herschel first showed the sidereal universe as accessible to investigation, and thereby offered to science new worlds— majestic, manifold, "infinitely infinite" to our apprehension in number, variety, and extent—for future conquest. Their gradual appropriation has absorbed, and will long continue to absorb, the powers which it has served to develop.

But this is not the only direction in which astronomy has enlarged, or rather levelled, its boundaries. The unification of the physical sciences is perhaps the greatest intellectual feat of recent times. The process has included astronomy; so that, like Bacon, she may now be said to have "taken all knowledge" (of that kind) "for her province." In return, she proffers potent aid for its increase. Every comet that approaches the sun is the scene of experiments in the electrical illumination of rarefied matter, performed on a huge scale for our benefit. The sun, stars, and nebulae form so many celestial laboratories, where the nature and mutual relations of the chemical "elements" may be tried by more stringent tests than sublunary conditions afford. The laws of terrestrial magnetism can be completely investigated only with the aid of a concurrent study of the face of the sun. The positions of the planets will perhaps one day tell us something of impending droughts, famines, and cyclones.

Astronomy generalizes the results of the other sciences. She exhibits the laws of Nature working over a wider area, and under more varied conditions, than ordinary experience presents. Ordinary experience, on the other hand, has become indispensable to her progress. She takes in at one view the indefinitely great and the indefinitely little. The mutual revolutions of the stellar multitude during tracts of time which seem to lengthen out to eternity as the mind attempts to traverse them, she does not admit to be beyond her ken; nor is she indifferent to the constitution of the minutest atom of matter that thrills the ether into light.

Agnes Clerke's ability to chart progress without condescension towards her scientific predecessors is nicely illustrated by her discussion of Alexander Wilson's and William Herschel's constructively mistaken ideas concerning the nature of the sun. Her reflections elegantly anticipate the vocabulary of models, "useful fictions," and "paradigm shifts" that became the stuff of much theoretical discourse in the last half of the twentieth century.

For 164 years after Galileo first levelled his telescope at the setting sun, next to nothing was learned as to its nature; and the facts immediately ascertained of its rotation on an axis nearly erect to the plane of the ecliptic, in a period of between twenty-five and twenty-six days, and of the virtual limitation of the spots to a so-called "royal" zone extending some thirty degrees north and south of the solar equator, gained little either in precision or development from five generations of astronomers.

But in November 1769 a spot of extraordinary size engaged the attention of Alexander Wilson, professor of astronomy in the University of Glasgow. He watched it day by day, and to good purpose. As the great globe slowly revolved, carrying the spot towards its western edge, he was struck with the gradual contraction and final disappearance of the penumbra *on the side next the center of the disc;* and when, on the 6th of December, the same spot re-emerged on the eastern limb, he perceived, as he had anticipated, that the shady zone was now deficient *on the opposite side,* and resumed its original completeness as it returned to a central position. Similar perspective effects were visible in numerous other spots subsequently examined by him, and he was thus in 1774 able to prove by strict geometrical reasoning that such appearances were, as a matter of fact, produced by vast excavations in the sun's substance. . . . Wilson's demonstration came with all the surprise of novelty, as well as with all the force of truth.

The general theory by which it was accompanied rested on a very different footing. It was avowedly tentative, and was set forth in the modest shape of an interrogatory. "Is it not reasonable to think," he asks, "that the great and stupendous body of the sun is made up of two kinds of matter, very different in their qualities; that by far the greater part is solid and dark, and that this immense and dark globe is encompassed with a thin covering of that resplendent substance from which the sun would seem to derive the whole of his vivifying heat and energy?" . . . From these hints, supplemented by his own diligent observations and sagacious reasonings, Herschel elaborated a scheme of solar constitution which held its ground until the physics of the sun were revolutionized by the spectroscope.

A cool, dark, solid globe, its surface diversified with mountains and valleys, clothed with luxuriant vegetation, and "richly stored with inhabitants," protected by a heavy cloud-canopy from the intolerable glare of the upper luminous region, where the dazzling coruscations of a solar aurora some thousands of miles in depth evolved the stores of light and heat which vivify our world—such was the central luminary which Herschel constructed with his wonted ingenuity, and described with his wonted eloquence. . . .

We smile at conclusions which our present knowledge condemns as extravagant and impossible, but such incidental flights of fancy in no way derogate from the high value of Herschel's contributions to solar science. The cloud-like character which he attributed to the radiant shell of the sun (first named by Schröter the "photosphere") is borne out by all recent investigations; he observed its mottled or corrugated aspect, resembling, as he described it, the roughness on the rind of an orange; showed that "faculae" are elevations or heaped-up ridges of the disturbed photospheric matter; and threw out the idea that spots may be caused by an excess of the ordinary luminous emissions. A certain "empyreal" gas was, he supposed (very much as Wilson had done), generated in the body of the sun, and rising everywhere by reason of its lightness, made for itself, when in moderate quantities, small openings or "pores," abundantly visible as dark points on the solar disc. But should an uncommon quantity be formed, "it will," he maintained, "burst through the planetary regions of clouds, and thus will produce great openings; then, spreading itself above them, it will occasion large shallows (penumbrae), and, mixing afterwards gradually with other superior gasses, it will promote the increase, and assist in the maintenance of the general luminous phenomena."

This partial anticipation of the modern view that the solar radiations are maintained by some process of circulation within the solar mass, was reached by Herschel through prolonged study of the phenomena in question. The novel and important idea contained in it, however, it was at that time premature to attempt to develop. But though many of the subtler suggestions of Herschel's genius passed unnoticed by his contemporaries, the main result of his solar researches was an unmistakable one. It was nothing less than the definitive introduction into astronomy of the paradoxical conception of the central fire and hearth of our system as a cold, dark, terrestrial mass, wrapped in a mantle of innocuous radiance—an earth, so to speak, within—a sun without.

Let us pause for a moment to consider the value of this remarkable innovation. It certainly was not a step in the direction of truth. On the contrary, the crude notions of Anaxagoras and Xeno approached more nearly to what we now know of the sun, than the complicated structure devised for the happiness of a nobler race of beings than our own, by the benevolence of eighteenth-century astronomers. And yet it undoubtedly constituted a very important advance in science. It was the first earnest attempt to bring solar phenomena within the compass of a rational system; to put together into a consistent whole the facts ascertained; to fabricate, in short, a solar machine that would in some fashion work. It is true that the materials were inadequate and the design faulty. The resulting construction has not proved strong enough to stand the wear and tear of time and discovery, but has had to be taken to pieces and remodelled on a totally different plan. But the work was not therefore done in vain. None of Bacon's aphorisms show a clearer insight into the relations between the human mind and the external world than that which declares "Truth to emerge sooner from error than from confusion." A definite theory (even if a false one) gives holding-ground to thought. Facts acquire a meaning with reference to it. It affords a motive for accumulating them and a means of coordinating them; it provides a framework for their arrangement and a receptacle for their preservation, until they become too strong and numerous to be any longer included within arbitrary limits, and shatter the vessel originally framed to contain them.

Such was the purpose subserved by Herschel's theory of the sun. It helped to *clarify* ideas on the subject. The turbid sense of groping and viewless ignorance gave place to the lucidity of a plausible scheme. The persuasion of knowledge is a keen incentive to its increase. Few men care to investigate what they are obliged to admit themselves entirely ignorant of; but once started on the road to knowledge (real or supposed), they are eager to pursue it. By the promulgation of a confident and consistent view regarding the nature of the sun, accordingly, research was encouraged, because it was rendered hopeful, and inquirers were shown a path leading indefinitely onwards where an impassable thicket had before seemed to bar the way.

SOURCE: Agnes Mary Clerke, *Popular History of Astronomy During the Nineteenth Century*, 2nd edition, London, 1893.

Part Five

Part Five

THE UNIVERSE RE-IMAGINED

54

The Peculiar Interest of Mars

Giovanni Schiaparelli and Percival Lowell

Of all the planets it is Mars that has held the most persistent and the firmest grip upon the human imagination—as the god of war, as the "red planet," and as the longest-standing candidate for receptacle of extraterrestrial life within the solar system. Moreover, the ability of Mars to fire the imagination has not generally frustrated science but furthered it, if often by a circuitous route. The most famous among these winding pathways is exemplified by the story—one that virtually fuses science and science fiction—regarding the "canals" of Mars.

The first member of the greatest pair of astronomers to put their stamp on Martian topography and speculation was Giovanni Virginio Schiaparelli (1835–1910), director of the Royal Observatory in Milan, who viewed Mars during a close "opposition" in the summer of 1877 and reported what appeared to be canali *across the surface of the planet. It has often been pointed out that Italian* canali *can be translated into English merely as "channels"; the word need not mean "canals." Nevertheless, Schiaparelli, in spite of his cautious scientific style, does not strenuously resist the connotation of design that accompanies "canal."*

All the vast extent of the continents [on Mars] is furrowed upon every side by a network of numerous lines or fine stripes of a more or less pronounced dark color, whose aspect is very variable. These traverse the planet for long distances in regular lines, that do not at all resemble the winding courses of our streams. Some of the shorter ones do not reach 500 kilometers (300 miles), others on the other hand extend for many thousands, occupying a quarter or sometimes even a third of a circumference of the planet. Some of these are very easy to see. . . . Others in turn are extremely difficult, and resemble the finest thread of spider's web drawn across the disc. They are subject also to great variations in their breadth, which may reach 200 or even 300 kilometers (120 to 180 miles) . . . whilst some are scarcely 30 kilometers (18 miles) broad.

These lines or stripes are the famous canals of Mars, of which so much has been said. As far as we have been able to observe them hitherto, they are certainly fixed configurations upon the planet. The Nilosyrtis has been seen in that place for nearly one hundred years, and some of the others for at least thirty years. Their length and arrangement are constant, or vary only between very narrow limits. Each of them always begins and ends between the same regions. But their appearance and their degree of visibility vary greatly, for all of them, from one opposition to another, and even from one week to another, and these variations do not take place simultaneously and according to the same laws for all, but in most cases happen apparently capriciously, or at least according to laws not sufficiently simple for us to be able to unravel. . . .

Every canal (for now we shall so call them), opens at its ends either into a sea, or into a lake, or into another canal, or else into the intersection of several other canals. None of them have yet been seen cut off in the middle of the continent, remaining without beginning or without end. This fact is of the highest importance.

Schiaparelli goes on to describe the periodic "gemination" (doubling) of the canals and to seek an explanation for the phenomenon.

Having regard then to the principle that in the explanation of natural phenomena it is universally agreed to begin with the simplest suppositions, the first hypotheses on the nature and cause of the geminations have for the most part put in operation only the laws of inorganic nature. Thus, the gemination is supposed to be due either to the effects of light in the atmosphere of Mars, or to optical illusions produced by vapors in various manners, or to glacial phenomena of a perpetual winter, to which it is known all the planets will be condemned, or to double cracks in its surface, or to single cracks of which the images are doubled by the effect of smoke issuing in long lines and blown laterally by the wind. The examination of these ingenious suppositions leads us to conclude that none of them seem to correspond entirely with the observed facts, either in whole or in part. Some of these hypotheses would not have been proposed, had their authors been able to examine the geminations with their own eyes. Since some of these may ask me directly—Can you suggest anything better? I must reply candidly, No.

It would be far more easy if we were willing to introduce the forces pertaining to organic nature. Here the field of plausible supposition is immense, being capable of making an infinite number of combinations capable of satisfying the appearances even with the smallest and simplest means. Changes of vegetation over a vast area, and the production of animals, also very small, but in enormous multitudes, may well be rendered visible at such a distance. An observer placed in the moon would be able to see such an appearance at the times in which agricultural operations are carried out upon one vast plain—the seed time and the gathering of the harvest. In such a manner also would the flowers of the plants of the great steppes of Europe and Asia be rendered visible at the distance of Mars—by a variety of coloring. A similar system of operations produced in that planet may thus certainly be rendered visible to us. But how difficult for the Lunarians and the Areans [i.e., Martians] to be able to imagine the true causes of such changes of appearances, without first at least some superficial knowledge of terrestrial nature! So also for us, who know so little of the physical state of Mars, and nothing of its organic world, the great liberty of possible supposition renders arbitrary all explanations of this sort, and constitutes the gravest obstacle to the acquisition of well founded notions. All that we may hope is that with time the uncertainty of the problem will gradually diminish, demonstrating, if not what the geminations are, at least what they can not be. We may also confide a little in what Galileo called "the courtesy of Nature," thanks to which, sometime from an unexpected source, a ray of light will illuminate an investigation at first believed inaccessible to our speculations, and of which we have a beautiful example in celestial chemistry. Let us therefore hope and study.

Schiaparelli's true successor, who did indeed combine enormous measures of hope and study, was Percival Lowell (1855–1916). Son of the famous Lowell family of Massachusetts, Percival, after a career travelling and writing, turned his attention to astronomy in the 1890s, founding his own observatory in the clear air atop a mesa near Flagstaff, Arizona. Equipped with this powerful means of sight to supplement his remarkable energy and eloquence, Lowell ignited the popular imagination concerning earth's red neighbor.

Once in about every fifteen years a startling visitant makes his appearance upon our midnight skies—a great red star that rises at sunset through the haze about the eastern horizon, and then, mounting higher with the deepening night, blazes forth against the dark background of space with a splendor that outshines Sirius and rivals the giant Jupiter himself. Startling for its size, the stranger looks the more fateful for being a fiery red. Small wonder that by many folk it is taken for a portent. Certainly, no one who had not followed in their courses what the Greeks so picturesquely called "the wanderers" (*ho planetoi*) would recognize in the apparition an orderly member of our own solar family. Nevertheless, one of the wanderers it is, for that star is the planet Mars, large because for the moment near, having in due course again been overtaken by the earth, in her swifter circling about the sun, at that point in space where his orbit and hers make their closest approach.

Although the apparent new-comer is neither new nor intrinsically great, he possesses for us an interest out of all proportion to his size or his relative importance in the universe; and this for two reasons: first, because he is our own cosmic kin; and secondly, because no other heavenly body, Venus and the moon alone excepted, ever approaches us so near. What is more, we see him at such times better than we ever do Venus, for the latter, contrary to what her name might lead one to expect, keeps herself so constantly cloaked in cloud that we are permitted only the most meagre peeps at her actual surface; while Mars, on the other hand, lets us see him as he is, no cloud-veil of his, as a rule, hiding him from view. He thus offers us effective opportunities for study at closer range than does any other body in the universe except the moon. And the moon balks inquiry at the outset. For that body, from which we might hope to learn much, appears upon inspection to be, cosmically speaking, dead. Upon her silent surface next to nothing now takes place save for the possible crumbling of a crater wall. For all practical purposes Mars is our nearest neighbor in space. Of all the orbs about us, therefore, he holds out most promise of response to that question which man instinctively makes as he gazes up at the stars: What goes on upon all those distant globes? Are they worlds, or are they mere masses of matter? Are physical forces alone at work there, or has evolution begotten something more complex, something not unakin to what we know on earth as life? It is in this that lies the peculiar interest of Mars.

Of course Mars has not at all lost its interest since the Mariner, Viking, and Sojourner expeditions have shown us pictures quite contrary to that envisaged by Lowell. But who in 1895 could fail to be enchanted by the persuasive and thoroughgoing image he constructed? Much of the force of the picture derives from its use of earthly analogy and comparison, even though Lowell clearly wishes to deny that earth, or humankind, should be the measure of all things.

[Let us review] the chain of reasoning by which we have been led to regard it probable that upon the surface of Mars we see the effects of local intelligence. We find, in the first place, that the broad physical conditions of the planet are not antagonistic to some form of life; secondly, that there is an apparent dearth of water upon the planet's surface, and therefore, if beings of sufficient intelligence inhabited it, they would have to resort to irrigation to support life; thirdly, that there turns out to be a network of markings covering the disk precisely counterparting what a system of irrigation would look like; and, lastly, that there is a set of spots placed where we should expect to find the lands thus artificially fertilized, and behaving as such constructed oases should. All this, of course, may be a set of coincidences, signifying nothing; but the probability points the other way. As to details of explanation, any we may adopt will undoubtedly be found, on closer acquaintance, to vary from the actual Martian state of things; for any Martian life must differ markedly from our own.

The fundamental fact in the matter is the dearth of water. If we keep this in mind, we shall see that many of the objections that spontaneously arise answer themselves. The supposed herculean task of constructing such canals disappears at once; for, if the canals be dug for irrigation purposes, it is evident that what we see, and call by ellipsis the canal, is not really the canal at all, but the strip of fertilized land bordering it—the thread of water in the midst of it, the canal itself, being far too small to be perceptible. In the case of an irrigation canal seen at a distance, it is always the strip of verdure, not the canal, that is visible, as we see in looking from afar upon irrigated country on the earth.

We may, perhaps, . . . consider for a moment how different in its details existence on Mars must be from existence on the earth. One point out of many bearing on the subject, the simplest and most certain of all, is the effect of mere size of habitat upon the size of the inhabitant; for geometrical conditions alone are most potent factors in the problem of life. Volume and mass determine the force of gravity upon the surface of a planet, and this is more far-reaching in its effects than might at first be thought. Gravity on the surface of Mars is only a little more than one third what it is on the surface of the earth. This would work in two ways to very different conditions of existence from those to which we are accustomed. To begin with, three times as much work, as for example, in digging a canal, could be done by the same expenditure of muscular force. If we were transported to Mars, we should be pleasingly surprised to find all our manual labor suddenly lightened threefold. But, indirectly, there might result a yet greater gain to our capabilities; for if Nature chose she could afford there to build her inhabitants on three times the scale she does on earth without their ever finding it out except by planetary comparison. Let us see how.

As we all know, a large man is more unwieldy than a small one. An elephant refuses to hop like a flea, not because he considers the act undignified, but simply because he cannot bring it about. If we could, we should all jump straight across the street, instead of painfully paddling through the mud. Our inability to do so depends upon the size of the earth, not upon what it at first seems to depend on, the size of the street. . . .

Now [suppose] the possible inhabitant of Mars . . . to be constructed three times as large as a human being in every dimension. If he were on earth, he would weigh twenty-seven times as much, but on the surface of Mars, since gravity there is only about one third of what it is here, he would weigh but nine times as much. The cross-section of his muscles would be nine times as great. Therefore the ratio of his supporting power to the weight he must support would be the same as ours. Consequently, he would be able to stand with as little fatigue as we. Now consider the work he might be able to do. His muscles, having length, breadth, and thickness, would all be twenty-seven times as effective as ours. He would prove twenty-seven times as strong as we, and could accomplish twenty-seven times as much. But he would further work upon what required, owing to decreased gravity, but one third the effort to overcome. His effective force, therefore, would be eighty-one times as great as man's, whether in digging canals or in other bodily occupation. As gravity on the surface of Mars is really a little more than one third that at the surface of the earth, the true ratio is not eighty-one, but about fifty; that is, a Martian would be, physically, fifty-fold more efficient than man. . . .

Something more we may deduce about the characteristics of possible Martians, dependent upon Mars itself, a result of the age of the world they would live in.

A planet may in a very real sense be said to have a life of its own, of which what we call life may or may not be a subsequent detail. It is born, has its fiery youth, sobers into middle age, and just before this happens brings forth, if it be going to do so at all, the creatures on its surface which are, in a sense, its offspring. The speed with which it runs through its gamut of change prior to production depends upon its size; for the smaller the body the quicker it cools, and with it loss of heat means beginning of life for its offspring. It cools quicker because . . . it has relatively less inside for its outside, and it is through its outside that its inside cools. After it has thus become capable of bearing life, the sun quickens that life and supports it for we know not how long. But its duration is measured at the most by the sun's life. Now, inasmuch as time and space are not, as some philosophers have from their too mundane standpoint supposed, forms of our intellect, but essential attributes of the universe, the time taken by any process affects the character of the process itself, as does also the size of the body undergoing it. The changes brought about in a large planet by its cooling are not, therefore, the same as those brought about in a small one. Physically, chemically, and, to our present end, organically, the two results are quite diverse. So different, indeed, are they that unless the planet have at least a certain size it will never produce what we call life. . . .

Whatever the particular planet's line of development, however, in its own line, it proceeds to greater and greater degrees of evolution, till the process stops, dependent, probably, upon the sun. The point of development attained is, as regards its capabilities, measured by the planet's own age, since the one follows upon the other.

Now, in the special case of Mars, we have before us the spectacle of a world relatively well on in years, a world much older than the earth. To so much about his age Mars bears evidence on his face. He shows unmistakable signs of being old. Advancing planetary years have left their mark legible there. His continents are all smoothed down; his oceans have all dried up. . . . If once he had a chaotic youth, it has long since passed away. Although called after the most turbulent of the gods, he is at the present time, whatever he may have been once, one of the most peaceable of the heavenly host. . . .

Mars being thus old himself, we know that evolution on his surface must be similarly advanced. This only informs us of its condition relative to the planet's capabilities. Of its actual state our data are not definite enough to furnish much deduction. But from the fact that our own development has been comparatively a recent thing, and that a long time would be needed to bring even Mars to his present geological condition, we may judge any life he may support to be not only relatively, but really older than our own.

From the little we can see, such appears to be the case. The evidence of handicraft, if such it be, points to a highly intelligent mind behind it. Irrigation, unscientifically conducted, would not give us such truly wonderful mathematical fitness in the several parts to the whole as we there behold. A mind of no mean order would seem to have presided over the system we see—a mind certainly of considerably more comprehensiveness than that which presides over the various departments of our own public works. Party politics, at all events, have had no part in them; for the system is planet wide. Quite possibly, such Martian folk are possessed of inventions of which we have not dreamed, and with them electrophones and kinetoscopes are things of a bygone past, preserved with veneration in museums as relics of the clumsy contrivances of the simple childhood of the race. Certainly what we see hints at the existence of beings who are in advance of, not behind us, in the journey of life.

Lowell goes on to cultivate an antianthropocentric polemic in order to suggest that the burden of proof lies with those in the extraterrestrial life debate who take a skeptical position. He concludes:

One thing, however, we can do, and that speedily: look at things from a standpoint raised above our local point of view; free our minds at least from the shackles that of necessity tether our bodies; recognize the possibility of others in the same light that we do the certainty of ourselves. That we are the sum and substance of the capabilities of the cosmos is something so preposterous as to be exquisitely comic. We pride ourselves upon being men of the world, forgetting that this is but objectionable singularity unless we are, in some wise, men of more worlds than one. For, after all, we are but a link in a chain. Man is merely this earth's highest production to date. That he in any sense gauges the possibilities of the universe is humorous. He does not, as we can easily foresee, even gauge those of this planet. He has been steadily bettering from an immemorial past, and will apparently continue to improve through an incalculable future. Still less does he gauge the universe about him. He merely typifies in an imperfect way what is going on elsewhere, and what, to a mathematical certainty, is in some corners of the cosmos indefinitely excelled.

If astronomy teaches anything, it teaches that man is but a detail in the evolution of the universe, and that resemblant though diverse details are inevitably to be expected in the host of orbs around him. He learns that, though he will probably never find his double anywhere, he is destined to discover any number of cousins scattered through space.

As John Noble Wilford puts it, "Lowell was inspired, enthusiastic, eloquent, and, of course, absolutely wrong about Mars." And yet the curiosity and enthusiasm he generated, in Carl Sagan's words, "turned on all the eight-year-olds who came after him, and who eventually turned into the present generation of astronomers" (Wilford, Mars Beckons [New York: Knopf, 1990], 33, 35). Lowell went on to notice and record perturbations in the orbits of Uranus and Neptune which led him to predict, more than a decade before the actual discovery of Pluto, the existence of a "trans-Neptunian planet." His greatest accomplishment, however, was one more of imagination than of observation; and, as with so many cosmological theories across the centuries, its wrongness did not keep it from being inspiring and significant.

SOURCES: Giovanni Schiaparelli, "The Planet Mars," trans. W. H. Pickering, *Astronomy and Astrophysics*, 13 (1894); Percival Lowell, *Mars*, Boston, 1895.

55

Cosmical Evolution

G. H. Darwin

G. H. Darwin (1845–1912), the second son of Charles Darwin, extended Darwinism from biology to cosmology. For the twentieth century, his cosmic evolutionism became a dominant legacy (see, for example, Chapter 78). His principal scientific achievement was an analysis of "tidal" effects—and extended processes—among bodies such as the sun, moon, and earth. That many of his conclusions have been refuted does not diminish the fascination or importance of Darwin's speculation.

In the following selection from an address entitled "Cosmical Evolution," which Darwin delivered in Cape Town as president of the British Association in 1905, we discern his awareness of himself as torchbearer of Darwinism into new realms, realms which must be explored by means of a combination of science and imagination.

We do not know whether the last hundred years will be regarded for ever as the *saeculum mirabile* of discovery, or whether it is but the prelude to yet more marvelous centuries. To us living men, who scarcely pass a year of our lives without witnessing some new marvel of discovery or invention, the rate at which the development of knowledge proceeds is truly astonishing; but from a wider point of view the scale of time is relatively unimportant, for the universe is leisurely in its procedure. Whether the changes which we witness be fast or slow, they form a part of a long sequence of events which begin in some past of immeasurable remoteness and tend to some end which we cannot foresee. It must always be profoundly interesting to the mind of man to trace successive cause and effect in the chain of events which make up the history of the earth and all that lives on it, and to speculate on the origin and future fate of animals, and of planets, suns, and stars. I shall try, then, to set forth in my address some of the attempts which have been made to formulate evolutionary speculation. This choice of subject has moreover been almost forced on me by the scope of my own scientific work, and it is, I think, justified by the name which I bear. . . .

The man who propounds a theory of evolution is attempting to reconstruct the history of the past by means of the circumstantial evidence afforded by the present. The historian of man, on the other hand, has the advantage over the evolutionist in that he has the written records of the past on which to rely. The discrimination of the truth from amongst discordant records is frequently a work demanding the highest qualities of judgment; yet when this end is attained it remains for the historian to convert the arid skeleton of facts into a living whole by clothing it with the flesh of human motives and impulses. For this part of his task he needs much of that power of entering into the spirit of other men's lives which goes to the making of a poet. Thus the historian should possess not only the patience of the man of science in the analysis of facts, but also the imagination of the poet to grasp what the facts have meant. . . .

The evolutionist is spared the surpassing difficulty of the human element, yet he also needs imagination, although of a different character from that of the historian. In its lowest form his imagination is that of the detective who reconstructs the story of a crime; in its highest it demands the power of breaking loose from all the trammels of convention and education, and of imagining something which has never occurred to the mind of man before.

In his work The Tides and Kindred Phenomena in the Solar System *(1898), Darwin attempts to account for relationships between the earth and the moon, and for lunar genesis itself, in terms of gravitational—or more specifically tidal—effects acting over time. Later in the same speech on "Cosmical Evolution," and using vocabulary such as "crisis" and transitional forms between "species," he sketches the outlines of his hypothesis. It is worth noting too that Darwin's cosmic evolutionism operates within a purely Newtonian physics.*

If a planet is covered with oceans of water and air, or if it is formed of plastic molten rock, tidal oscillations must be generated in its mobile parts by attractions of its satellites and of the sun. Such movements must be subject to frictional resistance, and the planet's rotation will be slowly retarded by tidal friction in much the same way that a fly-wheel is gradually stopped by any external cause of friction. Since action and reaction are equal and opposite, the action of the satellites on the planet, which causes the tidal friction of which I speak, must correspond to a reaction of the planet on the motion of the satellites.

At any moment of time we may regard the system composed of the rotating planet with its attendant satellite as a stable species of motion, but the friction of the tides introduces forces which produce a continuous, although slow, transformation in the configuration. It is, then, clearly of interest to trace backwards in time the changes produced by such a continuously acting cause, and to determine the initial condition from which the system of planet and satellite must have been slowly degrading. We may also look forward, and discover whither the transformation tends.

Let us consider, then, the motion of the earth and moon revolving in company round the sun, on the supposition that the friction of the tides in the earth is the only effective cause of change. . . . This is not the time to attempt a complete exposition of the manner in which tidal friction gives rise to the action and reaction between planet and satellite. . . . It must suffice to set forth the results in their main outlines, and, as in connection with the topic of evolution retrospect is perhaps of greater interest than prophecy, I shall begin with the consideration of the past.

At the present time the moon, moving at a distance of 240,000 miles from the earth, completes her circuit in twenty-seven days. Since a day is the time of one rotation of the earth on its axis, the angular motion of the earth is twenty-seven times as rapid as that of the moon.

Tidal friction acts as a brake on the earth, and therefore we look back in retrospect to times when the day was successively twenty-three, twenty-two, twenty-one of our present hours in length, and so on backward to still shorter days. But during all this time the reaction on the moon was at work, and it appears that its effect must have been such that the moon also revolved round the earth in a shorter period than it does now; thus the month also was shorter in absolute time than it now is. These conclusions are absolutely certain, although the effects on the motions of the earth and of the moon are so gradual that they can only doubtfully be detected by the most refined astronomical measurements.

We take the "day," regarding it as a period of variable length, to mean the time occupied by a single rotation of the earth on its axis; and the "month," likewise variable in absolute length, to mean the time occupied by the moon in a single revolution round the earth. Then, although there are now twenty-seven days in a month, and although both day and month were shorter in the past, yet there is, so far, nothing to tell us whether there were more or less days in the month in the past. . . .

Now it appears [in fact] from mathematical calculation that the day must now be suffering a greater degree of prolongation than the month, and accordingly in retrospect we look back to a time when there were more days in the month than at present. That number was once twenty-nine, in place of the present twenty-seven; but the epoch of twenty-nine days in the month is a sort of crisis in the history of moon and earth, for yet earlier the day was shortening less rapidly than the month. Hence, earlier than the time when there were twenty-nine days in the month, there was a time when there was a reversion to the present smaller number of days.

We thus arrive at the curious conclusion that there is a certain number of days to the month, namely twenty-nine, which can never have been exceeded, and we find that this crisis was passed through by the earth and moon recently; but, of course, a recent event in such a long history may be one which happened some millions of years ago.

Continuing our retrospect beyond this crisis, both day and month are found continuously shortening, and the number of days in the month continues to fall. No change in conditions which we need pause to consider now supervenes, and we may ask at once, what is the initial stage to which the gradual transformation points? I say, then, that on following the argument to its end the system may be traced back to a time when the day and month were identical in length, and were both only about four or five of our present hours. The identity of day and month means that the moon was always opposite to the same side of the earth; thus at the beginning the earth always presented the same face to the moon, just as the moon now always shows the same face to us. Moreover, when the month was only some four or five of our present hours in length the moon must have been only a few thousand miles from the earth's surface—a great contrast with the present distance of 240,000 miles.

It might well be argued from this conclusion alone that the moon separated from the earth more or less as a single portion of matter at a time immediately antecedent to the initial stage to which she has been traced. But there exists a yet more weighty argument favorable to this view, for it appears that the initial stage is one in which the stability of the species of motion is tottering, so that the system presents the characteristics of a transitional form, which we have seen to denote a change of type or species. . . .

Darwin's hypothesis, even if it is false, is compelling as a common-sense model. Anyone who has flipped a water-filled balloon into the air has observed how its contents tends to form into two globes within its elastic confines. Alternatively, consider the figure skater who twirls in a tight spin: As he extends his arms, the spin decreases in rapidity. But imagine (contrary to fact and all decorum!) that the skater's arms detach themselves altogether from his body but remain in orbit around him. In such a case the arms alone would slow down while the skater's body continued its rapid spin. Such is the case of the earth and the moon, in Darwin's hypothesis. Yet because the satellite and the body continue to influence each other tidally, the rate of revolution of the satellite decreases in keeping with the slower rate of rotation of the main body, which in turn is partly speeded up by the reciprocated (but proportionately weaker) tidal influence of the satellite.

The moon (the satellite under discussion), however, is not only revolving but also rotating, and Darwin sees that rotation too as having been brought to its present state by the effects of tidal friction.

We have pursued the changes into the past, and I will refer but shortly to the future. The day and month are both now lengthening, but the day changes more quickly than the month. Thus the two periods tend again to become equal to one another, and it appears that when that goal is reached both day and month will be as long as fifty-five of our present days. The earth will then always show the same face to the moon, just as it did in the remotest past. But there is a great contrast between the ultimate and initial conditions, for the ultimate stage, with day and month both equal to fifty-five of our present days, is one of great stability in contradistinction to the vanishing stability which we found in the initial stage.

Since the relationship between the moon and earth is a mutual one, the earth may be regarded as a satellite of the moon, and if the moon rotated rapidly on her axis, as was probably once the case, the earth must at that time have produced tides in the moon. The mass of the moon is relatively small, and the tides produced by the earth would be large; accordingly, the moon would pass through the several stages of her history much more rapidly than the earth. Hence it is that the moon has already advanced to that condition which we foresee as the future fate of the earth, and now always shows to us the same face.

If the earth and moon were the only bodies in existence, this ultimate stage when the day and month were again identical in length, would be one of absolute stability, and therefore eternal; but the presence of the sun introduces a cause for yet further changes. I do not, however, propose to pursue the history to this yet remoter futurity, because our system must contain other seeds of decay which will probably bear fruit before these further transformations could take effect.

SOURCE: George Howard Darwin, *Scientific Papers*, vol. 4, Cambridge: Cambridge UP, 1911.

56

Cosmos Without Peer and Without Price

G. K. Chesterton

G. K. Chesterton (1874–1936) is famous as a political and religious contro-
versialist, writer of detective stories, and wielder of paradoxes, not as a cos-
mologist. But his controversial verve and skill at penetrating riddles, as well
as his sweeping love of literature, powerfully befit a meditation on the
uniqueness of the universe.

The one thing [modern thought] loved to talk about was expansion and
largeness. Herbert Spencer . . . popularized this contemptible notion that the
size of the solar system ought to overawe the spiritual dogma of man. Why
should a man surrender his dignity to the solar system any more than to a
whale? If mere size proves that man is not the image of God, then a whale
may be the image of God. . . . It is quite futile to argue that man is small compared to the cosmos; for man was always small compared to the nearest tree.
But Herbert Spencer, in his headlong imperialism, would insist that we had in
some way been conquered and annexed by the astronomical universe. He
spoke about men and their ideals exactly as the most insolent Unionist talks
about the Irish and their ideals. He turned mankind into a small nationality.
And his evil influence can be seen even in the most spirited and honorable of
later scientific authors, notably in the early romances of Mr. H. G. Wells.
Many moralists have in an exaggerated way represented the earth as wicked.
But Mr. Wells and his school made the heavens wicked. We should lift up our
eyes to the stars from whence would come our ruin.

But the expansion of which I speak was much more evil than all this. I have remarked that the materialist, like the madman, is in prison, in the prison of one thought. These people seemed to think it singularly inspiring to keep on saying that the prison was very large. The size of this scientific universe gave one no novelty, no relief. The cosmos went on for ever, but not in its wildest constellation could there be anything really interesting, anything, for instance, such as forgiveness or free will. The grandeur or infinity of the secret of its cosmos added nothing to it. It was like telling a prisoner in Reading jail that he would be glad to hear that the jail now covered half the country. The warder would have nothing to show the man except more and more long corridors of stone lit by ghastly lights and empty of all that is human. So these expanders of the universe had nothing to show us except more and more infinite corridors of space lit by ghastly suns and empty of all that is divine.

In fairyland there had been a real law, a law that could be broken, for the definition of a law is something that can be broken. But the machinery of this cosmic prison was something that could not be broken, for we ourselves were only part of its machinery. We were either unable to do things or we were destined to do them. The idea of the mystical condition quite disappeared; one can neither have the firmness of keeping laws nor the fun of breaking them. The largeness of this universe had nothing of that freshness and airy outlook which we have praised in the universe of the poet. This modern universe is literally an empire; that is, it was vast, but it is not free. One went into larger and larger windowless rooms, rooms big with Babylonian perspective; but one never found the smallest window or a whisper of outer air. . . .

According to [the materialists] the cosmos was one thing since it had one unbroken rule. Only (they would say) while it is one thing it is also the only thing there is. Why, then, should one worry particularly to call it large? There is nothing to compare it with. It would be just as sensible to call it small. A man may say, "I like this vast cosmos, with its throng of stars and its crowd of varied creatures." But if it comes to that, why should not a man say, "I like this cosy little cosmos, with its decent number of stars and as neat a provision of live stock as I wish to see"? . . .

I may express this . . . feeling of cosmic cosiness by allusion to another book always read in boyhood, *Robinson Crusoe*, which . . . owes its eternal vivacity to the fact that it celebrates the poetry of limits, nay, even the wild romance of prudence. Crusoe is a man on a small rock with a few comforts just snatched from the sea: the best thing in the book is simply the list of things saved from the wreck. The greatest of poems is an inventory. Every kitchen tool becomes ideal because Crusoe might have dropped it in the sea.

It is a good exercise, in empty or ugly hours of the day, to look at anything, the coal-scuttle or the bookcase, and think how happy one could be to have brought it out of the sinking ship on to the solitary island. But it is a better exercise still to remember how all things have had this hair-breadth escape: everything has been saved from a wreck. Every man has had one horrible adventure: as a hidden untimely birth he had not been, as infants that never see the light. Men spoke much in my boyhood of restricted or ruined men of genius, and it was common to say that many a man was a Great Might-Have-Been. To me it is a more solid and startling fact that any man in the street is a Great Might-Not-Have-Been.

But I really felt (the fancy may seem foolish) as if all the order and number of things were the romantic remnant of Crusoe's ship. That there are two sexes and one sun was like the fact that there were two guns and one axe. It was poignantly urgent that none should be lost; but somehow, it was rather fun that none could be added. The trees and the planets seemed like things saved from the wreck; and when I saw the Matterhorn I was glad that it had not been overlooked in the confusion. I felt economical about the stars as if they were sapphires (they are called so in Milton's Eden); I hoarded the hills. For the universe is a single jewel, and while it is a natural cant to talk of a jewel as peerless and priceless, of this jewel it is literally true. This cosmos is indeed without peer and without price. For there cannot be another one.

SOURCE: Gilbert Keith Chesterton, *Orthodoxy*, London, 1908.

57

Curved Space and Poetry of the Universe

Robert Osserman

Robert Osserman (b. 1926) has since 1955 been professor of mathematics at Stanford University in California. His (in the best sense of the word) popular book Poetry of the Universe *introduces to non-mathematicians some of the fundamental mathematical notions without which post-Einsteinian work in physics and cosmology must remain all but inaccessible. Throughout the book Osserman uncondescendingly celebrates imagination and points to its connections with, rather than alienation from, mathematical thinking. He writes in his preface:*

There are few people left who believe in a flat earth, but most of the world's population continues to think in terms of a flat universe. Just as everyday experience led us to think of the earth as flat or planar rather than curved, so does our perception of the world around us lead us to view space as flat or "euclidean." It requires as great an effort of the imagination [today] to conceive of curved space as it did a thousand years ago to conceive of the earth as a gigantic ball somehow suspended or floating freely in an even more gigantic expanse of space. Nevertheless, the evidence is overwhelming that space is indeed curved.

Osserman expounds some of the major events in the history of mathematics, in particular Bernhard Riemann's development of non-Euclidean geometry, which begins with defining "plane" as the surface of a sphere. Riemannian geometry is important not only because it helps us understand the curvature of space according to relativistic physics, but also because it was essential to the formulation of relativistic physics in the first place. In his admirably lucid prose Osserman works to keep mathematics and the imagination pulling in the same direction.

There are two common misconceptions about the curvature of space. The first is that curvature is a rather vague or qualitative concept to contrast with flatness. It is in fact very precise, assigning to each point in space and each direction at that point an exact number, determined by the shape of space near the specific location. The second misconception is that in order to describe or picture curved space one has somehow to think of it as "curving" into the fourth dimension. That image can be a useful aid in visualizing curved space for those who have familiarized themselves with what mathematicians call "four-dimensional euclidean space." Unfortunately, the concept of four-dimensional space is one that science popularizers and science fiction writers have often laced with mystical overtones. For someone who has not worked through the mathematical details, it is more likely to cloud than to clarify one's understanding. Once again the point is simply that when measurements are made in ordinary three-dimensional space, they may well not jibe with the results that are embodied in euclidean geometry; what "curvature" measures is the degree and kind of deviation from the euclidean model.

Our experience confirms that euclidean geometry offers a good description of space on a small scale. But we have absolutely no reason to assume that it also holds on a larger intergalactic scale. It is that extrapolation, from small to large scale, that makes us think in terms of a flat universe exactly as an earlier age believed in a flat earth.

Our attempts to create an accurate map of the earth offer a good analogy to the dilemma of "mapping" the universe. Maps of towns or local regions may be seemingly accurate on a small scale, but become wildly distorted on a larger scale, because no flat map can accurately depict the curve of a sphere. The same may be true of the universe. While we have learned to overcome our flat-earth mentality in measurements on a global scale, we have not yet overcome our tendency to think in terms of a flat universe.

Riemann not only invented the idea of curved space and explained how to actually compute its curvature; he also proposed a radically different model from the usual euclidean one for the entire universe. Specifically, he provided a description of the universe if it turned out to have the shape of a "spherical space." That would be the case if space had *constant positive curvature*.

We can most easily describe Riemann's universe by recalling the map of the earth depicted in the form of two separate hemispheres, each showing one

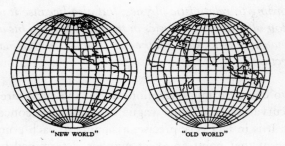

The two hemispheres of the earth.

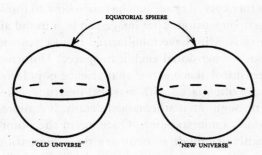

Riemann's universe.

"side" of the earth. Riemann's universe can be depicted in much the same manner.

Picture the earth at the center of the sphere on the left, and picture the inside of that sphere as representing all that we can currently see of the universe through our largest telescopes. Now picture a civilization far beyond the range of those telescopes, situated at the center of the right-hand sphere, peering through their own telescopes, whose range includes everything inside the right-hand sphere.

It is easy to conceive of various possibilities: the two spheres might be very far apart with lots of the universe between them; or they might overlap, with certain galaxies visible from both civilizations. Riemann suggests a third possibility. They might not overlap *and* they might together constitute the entire universe.

In other words, the part of the universe that we can reach with our telescopes lies inside a great sphere whose outer boundary may also be the outer boundary *from the other side* of another civilization. That outer boundary would be the "equatorial sphere" dividing the universe into two parts: the "Old Universe" that we know and the "New Universe" that a twenty-first-century space Columbus may set out to explore.

One reason that such a depiction of the universe seems to be artificial, if not impossible, is the same reason that attempts to draw accurate maps of the earth are doomed to failure: because of the curvature of the earth, the maps we draw are distortions of reality. In our pictures of the two hemispheres, we can draw our maps so that distances from the center are accurate, but then lengths of concentric circles will become more and more distorted away from the center, since the (positive) curvature of the earth's surface results in circles that are smaller than would be indicated by the corresponding circles on our flat map. On a flat map, those circles keep getting bigger and bigger, while in reality, the true circles on earth get bigger until they reach a maximum length—a great circle (corresponding to the boundary circle on the map)—and then start contracting back down toward the antipodal point.

If space had a fixed positive curvature in Riemann's sense, then the "circles of light" in our thought experiment would behave in exactly the same fashion. They would initially get longer and longer, although *less quickly* than in flat euclidean space, and eventually they would reach a maximum circumference at the outer shell of our part of the universe. They would then start to get smaller, and eventually contract down to a point at the "opposite end" of the universe, which on our map is the center of the right-hand sphere—the "antipodal point" to us in the universe. That would be the point of the universe farthest away from us. If we traveled on a spaceship, continuing "straight ahead" in any direction, we would eventually reach the antipodal point. If we kept on going past that point we would end up back where we started.

One feature of this model of the universe that particularly pleased Riemann is that it solved the age-old problem of the "edge" of the universe. Some philosophers had speculated that the universe was infinite in extent, going on forever in all directions. That theory was considered and rejected as implausible by many of those who pondered such questions seriously, from Plato and Aristotle to Newton and Leibniz. But the alternative seemed equally dubious: if it did not go on forever, then—like the flat earth—it would end somewhere; and what was beyond that?

Riemann's model resolved that paradox, which is rooted in the assumption that the universe is flat, or euclidean. If instead it is positively curved and Riemannian, then it can be finite in extent and still not have any "edge" or "boundary." In Riemann's model, every part of the universe looks just like every other part, as far as shapes and measurements go.

At the time that Riemann first presented this picture as a possible description of the real world, it must have seemed little more than a creation of his "gloriously fertile" imagination. And even today—a century and a half later—it stretches our imaginations to the limit to encompass Riemann's vision.

There is a much quoted story about David Hilbert, who one day noticed that a certain student had stopped attending class. When told that the student had decided to drop mathematics to become a poet, Hilbert replied, "Good—he did not have enough imagination to be a mathematician."

In a rare confluence of the poetic and mathematical imagination, the poet Dante arrived at a view of the universe with striking similarities to that of Riemann. In the *Divine Comedy*, Dante describes the universe as consisting of two parts. One part has its center at the earth, surrounded by larger and larger spheres on which move the moon, the sun, successive planets, and the fixed stars. The outer sphere, bounding all the visible universe, is called the *Primum Mobile*. What lies beyond is the "Empyrean," which Dante pictures as another sphere, with various orders of angels circling on concentric spheres about a center where a point of light radiates with almost blinding intensity.

The poet is led by Beatrice from the surface of the earth, through the various spheres of the visible universe, and all the way to the *Primum Mobile*. Looking out from there, he finds himself looking *in* to the sphere of the Empyrean. There is no indication that one must choose a particular point on the *Primum Mobile*; presumably, looking out at any point would give a view into the Empyrean. In other words, we are to think of the Empyrean as somehow both surrounding the visible universe and adjacent to it. If that is the case, then the universe according to Dante would coincide exactly with the universe according to Riemann; they would differ only in the labels.

Riemann's vision is of course more "scientific" than Dante's, in that it is quantitative as well as qualitative—Riemann gives formulas from which one can derive the area of concentric spheres, the circumference of circles, and so on.

The shape of the Dante-Riemannn universe is what mathematicians call *spherical space* or a *hypersphere*. It is like the ordinary sphere—elevated to a higher dimension. The analogies are clear; concentric circles on an ordinary sphere initially get larger, reach a maximum size and then start to get smaller. On a hypersphere, concentric (ordinary) *spheres* start by getting larger, reach a maximum size, and then contract down. On both the sphere and the hypersphere, starting at any point in any direction and continuing "straight ahead" will eventually lead back to the starting point. Furthermore, the total distance traveled will be the same, no matter what the starting point and direction.

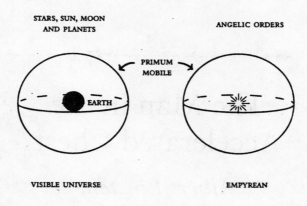

STARS, SUN, MOON
AND PLANETS

ANGELIC ORDERS

PRIMUM
MOBILE

EARTH

VISIBLE UNIVERSE

EMPYREAN

Dante's universe.

A sphere and a hypersphere can be of any size; the size is determined by the total length of a round trip from any point. The length of the round trip also determines the curvature: the longer the round trip, the smaller the curvature and the closer the geometry is to euclidean.

Riemann's conception of a spherical space, together with his suggestion that such a space may describe the actual shape of our universe, constitutes one of the most original and radical departures from the standard world-view in the history of science. A leading twentieth-century physicist, Max Born, has said: "This suggestion of a finite, but unbounded space is one of the greatest ideas about the nature of the world which ever has been conceived." Ironically, Born thought he was referring to an idea of Einstein, since Einstein incorporated Riemann's spherical space into his work on cosmology, along with two other fundamental ideas of Riemann: the curvature of space and the description of a curved space of four dimensions. Riemann invented all these concepts as well as an important alternative to spherical space: "hyperbolic space"—of equal interest to modern cosmologists—while still in his twenties. He presented them at Göttingen in 1854, at the age of twenty-eight—in a lecture that we now see clearly in retrospect as marking the birth of modern cosmology.

As it turned out, Riemann was still missing one key element for a complete picture of the universe. For that the world would have to wait another half century, for the birth of Albert Einstein.

SOURCE: Robert Osserman, *Poetry of the Universe: A Mathematical Exploration of the Cosmos*, New York: Anchor Books, 1995.

58

The Man in the
Accelerated Chest

Albert Einstein

It would be interesting to know, in an age whose physics takes its name from that of Albert Einstein (1879–1955), how many have ever read any of his writings. The non-scientist can be heartened, however, that in 1916 Einstein did pen a little book for the laity in which he sketches the outlines of both the special and the general theories of relativity. The book, entitled simply Relativity, *does not pretend that understanding this topic will be easy. Einstein's recommendation that the reader be prepared to exercise "a fair amount of patience and force of will" is worth repeating here. Einstein also declares (less seriously, one suspects) that he has not paid "the slightest attention to the elegance of the presentation, . . . matters of elegance [being best] left to the tailor and the cobbler." Nevertheless, having wished the reader "a few happy hours of suggestive thought," Einstein goes on to present a remarkably elegant and accessible survey of relativity.*

Before turning to the heart of Einstein's discussion, we do well to review his summary of the Cartesian system of coordinates.

[The Cartesian system of coordinates] consists of three plane surfaces perpendicular to each other and rigidly attached to a rigid body. Referred to a system of coordinates, the scene of any event will be determined (for the main part) by the specification of the lengths of the three perpendiculars or coordinates (x, y, z) which can be dropped from the scene of the event to those three plane surfaces. The lengths of these three perpendiculars can be determined by a series of manipulations with rigid measuring-rods performed according to the rules and methods laid down by Euclidean geometry.

In practice, the rigid surfaces which constitute the coordinates are generally not available; furthermore, the magnitudes of the coordinates are not actually determined by constructions with rigid rods, but by indirect means. If the results of physics and astronomy are to maintain their clearness, the physical meaning of specifications of position must always be sought in accordance with the above considerations.

We thus obtain the following result: Every description of events in space involves the use of a rigid body to which such events have to be referred. The resulting relationship takes for granted that the laws of Euclidean geometry hold for "distances," the "distance" being represented physically by means of the convention of two marks on a rigid body.

This use of coordinates and of Euclidean geometry, of course, is something from which Einstein later departs in his general theory of relativity. But the general is linked with the specific, or special; and the special theory of relativity arises against the backdrop of this "Galileian"—or Newtonian, or "classical"—version of mechanics.

Einstein's discussion employs a species of thought experiment—whose use goes back to Newton, Copernicus, Oresme, and Ptolemy, and is adapted again later by Wheeler (see Chapter 60)—that involves imagining physical motion within an actual self-contained frame of reference such as a ship or, as Einstein prefers, a train.

SPACE AND TIME IN CLASSICAL MECHANICS

"The purpose of mechanics is to describe how bodies change their position in space with time." I should load my conscience with grave sins against the sacred spirit of lucidity were I to formulate the aims of mechanics in this way, without serious reflection and detailed explanations. Let us proceed to disclose these sins.

It is not clear what is to be understood here by "position" and "space." I stand at the window of a railway carriage which is travelling uniformly, and drop a stone on the embankment, without throwing it. Then, disregarding the influence of the air resistance, I see the stone descend in a straight line. A pedestrian who observes the misdeed from the footpath notices that the stone falls to earth in a parabolic curve. I now ask: Do the "positions" traversed by the stone lie "in reality" on a straight line or on a parabola? Moreover, what is meant here by motion "in space"? From the considerations of the previous section the answer is self-evident. In the first place, we entirely shun the vague word "space," of which, we must honestly acknowledge, we cannot form the slightest conception, and we replace it by "motion relative to a practically rigid body of reference." . . . If instead of "body of reference" we insert "system of coordinates," which is a useful idea for mathematical description, we are in a position to say: The stone traverses a straight line relative to a system of coordinates rigidly attached to the carriage, but relative to a system of coordinates rigidly attached to the ground (embankment) it describes a parabola. With the aid of this example it is clearly seen that there is no such thing as an independently existing trajectory (literally, "path-curve"), but only a trajectory relative to a particular body of reference.

In order to have a *complete* description of the motion, we must specify how the body alters its position *with time*; that is, for every point on the trajectory it must be stated at what time the body is situated there. These data must be supplemented by such a definition of time that, in virtue of this definition, these time-values can be regarded essentially as magnitudes (results of measurements) capable of observation. If we take our stand on the ground of classical mechanics, we can satisfy this requirement for our illustration in the following manner. We imagine two clocks of identical construction; the man at the railway-carriage window is holding one of them, and the man on the footpath the other. Each of the observers determines the position on his own reference-body occupied by the stone at each tick of the clock he is holding in his hand. In this connection we have not taken account of the inaccuracy involved by the finiteness of the velocity of light. . . .

As becomes clear elsewhere in Einstein's work, the finiteness and constancy of the velocity of light is an "absolute" at the foundation of the relativity of space and time. At this point in the discussion, however, Einstein pauses to remind the reader of one of the main tenets of the system of physics he is working to undermine.

THE GALILEIAN SYSTEM OF COORDINATES

As is well known, the fundamental law of the mechanics of Galilei-Newton, which is known as the *law of inertia*, can be stated thus: A body removed sufficiently far from other bodies continues in a state of rest or of uniform motion in a straight line. This law not only says something about the motion of bodies, but it also indicates the reference-bodies or systems of coordinates, permissible in mechanics, which can be used in mechanical description. The visible fixed stars are bodies for which the law of inertia certainly holds to a high degree of approximation. Now if we use a system of coordinates which is rigidly attached to the earth, then, relative to this system, every fixed star describes a circle of immense radius in the course of an astronomical day, a result which is opposed to the statement of the law of inertia. So that if we adhere to this law we must refer these motions only to systems of coordinates relative to which the fixed stars do not move in a circle. A system of coordinates of which the state of motion is such that the law of inertia holds relative to it is called a "Galileian system of coordinates." The laws of the mechanics of Galilei-Newton can be regarded as valid only for a Galileian system of coordinates.

In the second part of his book Einstein makes the transition from the special to the general theory of relativity, which is our main interest insofar as we are concerned with physical and mathematical description of the cosmos as a whole and in particular with the nature of gravity.

SPECIAL AND GENERAL PRINCIPLE OF RELATIVITY

The basal principle, which was the pivot of all our previous considerations, was the *special* principle of relativity, i.e., the principle of the physical relativity of all *uniform* motion. Let us once more analyze its meaning carefully.

It was at all times clear that, from the point of view of the idea it conveys to us, every motion must only be considered as a relative motion. Returning to the illustration we have frequently used of the embankment and the railway carriage, we can express the fact of the motion here taking place in the following two forms, both of which are equally justifiable:

(a) The carriage is in motion relative to the embankment.
(b) The embankment is in motion relative to the carriage.

In (a) the embankment, in (b) the carriage, serves as the body of reference in our statement of the motion taking place. If it is simply a question of detaching or of describing the motion involved, it is in principle immaterial to what reference-body we refer the motion. As already mentioned, this is self-evident, but it must not be confused with the much more comprehensive statement called "the principle of relativity," which we have taken as the basis of our investigations.

The principle we have made use of not only maintains that we may equally well choose the carriage or the embankment as our reference-body for the description of any event (for this too is self-evident). Our principle rather asserts what follows: If we formulate the general laws of nature as they are obtained from experience, by making use of

    (a) the embankment as reference-body,

    (b) the railway carriage as reference-body,

then these general laws of nature (for example, the laws of mechanics or the law of the propagation of light in a vacuum) have exactly the same form in both cases. This can be expressed as follows: For the *physical* description of natural processes, neither of the reference-bodies K, K' is unique (literally, "specially marked out") as compared with the other. Unlike the first, this latter statement need not of necessity hold a priori; it is not contained in the conceptions of "motion" and "reference-body" nor derivable from them; only *experience* can decide as to its correctness or incorrectness.

Up to the present, however, we have by no means maintained the equivalence of *all* bodies of reference K in connection with the formulation of natural laws. Our course was more on the following lines. In the first place, we started out from the assumption that there exists a reference-body K, whose condition of motion is such that the Galileian law holds with respect to it: A particle left to itself and sufficiently far removed from all other particles moves uniformly in a straight line. With reference to K (Galileian reference-body) the laws of nature were to be as simple as possible. But in addition to K, all bodies of reference K' would be given preference in this sense, and they should be exactly equivalent to K for the formulation of natural laws, provided that they are in a state of *uniform rectilinear and non-rotary motion* with respect to K; all these bodies of reference are to be regarded as Galileian reference-bodies. The validity of the principle of relativity was assumed only for these reference-bodies, but not for others (for example, those possessing motion of a different kind). In this sense we speak of the *special* principle of relativity, or special theory of relativity.

In contrast to this we wish to understand by the "general principle of relativity" the following statement: All bodies of reference K, K', etc., are equivalent for the description of natural phenomena (formulation of the general laws of nature), whatever may be their state of motion. But before proceeding farther, it ought to be pointed out that this formulation must be replaced later by a more abstract one, for reasons which will become evident at a later stage.

Since the introduction of the special principle of relativity has been justified, every intellect which strives after generalization must feel the temptation to venture the step towards the general principle of relativity. But a simple and apparently quite reliable consideration seems to suggest that, for the present at any rate, there is little hope of success in such an attempt. Let us imagine ourselves transferred to our old friend the railway carriage, which is travelling at a uniform rate. As long as it is moving uniformly, the occupant of the carriage is not sensible of its motion, and it is for this reason that he can without reluctance interpret the facts of the case as indicating that the carriage is at rest, but the embankment in motion. Moreover, according to the special principle of relativity, this interpretation is quite justified also from a physical point of view.

If the motion of the carriage is now changed into a non-uniform motion, as for instance by a powerful application of the brakes, then the occupant of the carriage experiences a correspondingly powerful jerk forwards. The retarded motion is manifested in the mechanical behavior of bodies relative to the person in the railway carriage. The mechanical behavior is different from that of the case previously considered, and for this reason it would appear to be impossible that the same mechanical laws hold relatively to the non-uniformly moving carriage, as hold with reference to the carriage when at rest or in uniform motion. At all events it is clear that the Galileian law does not hold with respect to the non-uniformly moving carriage. Because of this, we feel compelled at the present juncture to grant a kind of absolute physical reality to non-uniform motion, in opposition to the general principle of relativity. But in what follows we shall soon see that this conclusion cannot be maintained.

The Gravitational Field

"If we pick up a stone and then let it go, why does it fall to the ground?" The usual answer to this question is: "Because it is attracted to the earth." Modern physics formulates the answer rather differently for the following reason. As a result of the more careful study of electromagnetic phenomena, we have come to regard action at a distance as a process impossible without the intervention of some intermediary medium. If, for instance, a magnet attracts a piece of iron, we cannot be content to regard this as meaning that the magnet acts directly on the iron through the intermediate empty space, but we are constrained to imagine—after the manner of Faraday—that the magnet always calls into being something physically real in the space around it, that something being what we call a "magnetic field." In its turn this magnetic field operates on the piece of iron, so that the latter strives to move towards the magnet. We shall not discuss here the justification for this incidental conception, which is indeed a somewhat arbitrary one. We shall only mention that with its aid electromagnetic phenomena can be theoretically represented much more satisfactorily than without it, and this applies particularly to the transmission of electromagnetic waves. The effects of gravitation also are regarded in an analogous manner.

The action of the earth on the stone takes place indirectly. The earth produces in its surroundings a gravitational field, which acts on the stone and produces its motion of fall. As we know from experience, the intensity of the action on a body diminishes according to a quite definite law, as we proceed farther and farther away from the earth. From our point of view this means: The law governing the properties of the gravitational field in space must be a perfectly definite one, in order correctly to represent the diminution of gravitational action with the distance from operative bodies. It is something like this: The body (for example, the earth) produces a field in its immediate neighborhood directly; the intensity and direction of the field at points farther removed from the body are thence determined by the law which governs the properties in space of the gravitational fields themselves.

In contrast to electric and magnetic fields, the gravitational field exhibits a most remarkable property, which is of fundamental importance for what follows. Bodies which are moving under the sole influence of a gravitational field receive an acceleration, *which does not in the least depend either on the material or on the physical state of the body*. For instance, a piece of lead and a piece of wood fall in exactly the same manner in a gravitational field (in a vacuum), when they start off from rest or with the same initial velocity. This law, which holds most accurately, can be expressed in a different form in light of the following consideration.

According to Newton's law of motion, we have

$$(\text{Force}) = (\text{inertial mass}) \times (\text{acceleration}),$$

where the "inertial mass" is a characteristic constant of the accelerated body. If now gravitation is the cause of the acceleration, we then have

$$(\text{Force}) = \frac{(\text{gravitational mass})}{(\text{inertial mass})} \times (\text{intensity of the gravitational field})$$

If now, as we find from experience, the acceleration is to be independent of the nature and the condition of the body and always the same for a given gravitational field, then the ratio of the gravitational to the inertial mass must likewise be the same for all bodies. By a suitable choice of units we can thus make this ratio equal to unity. We then have the following law: The *gravitational* mass of a body is equal to its *inertial* mass.

It is true that this important law had hitherto been recorded in mechanics, but it had not been *interpreted*. A satisfactory interpretation can be obtained only if we recognize the following fact: *The same* quality of a body manifests itself according to circumstances as "inertia" or as "weight" (literally, "heaviness"). In the following section we shall show to what extent this is actually the case, and how this question is connected with the general postulate of relativity.

THE EQUALITY OF INERTIAL
AND GRAVITATIONAL MASS AS AN ARGUMENT FOR
THE GENERAL POSTULATE OF RELATIVITY

We imagine a large portion of empty space, so far removed from stars and other appreciable masses that we have before us approximately the conditions required by the fundamental law of Galilei. It is then possible to choose a Galileian reference-body for this part of space (world), relative to which points at rest remain at rest and points in motion continue permanently in uniform rectilinear motion. As reference-body let us imagine a spacious chest resembling a room with an observer inside who is equipped with apparatus. Gravitation naturally does not exist for this observer. He must fasten himself with strings to the floor; otherwise the slightest impact against the floor will cause him to rise slowly towards the ceiling of the room.

To the middle of the lid of the chest is fixed externally a hook with rope attached, and now a "being" (what kind of being is immaterial to us) begins pulling at this with a constant force. The chest together with the observer then begin to move "upwards" with a uniformly accelerated motion. In course of time their velocity will reach unheard-of values—provided that we are viewing all this from another reference-body which is not being pulled with a rope.

But how does the man in the chest regard the process? The acceleration of the chest will be transmitted to him by the reaction of the floor of the chest. He must therefore take up this pressure by means of his legs if he does not wish to be laid out full length on the floor. He is then standing in the chest in exactly the same way as anyone stands in a room of a house on our earth. If he releases a body which he previously had in his hand, the acceleration of the chest will no longer be transmitted to this body, and for this reason the body will approach the floor of the chest with an accelerated relative motion. The observer will further convince himself *that the acceleration of the body towards the floor of the chest is always of the same magnitude, whatever kind of body he may happen to use for the experiment.*

Relying on his knowledge of the gravitational field (as was discussed in the preceding section), the man in the chest will thus come to the conclusion that he and the chest are in a gravitational field which is constant with regard to time. Of course he will be puzzled for a moment as to why the chest does not fall in this gravitational field. Just then, however, he discovers the hook in the middle of the lid of the chest and the rope which is attached to it, and he consequently comes to the conclusion that the chest is suspended at rest in the gravitational field.

Ought we to smile at the man and say that he errs in his conclusion? I do not believe we ought to if we wish to remain consistent; we must rather admit that his mode of grasping the situation violates neither reason nor known mechanical laws. Even though it is being accelerated with respect to the "Galileian space" first considered, we can nevertheless regard the chest as being at rest. We have thus good grounds for extending the principle of relativity to include bodies of reference which are accelerated with respect to each other, and as a result we have gained a powerful argument for a generalized postulate of relativity.

We must note carefully that the possibility of this mode of interpretation rests on the fundamental property of the gravitational field of giving all bodies the same acceleration, or, what comes to the same thing, on the law of the equality of inertial and gravitational mass. If this natural law did not exist, the man in the accelerated chest would not be able to interpret the behavior of the bodies around him on the supposition of a gravitational field, and he would not be justified on the grounds of experience in supposing his reference-body to be "at rest."

Suppose that the man in the chest fixes a rope to the inner side of the lid, and that he attaches a body to the free end of the rope. The result of this will be to stretch the rope so that it will hang "vertically" downwards. If we ask for an opinion of the cause of tension in the rope, the man in the chest will say: "The suspended body experiences a downward force in the gravitational field, and this is neutralized by the tension of the rope; what determines the magnitude of the tension of the rope is the *gravitational mass* of the suspended body." On the other hand, an observer who is poised freely in space will interpret the condition of things thus: "The rope must perforce take part in the accelerated motion of the chest, and it transmits this motion to the body attached to it. The tension of the rope is just great enough to effect the acceleration of the body. That which determines the magnitude of the tension of the rope is the *inertial mass* of the body." Guided by this example, we see that our extension of the principle of relativity implies the *necessity* of the law of the equality of inertial and gravitational mass. Thus we have obtained a physical interpretation of this law.

From our consideration of the accelerated chest we see that a general theory of relativity must yield important results on the laws of gravitation. In point of fact, the systematic pursuit of the general idea of relativity has supplied the laws satisfied by the gravitational field. Before proceeding farther, however, I must warn the reader against a misconception suggested by these considerations. A gravitational field exists for the man in the chest, despite the fact that there was no such field for the coordinate system first chosen. Now we might easily suppose that the existence of a gravitational field is always only an *apparent* one. We might also think that, regardless of the kind of gravitational field which may be present, we could always choose another reference-body such that *no* gravitational field exists with reference to it. This is by no means true for all gravitational fields, but only for those of quite special form. It is, for instance, impossible to choose a body of reference such that, as judged from it, the gravitational field of the earth (in its entirety) vanishes.

We can now appreciate why that argument is not convincing which we brought forward against the general principle of relativity [earlier]. It is certainly true that the observer in the railway carriage experiences a jerk forwards as a result of the application of the brake, and that he recognizes in this the non-uniformity of motion (retardation) of the carriage. But he is compelled by nobody to refer this jerk to a "real" acceleration (retardation) of the carriage. He might also interpret his experience thus: "My body of reference (the carriage) remains permanently at rest. With reference to it, however, there exists (during the period of the application of the brakes) a gravitational field which is directed forwards and which is variable with respect to time. Under the influence of this field, the embankment together with the earth moves non-uniformly in such a manner that their original velocity in the backwards direction is continuously reduced."

SOURCE: Albert Einstein, *Relativity: The Special and General Theory,* trans. Robert W. Lawson, New York: Henry Holt, 1920.

59

It Is Not True That "All Is Relative"

Richard Feynman

Richard Phillips Feynman (1918–1988) had a colorful and public career—from the Manhattan Project, to a Nobel Prize for Physics in 1965, to the commission investigating the Challenger disaster not long before his death. By the general public, Feynman is best known as a forceful and imaginative teacher, a showman among scientists—"showman" here having a sense not derogatory but rather in keeping with the nature of the science he professed, which involved empirical demonstration. Even while explaining relativity theory, which to many still seems such an abstraction, Feynman insists on the empirical nature of his subject and, as in his renowned "Lectures on Physics" (1961–63), aims his satire at those who would reduce it to either woolly, trivializing "philosophy" or lazy paradox. His discussion is important not merely for its satire, however, but for its clarification (1) of the empirical nature of relativity and (2) of what conclusions may and may not be drawn from it.

[Jules Henri] Poincaré [1854–1912] made the following statement of the principle of relativity: "According to the principle of relativity, the laws of physical phenomena must be the same for a fixed observer as for an observer who has a uniform motion of translation relative to him, so that we have not, nor can we possibly have, any means of discerning whether or not we are carried along in such a motion."

When this idea descended upon the world, it caused a great stir among philosophers, particularly the "cocktail-party philosophers," who say, "Oh, it is very simple: Einstein's theory says all is relative!" In fact, a surprisingly large number of philosophers, not only those found at cocktail parties (but rather than embarrass them, we shall just call them "cocktail-party philosophers"), will say, "That all is relative is a consequence of Einstein, and it has profound influences on our ideas." In addition, they say "It has been demonstrated in physics that phenomena depend upon your frame of reference." We hear that a great deal, but it is difficult to find out what it means. Probably the frames of reference that were originally referred to were the coordinate systems which we use in the analysis of the theory of relativity. So the fact that "things depend upon your frame of reference" is supposed to have had a profound effect on modern thought. One might well wonder why, because, after all, that things depend upon one's point of view is so simple an idea that it certainly cannot have been necessary to go to all the trouble of the physical relativity theory in order to discover it. That what one sees depends on his frame of reference is certainly known to anybody who walks around, because he sees an approaching pedestrian first from the front and then from the back; there is nothing deeper in most of the philosophy which is said to have come from the theory of relativity than the remark that "A person looks different from the front than from the back." The old story about the elephant that several blind men describe in different ways is another example, perhaps, of the theory of relativity from the philosopher's point of view.

But certainly there must be deeper things in the theory of relativity than just this simple remark that "A person looks different from the front than from the back." Of course relativity is deeper than this, because *we can make definite predictions with it.* It certainly would be rather remarkable if we could predict the behavior of nature from such a simple observation alone.

There is another school of philosophers who feel very uncomfortable about the theory of relativity, which asserts that we cannot determine our absolute velocity without looking at something outside, and who would say, "It is obvious that one cannot measure his velocity without looking outside. It is self-evident that it is *meaningless* to talk about the velocity of a thing without looking outside; the physicists are rather stupid for having thought otherwise, but it has just dawned on them that this is the case. If only we philosophers had realized what the problems were that the physicists had, we could have decided immediately by brainwork that it is impossible to tell how fast one is moving without looking outside, and we could have made an enormous contribution to physics." These philosophers are always with us, struggling in the periphery to try to tell us something, but they never really understand the subtleties and depths of the problem.

Our inability to detect absolute motion is a result of *experiment* and not a result of plain thought, as we can easily illustrate. In the first place, Newton believed that it was true that one could not tell how fast he is going if he is moving with uniform velocity in a straight line. In fact, Newton first stated the principle of relativity. . . . Why then did the philosophers not make all this fuss about "all is relative," or whatever, in Newton's time? Because it was not until Maxwell's theory of electrodynamics was developed that there were physical laws that suggested that one *could* measure his velocity without looking outside; soon it was found *experimentally* that one could *not*.

Now, *is* it absolutely, definitely, philosophically *necessary* that one should not be able to tell how fast he is moving without looking outside? One of the consequences of relativity was the development of a philosophy which said, "You can only define what you can measure! Since it is self-evident that one cannot measure a velocity without seeing what he is measuring it relative to, therefore it is clear that there is no *meaning* to absolute velocity. The physicists should have realized that they can talk only about what they can measure." But *that is the whole problem*: whether or not one *can define* absolute velocity is the same as the problem of whether or not one *can detect in an experiment*, without looking outside, whether he is moving. In other words, whether or not a thing is measurable is not something to be decided *a priori* by thought alone, but something that can be decided only by experiment. Given the fact that the velocity of light is 186,000 mi/sec, one will find few philosophers who will calmly state that it is self-evident that if light goes 186,000 mi/sec inside a car, and the car is going 100,000 mi/sec, that the light also goes 186,000 mi/sec past an observer on the ground. That is a shocking fact to them; the very ones who claim it is obvious find, when you give them a specific fact, that it is not obvious.

Feynman concludes his swipe at the cocktail-party philosophers by recurring to problems such as Newton's experiment with the bucket or the astronaut's free-float backward somersault that leaves him wondering whether he or the space capsule is rotating (see Chapter 34). In short, either the astronaut or Feynman's straw philosopher could discover empirically, and only empirically, which body was actually moving.

Finally, there is even a philosophy which says that one cannot detect *any* motion except by looking outside. It is simply not true in physics. True, one cannot perceive a *uniform* motion in a *straight line*, but if the whole room were *rotating* we would certainly know it, for everybody would be thrown to the wall—there would be all kinds of "centrifugal" effects. That the earth is turning on its axis can be determined without looking at the stars, by means of the so-called Foucault pendulum, for example. Therefore it is not true that "all is relative"; it is only *uniform velocity* that cannot be detected without looking outside. Uniform *rotation* about a fixed axis *can* be. When this is told to a philosopher, he is very upset that he did not really understand it, because to him it seems impossible that one should be able to determine rotation about an axis without looking outside. If the philosopher is good enough, after some time he may come back and say, "I understand. We really do not have such a thing as absolute rotation; we are really rotating *relative to the stars*, you see. And so some influence exerted by the stars on the object must cause the centrifugal force."

Now, for all we know, that is true; we have no way, at the present time, of telling whether there would have been centrifugal force if there were no stars and nebulae around. We have not been able to do the experiment of removing all the nebulae and then measuring our rotation, so we simply do not know. We must admit that the philosopher may be right. He comes back, therefore, in delight and says, "It is absolutely necessary that the world ultimately turn out to be this way: *absolute* rotation means nothing; it is only *relative* to the nebulae." Then we say to him, "*Now*, my friend, is it or is it not obvious that uniform velocity in a straight line, *relative to the nebulae*, should produce no effects inside a car?" Now that the motion is no longer absolute, but is a motion *relative to the nebulae*, it becomes a mysterious question, and a question that can be answered only by experiment.

Finally, for our purposes, Feynman pours his scorn on the "twin paradox" supposedly created by relativity, dissolving it by means of appeal once more to experiment and experience.

To continue our discussion of the Lorentz transformation [the equation that describes the foreshortening or contraction of a body in the direction of its travel] and relativistic effects, we consider a famous so-called "paradox" of Peter and Paul, who are supposed to be twins, born at the same time. When they are old enough to drive a space ship, Paul flies away at very high speed. Because Peter, who is left on the ground, sees Paul going so fast, all of Paul's clocks appear to go slower, his heart beats go slower, his thoughts go slower, everything goes slower, from Peter's point of view. Of course, Paul notices nothing unusual, but if he travels around and about for a while and then comes back, he will be younger than Peter, the man on the ground! That is actually right; it is one of the consequences of the theory of relativity which has been clearly demonstrated. Just as the mu-mesons last longer when they are moving, so also will Paul last longer when he is moving. This is called a "paradox" only by the people who believe that the principle of relativity means that *all motion* is relative; they say, "Heh, heh, heh, from the point of view of Paul, can't we say that *Peter* was moving and should therefore appear to age more slowly? By symmetry, the only possible result is that both should be the same age when they meet." But in order for them to come back together and make the comparison, Paul must either stop at the end of the trip and make a comparison of clocks or, more simply, he has to come back, and the one who comes back must be the man who was moving, and he knows this, because he had to turn around. When he turned around, all kinds of unusual things happened to his space ship—the rockets went off, things jammed up against one wall, and so on—while Peter felt nothing.

So the way to state the rule is to say that *the man who has felt the accelerations,* who has seen things fall against the walls, and so on, is the one who would be the younger; that is the difference between them in an "absolute" sense, and it is certainly correct.

SOURCE: Richard P. Feynman, *Six Not-So-Easy Pieces*, Reading, Mass.: Addison-Wesley/Helix Books, 1997.

60

Spacetime Tells Matter How to Move

John Archibald Wheeler

John Archibald Wheeler (b. 1911), a much-decorated physical relativist, held chairs at both Princeton and the University of Texas and has continued to publish actively since his retirement in 1986. To the nonspecialist, his writings reveal not only lucidity (in spite of the fact that it is he who coined the term "black hole"!), but also an engaging philosophical playfulness.

In his A Journey into Gravity and Spacetime, Wheeler recounts the story of how the general theory of relativity began to emerge in Einstein's thinking one day in 1908. Einstein heard that a certain painter had fallen off a roof, and he sought the painter out in order to ask him how it felt to fall. The painter replied that he had felt nothing at all (presumably until he hit the ground). Einstein later described this as the beginning of "the greatest idea of my life," namely, that "in a gravitational field (of small spatial extension) things behave as they do in a space free of gravitation."

Two ideas are central to Einstein's conception of gravity. The first is free float. The second is spacetime curvature. What do these terms mean, and how do these ideas fit together?

Einstein's painter was right to believe that he was weightless as he fell. We who stand securely see things the wrong way around because the ground beneath our feet is all the time pushing us away from a natural state of motion. That natural state of motion is free fall, or, better said, free float. To see what free float really means, imagine that the painter falls directly into a shaft that extends straight through the Earth. The painter is never forced away from his natural state of motion by contact with the Earth. Where does the painter get his moving orders for his long, never-ending track through space and time? According to Einstein, the moving orders do not come from some mysterious "gravity" acting in some mysterious way from the center of the Earth or the center of the Sun. Instead, the moving orders come from the geometry of space and time right where the painter is located.

Mass thus gets its moving orders locally. Objects in free float that start off the same—at the same place and time, with the same speed and direction— pursue the same history. They do so because they get their moving orders from the same spacetime. In brief, *spacetime grips mass, telling it how to move. . . .*

If spacetime grips matter, telling it how to move, then it is not surprising to discover that matter grips spacetime, telling it how to curve. To understand this corollary notion, let's imagine what free-float spacetime-driven motion would look like if spacetime were not curved. Every object in free float would move in a straight line with uniform velocity forever and ever. The Earth and the other planets would not enjoy the companionship of the Sun. Each would float away on its own proud, disregarding course. Conceivable though such a universe is, it is not the universe that we know. Faced with this difficulty, we could give up the idea that spacetime tells mass how to move. But if we want to retain this idea, despite the observed curvature of planetary orbits and the identical curvature of the tracks of a ball and a bullet through spacetime, we will say, with Einstein, that spacetime itself is curved. Moreover, this curvature is greater at and within the Earth than it is far away from the Earth. In brief, *mass grips spacetime, telling it how to curve.*

Wheeler uses analogies and thought-experiments to illustrate the notion of spacetime curvature. One such analogy is provided by a trampoline, which of course consists of an elastic membrane stretched flat until deflected from a planar state by some weight. This surface provides an analogy in three dimensions to spacetime in four dimensions.

To imagine how spacetime is curved by mass, picture a cannon ball at rest in the middle of the trampoline. The fabric membrane of the trampoline, where it makes contact with the cannon ball, displays a constant semi-spherical curvature like that of the cannon ball itself. Beyond the circle of contact the curvature of the fabric is negative—it bends back the other way; and the farther away from the cannon ball one looks, the more the surface approaches zero curvature.

The cannon ball on the trampoline is analogous to a massive body such as the earth in spacetime. The area where the cannon ball makes contact with the fabric of the trampoline is like the volume of space that is bounded by the earth's surface: here spacetime has positive constant curvature proportionate to the mass of the earth. Beyond the earth's surface, spacetime curvature is negative, and diminishes more and more the farther from earth one goes.

The trampoline analogy is useful in three main ways. First, it generally introduces the notion of spacetime as having or being a "fabric." It is not mere nothing, not a mere vacuity; it is not absolute space. It interacts with mass.

Second, it illustrates Einstein's initially difficult notion that all interaction between spacetime and mass is local. Whereas for Newton gravitation is action at a distance, in relativity theory there is no such thing. All action is local action, just as the bending and stretching of the surface of the trampoline in any given small area is a function of the bending and stretching of the surface that is immediately adjacent to that area. Where the cannon ball sits, the curvature is congruent to the shape of the cannon ball. Beyond the circle of the cannon ball's resting place, each local area of the trampoline is stretched and bent by the local area of the trampoline which it adjoins; and that stretching and bending are "dissipated" the farther out from the cannon ball any given area of trampoline fabric is located.

Third, the curvature of the trampoline fabric can help us understand the mechanics of the motion of bodies that come within the cannon ball's "gravitational field." Imagine a small marble which we place on the surface of the trampoline (for the sake of simplicity we assume the marble does not itself depress the trampoline's fabric). If near the edge of the trampoline we release the marble, it will start to roll towards the cannon ball, and it will do so not because the cannon ball itself attracts the marble, but because the surface of the trampoline curves towards the place where the cannon ball sits. Moreover, the velocity of the marble will increase—it will accelerate—as it approaches the cannon ball, and this acceleration will itself increase because of the increasing negative curvature of the fabric, not directly because of the cannon ball itself.

Of course, the marble will finally smack into the cannon ball and come to rest "on its surface," just as objects that fall to earth come to rest on its surface. However, imagine for a moment (contrary to fact) that we can remove the cannon ball but keep the curvature of the trampoline's surface just as it was when the cannon ball was on it. Now what will the marble do? As before, it will accelerate, and at an increasing rate, as it approaches the circular area where the cannon ball sat. But as it passes into that circle, which bounds an area of constant positive curvature, its acceleration (but not yet its velocity!) will decrease, this decrease remaining obediently constant as long as the marble is rolling down towards the bottom of the wok-shaped curved area defined by the cannon ball. As the marble passes the center, its velocity will *begin to decrease in keeping with its decreasing acceleration. In other words, at this low point the marble's acceleration is zero. Past this point its acceleration is less than zero—that is to say, acceleration has become deceleration—though the rate of this deceleration remains constant until the marble passes up out of this circle of constant curvature, whereupon the deceleration will decrease in accordance with the now negative and decreasing curvature of the trampoline's fabric.*

This analogy to the gravitational field of the earth thus illustrates why, given the curvature of spacetime gripped by mass, an object will accelerate towards the earth at an increasing rate, reaching "one G" or approximately ten meters per second per second at the earth's surface. Then, of course, the object will smack into the earth and come to rest, unless it is lucky enough to arrive where there is a shaft straight through the center of the earth and out the other side (this shaft being an imaginary counterfactual equivalent of our removal of the cannon ball from the trampoline while assuming the fabric of the trampoline retains the same curvature as when the cannon ball was there). If the "falling" object enters such a shaft, its acceleration will decrease at a constant rate, passing zero acceleration—that is, starting to decelerate—as the object passes the center of the earth. When it passes out the other side of the earth, its deceleration will continue but at a decreasing rate in accordance with the now negative and decreasing curvature of the fabric of local spacetime.

Finally, if we add a second identical marble to our thought experiment we can better imagine what Wheeler calls contractile *and* noncontractile *spacetime curvature. If from the edge of the trampoline we release the two marbles, say, one second apart, then the distance between them will initially increase. It will increase because of the increasing rate of acceleration that each marble will experience, this in turn because of the negative curvature of the trampoline's fabric—that is, until* they cross into the circle *bounded by the cannon ball, whereupon, because of their decreasing rate of acceleration (followed by identically increasing rate of deceleration), they will come closer together and finally either collide or pass. Thus negative curvature is noncontractile (the marbles get farther apart). But by contrast, positive curvature such as that within the cannon ball's circular place of rest on the trampoline, and like the earth's spherical place within spacetime, is contractile, for within these bounds the rolling marbles' constantly and positively curved "trajectories" approach each other.*

To simplify and enliven the discussion of "free float" and of positive spacetime curvature, Wheeler elaborates the supposition of a shaft through the center of the earth and proposes the following thought experiment:

The idea is old of a shaft drilled through the center of the Earth, a direct route from west to east. The brave traveler releases his fingerhold on the western rim of the terrifying hole and 42 minutes later he coasts to a safe landing at the eastern rim on the opposite side of the Earth. Is there a variant of this story in which the traveler fails to grasp the eastern rim and ends up, after another 42 minutes, back at his starting point, to win from his joking friends the title of "Boomeranger"? . . .

Boomeranging. No better word is there to describe the imaginary journey of one who falls through the Earth with never an obstacle to bang into, and so come back to his starting point. As we begin planning our Boomerang Project, we soon realize that an enterprise so vast, so novel—even if only in the imagination—requires the very best minds. We quickly assemble the Boomerang Project Organizing Committee consisting of experts from around the world. We explain the objectives of the project to the committee and why we have named it the Boomerang Project. In their first decision, the committee members vote unanimously that to honor the Land of the Boomerang, one end of the shaft should be located in Australia.

Getting down to their deliberations, the committee members discover that lines running through the center of the Earth from various points on the coastline of Australia stake out an area filled with heaving North Atlantic waves, with not one bit of land anywhere in it. Consternation! No place for the other end of the shaft to exit—at least not above water. Imagine the relief of the committee members at finding in all that watery waste one point of promise, a point on the Midatlantic Ridge called the Atlantis Seamount. Because this point lies only 82 fathoms—about 150 meters—below the surface, the committee agrees that a watertight chamber, a caisson, can be built from the level plain of the seamount to the ocean's surface, thus creating an artificial island. . . .

The Boomerang Project begins! The caisson is built, the water is pumped out, and the proud flag of the new principality of Atlantis is planted by committee members at its highest point. Next, with the most modern machinery available, drilling of the shaft begins simultaneously from Atlantis and from its antipodal point, the little town of Taralga in New South Wales. A superstrong lining is installed in the Earth's interior; insulation also is installed to protect against the great temperatures there.

Various technical problems are solved, and the project proceeds:

All is ready. The maiden voyage of the earthship—dubbed Boomer I—is about to begin with Rob at the controls. We brief him on what to expect. He and his earthship will zoom effortlessly back and forth through the darkness of the evacuated boomerang shaft, Atlantis to Taralga, Taralga to Atlantis. . . . Mile after mile, hour after hour, he and Boomer I will float free. He will see no weight, feel no weight, have no weight. If he bounces a ping-pong ball against the wall of the brightly lit craft, its path will reveal to him no fall, no curve, no sign of gravity. Spacetime, master of motion, will appear to be as flat here as it is in the darkness of space far from planet, sun, or galaxy.

In dismay we realize that Rob's boomeranging will not show us anything about spacetime curvature. Is the whole Boomerang Project a bust? No, I assure everyone. We must look at the motion of two nearby boomerangers relative to each other, rather than at the motion of a single earthship relative to the Earth. In this way we can focus on local physics, as Einstein advises, not action-at-a-distance physics, Newton's original doorway to gravity. The wisdom of the organizing committee in planning a double boomerang shaft, like the tunnel of a major highway under a river, is now obvious. Not for counter-current traffic, however, do we need the second Atlantis-Taralga tube, but for a second earthship—Boomer II. Alix eagerly volunteers to pilot this craft, which will start off from Atlantis 2 seconds later than Rob does in BI. In consequence, Alix and BII initially will be 20 meters behind Rob and BI.

(Note that Wheeler here assumes G at 10 m/sec.^{2}. Thus, beginning from rest, during the first second, BI covers 5 meters, during the second second, 15 meters, during the third second, 25 meters, and so on. These assumptions would soon break down, however, precisely because the rate of acceleration changes as BI approaches the center of the earth.)

We've designed both earthcraft to be too light in mass to influence one another. They respond to the gravity of the far larger mass of the Earth without detectably influencing it or each other. In this sense they serve as *test masses*.

As soon as the pre-boomerang briefing is completed, we all assemble at the Atlantis passenger platform. At exactly 1 second before 11:18 a.m., Rob frees BI from its mooring and begins zooming past feature after feature of the Earth's interior. Alix departs 2 seconds later and passes the same landmarks, always 2 seconds after Rob does.

One second before noon Rob is at rest for a fleeting instant at the level of the Taralga passenger platform. Without a jolt he could then and there lock his craft to the wall of the shaft if he so chooses. He doesn't. He wants to keep on longer in free-float travel. At this instant Alix is 20 meters below the platform and still rising fast.

She finds her rise to be half as fast at noon itself. Alix is in the final slowdown of her approach to Taralga. At this point, only 5 meters below the platform, rising Alix meets descending Rob. He has already covered the first 5 meters of his return trip to Atlantis.

One second past noon Alix is at the summit of her flight. Rob is descending at twice the speed he had a second ago. He is already 20 meters down.

Two seconds past noon Alix herself is 5 meters down, well started on her return trip to Atlantis.

A minuet is in progress. The two boomerangers execute its next movement at 12:42, near Atlantis. There they encounter each other again; Alix slowing as she rises, Rob gaining speed as he starts downward on a new trip to Taralga.

The passage of yet another 42 minutes finds them passing once more, near Taralga.

The minuet goes on and on, ever maintaining its 42-minute rhythm, even as the two masses, in free float, zoom through the center of the Earth at the fantastic speed of 7.9 kilometers a second or 18,000 miles per hour.

(Note that this velocity is less than it would be if the acceleration of the "falling" objects remained 10 m/sec.². At a constant 1G, twenty-one minutes (1,260 seconds) would produce a velocity of 12.6 kilometers per second. However, as BI or BII approaches the center of the mass of the earth, its acceleration decreases and approaches zero.)

By late afternoon, it is clear that the Boomerang Project is a great success. Alix and Rob are fine. The earthships have survived intact. On their next approach to the Atlantis platform, the two boomerangers lock into the moorings and disembark, to the cheers of all, from BI and BII.

The amazement of it! We have seen two test masses, both in free float. They were together at high noon, at the point where their paths cross 5 meters below Taralga. As they left the crossing point, they floated apart in the darkness of the shaft. The separation between them every second grew 20 meters greater. Imagine Rob and Alix thus to be parting company in some faraway region of the cosmos where spacetime is essentially flat. Straight as arrows, their tracks would carry them to ever greater remoteness. Goodbye forever! Within and near the Earth, however, spacetime is not flat, nor is it so within the Atlantis-Taralga shaft. Not forever do Alix and Rob continue to increase their separation by an additional 20 meters with each passing second. As the metronome of the minuet prepared to boom out the 42-minute beat, Alix and Rob were both coming up on Atlantis. What is more, Alix was shortening Rob's lead. Each second then, the distance between them was not 20 meters greater, but 20 meters less!

The Alix-Rob separation increased fast in the first phase of the Taralga-to-Atlantis trip. It decreased fast in the final phase of that journey. Clearly at some place between the two terminals a changeover occurs from slow increase to slow decrease in separation, a place where that separation isn't changing at all—not changing, because it has crested. Crested where? Obviously at the center of the Earth. Crested at a separation how great? Sixteen kilometers, according to a bit of figuring we can skip.

Sixteen kilometers. Roughly only a thousandth of the diameter of the Earth. A peanut distance! Thus we're talking about *local* physics when we're analyzing the Alix-Rob separation. And Einstein tells us that physics only looks simple when we analyze it in local terms. So no more mention of Atlantis, or Taralga, or the position of Alix and Rob relative to the Earth. Only the distance between the two earthships is relevant.

What can we learn from the Alix-Rob separation about the nature of spacetime in a local region? That spacetime geometry does not let two local test masses fly apart forever. Initially, to be sure, Rob and Alix each saw the motion of the other as a uniform speed of separation, a constant recession velocity. When their separation became larger, however, the recession velocity began to diminish from minute to minute. This slowing of the recession velocity takes place faster, the larger the separation itself. In other words, the decrease in the recession velocity in one second is directly proportional to the average value of the separation itself during that second.

What happens when there is little or no actual separation, but a substantial speed of separation? Then little or no change in the recession velocity occurs in the course of a second. What happens when there is a larger separation, hundreds of meters, or even some kilometers? Then in the course of one second, the recession velocity decreases quite measurably. This decrease in the speed of separation depends only on the average value of the separation itself in that second, not at all on what the speed of recession happens to be.

As Alix and Rob boomeranged through the shaft, their separation, even at its greatest, was never more than a tiny fraction of the Earth's diameter. What brought free-float Alix and free-float Rob back together again time after time, when they would have separated forever if they were traveling through the flatness of faraway space? Ask this question as if freshly arrived in this magnificent universe. Cast off all school memories of a force acting at a distance, all misleading Newtonian overtones that go with the old word *gravity*, all dreams of some monstrous green puller hidden in the cracks and crannies of space. Instead, look for an answer, a local answer, a believable answer, at space and time right where the boomerangers are.

Spacetime geometry! Why invent anything more? The two earthships plough ahead free-float on the straightest world lines they possibly can. Every step of the way they acknowledge as guide and master the geometry of space right where they are and nothing more. But that geometry cannot be flat; otherwise Alix and Rob would separate forever. It must be curved. Only so can we understand in spacetime-geometry language why two worldlines, originally separating, and both conscientiously ploughing straight ahead nevertheless turn slowly, come back together, and cross. It is not the worldlines themselves that curve; it is the spacetime in which the earthships travel in free float that curves.

In addition to evidence of this spacetime curvature, we get a direct measure of the curvature from the motion of our two boomerangers. The ratio between the change in their speed of separation and their separation supplies a clean, clear measure of spacetime curvature right where the boomerangers are:

$$\text{spacetime curvature} = \frac{\text{decrease in separation speed per unit time}}{\text{separation}}$$

Because the initial separation between Rob and Alix was small, only 20 meters, their initial recession velocity of 20 meters per second decreased slowly. A second later, when the separation had increased to 40 meters, the separation itself was still increasing. However, the speed of separation was then diminishing twice as fast as it was an instant before. When the separation between Rob and Alix approached its peak value, 16 kilometers, the *speed* of separation between the two boomerangers dropped at its fastest pace. It dropped to zero, then dropped to negative—that is, changed from speed of separation to speed of approach.

Three very different separations, and three very different rates of change in separation speed. Amidst these differences, however, one pillar of constancy stands out: the ratio of the change in recession velocity to separation always has the same value. . . . only 1.73×10^{-23} per square meter. So small because the spacetime curvature imposed by matter [is] as "dilute" as the material of the Earth *is* small!

Spacetime curvature! That's gravity in brief.

SOURCE: John Archibald Wheeler, *A Journey into Gravity and Spacetime*, New York: Scientific American Library, 1990.

61

The Architecture of the Celestial Mansions

Annie Jump Cannon

Annie Jump Cannon (1863–1941) began work at the Harvard College Observatory in 1896 as one of a number of women whom E. C. Pickering (1846–1919), the director, employed as "computers." She rose to become the observatory's curator of astronomical photographs in 1911, and in 1921 she received a doctor of astronomy degree from Groningen University, the first woman to do so. In 1931, in his citation upon presenting Cannon with the Draper Medal of the National Academy of Science, Harlow Shapley (1885–1972) noted in his wry way that it was the first such medal "ever bestowed on a woman by the honorable body of fossils and one of the highest honors attainable by astronomers of any sex, race, religion, or political preference."

Cannon's work on the classification of stars, like much of that which lies at the foundations of astronomy, involved years of painstaking observation and the keeping of detailed records—in her case in the hundreds of thousands. Yet as her gently poetic and philosophical style attests, she also observed the relationship between the astronomical details and the cosmological big picture. The essay serves at the same time (complementing the selection from Huggins in Chapter 52) as a brief introduction to the astronomical uses of spectroscopy.

Sunlight and starlight are composed of waves of various lengths, which the eye, even aided by a telescope, is unable to separate. We must use more than a telescope. In order to sort out the component colors, the light must be dispersed by a prism, or split up by some other means. For instance, sunbeams, passing through rain drops, are transformed into the myriad-tinted rainbow. The familiar rainbow spanning the sky is Nature's most glorious demonstration that light is composed of many colors.

The very beginning of our knowledge of the nature of a star dates back to 1672, when Isaac Newton gave the world the results of his experiments on passing sunlight through a prism. To describe the beautiful band of rainbow tints, produced when sunlight was dispersed by his three-cornered piece of glass, he took from the Latin the word *spectrum*, meaning an appearance. The rainbow is the spectrum of the sun.

Hardly a more fascinating page in man's search for knowledge is to be found than this story of the analysis of light. May I rehearse just a few points of the story which are vital to the subject of classifying the stars? In 1814, more than a century after Newton, the spectrum of the sun was obtained in such purity that an amazing detail was seen and studied by the German optician Fraunhofer. He saw that the multiple spectral tints, ranging from delicate violet to deep red, were crossed by hundreds of fine dark lines. In other words, there were narrow gaps in the spectrum where certain shades were wholly blotted out.

For fifty years, many searched for a Rosetta-stone to solve the baffling mystery of these hieroglyphics traced by Nature's hand in the radiant sunbeams. The solution was actually found by studying the light of one of the best-known and commonest substances existing on our planet. It has sometimes happened that scientific men have been hardly able to earn their salt, but they have been able to prove that this omnipresent earthly substance, salt, or at least its sodium constituent, exists in the sun and distant stars.

We must remember that the word spectrum is applied not only to sunlight, but also to the light of any glowing substance when its rays are sorted out by a prism or a grating. Each substance thus treated sends out its own vibrations of particular wave lengths, which may be likened to singing its own song. Now the spectrum of salt, called sodium chloride by chemists, is very simple and includes two bright yellow lines. In the spectrum of the sun exactly the same shades of yellow are cut out by two black lines. Could there be any connection? Could the earthly yellow lines be made to change to black? Yes, it was found by experiment that they would do so instantly if a cooler vapor of salt were placed between the prism and a source of light that emits all wave lengths. Thus it was reasoned that some of the bright yellow light from the sun's hot surface was absorbed by cooler sodium vapors in the sun's atmosphere. Likewise two thousand black lines in the sun's spectrum were traced to iron, and indeed all the common substances, so familiar to us here on earth, have been found to exist in the sun by comparing its "absorption" spectrum with the bright line spectra given by these substances in laboratories.

It might have been expected that the sun, our parent, would contain the familiar earthly elements, as we were certainly, in a distant age, bone of his bone and flesh of his flesh. But what about the stars, so far away, apparently so faint? The sun outshines even the brilliant Sirius, the Dog Star, ten billion times. But the light of a star may be magnified several thousand fold by a telescope. Then, with a spectroscope attached to the telescope, we may behold a radiant and beautiful sight, for the twinkling starlight becomes a band showing all the rainbow colors, also crossed by the telltale dark lines. The stars then are suns.

Do the stars differ among themselves? If so, how may we learn about them? Although the human eye is such an admirable instrument, and, aided by the magnifying powers of a telescope, can penetrate far into space, it is not well adapted to observe the spectra of the stars. Fortunately, just at the time when astronomers were peering at starlight, chemists were as eagerly at work with compounds of bromides and silver, little dreaming that in their mixing-bowl lay the means of solving the riddle of the centuries concerning the great inverted bowl above their heads. The photographic method that the chemists developed has now completely superseded visual observations of stellar spectra. The film is more successful in registering faint light than the eye, and is sensitive also to rays which are too short or too long for the small gamut of human vision.

The Harvard Observatory was the first to undertake the photography of stellar spectra on a large scale. With a prism placed over the lens of a telescope, the spectrum of every star of sufficient brightness in the field of view can be photographed. We lose the beauty of the colors in the process, for they affect the film only as a background of light on which the dark lines are engraved. But stars can be classified from the position and strength of these lines.

Some photographs, eight by ten inches in size, covering a portion of the sky about twice as large as the bowl of the Big Dipper, show the spectra of four thousand stars. When we examine the spectra, we notice at once that many look alike. We may then select a few of the brighter stars as typical of a class. A simple system of designating the various classes of stars by the best known of all symbols, the alphabet, was originated by the Harvard Observatory, and has been adopted by the whole astronomical world. Let us see how some of the familiar stars are thus lettered.

When an astronomer speaks of a class A star, he refers to white stars like Sirius and Vega, in whose spectra we see a very strong series of dark lines caused by hydrogen in the atmosphere. For blue-white stars like Rigel we use the letter B. The gas, helium, so precious and so scarce here on earth, is very abundant in the atmosphere of these stars. The letter G is used for our own sun and other yellow stars; and for red stars like Betelgeuse, we use the letter M. Between A and G are the F stars; between G and M, the K stars. Other letters designate several rarer varieties. After a large number of stars had been classified, it was found that the letters B, A, F, G, K, M, stand for six divisions including the great majority of the stars. B must go before A in the astronomer's alphabet, because when it was too late to reverse the letters, the B stars were found to precede the A stars in life-history. Thus we have the so-called B A F classification, which is easy to remember, said an Irish astronomer, because B A F stands for Baffling.

A development of this system adds O stars at the head of the list; these are the bluest and hottest stars. Cannon provided the pons asinorum *that helped generations of astronomers remember the sequence, O B A F G K M: "Oh! Be a fine girl—kiss me."*

Cannon continues her exposition; and a preference for the passive voice ("a quarter of a million stars have already been arranged . . . ") should not disguise the fact that a large proportion of the research she refers to was her own.

Arranged in the given order, with intermediate divisions, the letters represent stars arranged in order of decreasing temperature and increasing redness. You have observed, I am sure, that stars differ in color. Do you not recall red Antares, the so-called Rival of Mars, and pure-white Spica in the summer sky? The color is a clue given us by Nature that stars differ in kind. But are all the stars growing colder and redder? No, the marvelous dark lines also tell the story of an ascending branch on a star's life tree. Stars pass first through the sequence from Class M to Class B, from red to blue; then later from B to M. Such is the life story of a star. In youth gigantic, rarefied, and red; in middle life very blue and hot and radiant; in old age shrunken, dense, and again red. This is not a fairy tale I am telling you, although the name of "giants" has been applied to stars in the youthful bloated stage, and the name of "dwarfs" to stars in the late condensed stage.

Twenty years ago it was assumed that stars differ in composition, that the A stars might have a monopoly of hydrogen, and the B stars, of helium, but the combined labors of chemists, physicists, and astronomers have pointed the way to the belief that the differences are mainly in temperature, and that the class of a star expresses the temperature which is required to produce the observed atmospheric conditions. The classifying of stars has moreover led to the belief that there is no new kind of matter in the universe, for the main features of the spectra of all stars can be accounted for in terms of substances known on earth.

A quarter of a million stars have already been arranged in forty classes from a study of the Harvard photographs. We now have much material to study the architecture of the celestial mansions and the streaming of the celestial tribes. Do stars of the same class, like birds of a feather, flock together? Why, yes, they often seem to. Nearly all of the B stars live in the Milky Way. They are extremely hot and brilliant, being in the very apex of star life. They are distant, too, for their great brilliancy enables them to shine through vast spaces. The A stars also seem to prefer the Galaxy. Stars like our own sun, of Class G, are scattered, however, over the whole sky, in all the highways and byways of the universe. If perchance any of them has had past adventures similar to that of the sun, and is now burdened with a family of planets, the inhabitants, if such there be, are blessed with a beneficent sunlight like our own. The very nearest star, Alpha Centauri, whose light takes but four and a third years to reach us, is of Class G.

The Great Nebula of Andromeda, which is faintly visible to the naked eye, looks something like a stationary comet. It is apparently not connected with our own universe of suns, but constitutes a separate system, to which the picturesque term Island Universe is sometimes applied. This system is estimated to be so far away that light, travelling 186,000 miles a second, requires nearly a million years to arrive; in other words, we see the nebula as it was a million years ago. Yet photographs exposed for many hours with large telescopes have revealed, in this Nebula of Andromeda, the existence of suns of Class G, like our own luminary.

Thus, peering into far-away spaces of the heavens, and looking back, as it were, into bygone epochs of time, we find stars composed of the same elements necessary to us today, vibrating in the same rhythm, sending out waves of the same lengths.

Classifying the stars has helped materially in all studies of the structure of the universe, than which no greater problem is presented to the human mind. While teaching man his relatively small sphere in creation, it also encourages him by its lessons of the unity of nature, and shows him that his power of comprehension allies him with the great intelligence that encompasses all.

SOURCE: "Classifying the Stars," in *The Universe of Stars*, ed. Harlow Shapley and Cecilia H. Payne, 2nd edition, Cambridge, Mass.: Harvard Observatory, 1929.

62

The Quickening Influence of the Universe

Cecilia Payne-Gaposchkin

Cecilia Payne-Gaposchkin (1900–1979) was a twentieth-century "Renaissance woman." Born in England, classically (and musically) educated, she acquired enough scientific and mathematical knowledge, largely through self-study, to enter Cambridge for a degree in botany, physics, and chemistry, though the second of these became her first love. Even before Cambridge, as she recounts in her absorbing autobiography, she had grasped the concept of relativity: "All motion, I had learned, was relative. Suddenly, as I was walking down a London street, I asked myself, 'relative to what?' The solid ground failed beneath my feet. With the familiar leaping of the heart I had my first sense of the Cosmos." Then at Cambridge, in 1919, came her next epiphany.

There was to be a lecture in the Great Hall of Trinity College. Professor [Arthur Stanley] Eddington was to announce the results of the eclipse expedition that he had led to Brazil in 1918. Four tickets for the lecture had been assigned to students at Newnham College and (almost by accident, for one of my friends was unable to go) a ticket fell to me.

The Great Hall was crowded. The speaker was a slender, dark young man with a trick of looking away from his audience and a manner of complete detachment. He gave an outline of the Theory of Relativity in popular language, as none could do better than he. He described the Lorenz-Fitzgerald contraction, the Michelson-Morley experiment and its consequences. He led up to the shift of the stellar images near the Sun as predicted by Einstein and described his verification of the prediction.

The result was a complete transformation of my world picture. I knew again the thunderclap that had come from the realization that all motion is relative. When I returned to my room I found that I could write down the lecture word for word. . . . For three nights, I think, I did not sleep. My world had been so shaken that I experienced something very like a nervous breakdown. The experience was so acute, so personal. . . .

The upshot was, perhaps, a foregone conclusion. I was done with biology, dedicated to physical science, forever.

After graduating from Cambridge in 1923, and believing there was no future for her as a scientist in England, Payne managed to obtain a position at Harvard College Observatory under the directorship of Harlow Shapley. In 1925, a mere two years after immigrating to the United States, she completed and published a Ph.D. dissertation, Stellar Atmospheres. *More than thirty years and hundreds of publications later, in 1956, after a career of struggling uphill professionally because of her sex, she became the first female full professor at Harvard. Yet near the end of her life she commented, with typical generosity: "All that I have done is respond to the quickening influence of the Universe. And during my whole scientific life I have never had a sense of being inferior, never been conscious that my fellow-scientists considered me inferior, on account of my sex."*

The passage that follows here is taken from "The Stuff Stars Are Made Of," one of a series of radio talks given by Shapley, Payne, and others and first published in 1926. It illustrates not only Payne's stylistic grace but also her keen awareness of how human theories "construct" the universe. In some more recent theorizing, including literary theorizing, this "construction" may get confused with the world "out there," rather as if (to echo one postmodern idea) the reader of a text becomes its author. Payne, however, while gently sustaining the metaphor of architects, builders, and building materials throughout her talk, does not finally confuse human theory with the deeper fabric of the cosmos. (One does well to remember that this talk was written before the advent of nuclear physics.)

The amount of matter in the stars is almost inconceivably great, and in discussing the substance of the stars we are virtually considering the chemistry of the universe. The astronomer has never been forced to doubt that a common chemistry and physics pervade the universe, and he has found in that useful assumption a consistent interpretation for a vast array of observed facts. The uniformity of the universe is indeed convincingly vindicated by the first step in the analysis of the stars.

In the practical task of making an object such as a building it is necessary to determine and obtain the materials, and to piece them together in a special way. If, in our minds, we presume to play the game of star building, we must determine and find the materials for it; moreover, the manner of setting them together is of extreme importance, but because our star building is an entirely theoretical and *post facto* business the leading stellar architects have not yet arrived at an agreement about the style, the plan, the foundations, or even the materials. For that reason it is impossible to avoid being something of a partisan in describing the material of the stars.

What do we mean when we talk of "what things are made of"? The earth beneath our feet is "made of" clay and other rocks, and more than half its surface is covered by seas, which are "made of" water. Our houses are of wood, or are built on a framework of steel. But these analyses are not ultimate. An electric current resolves water into hydrogen and oxygen, clay into silicon and aluminum. Burn a building or a living thing, and you find the atoms of carbon and iron, hydrogen, nitrogen, oxygen, and calcium.

Can we make for the stars the same analysis that we make for steel and clay? We cannot take a piece of the star Sirius and analyze it chemically. But there *is* something analyzable that we can get from Sirius, and from every other star we know. They are all pouring out light into space, and we can catch the light as it strikes the earth, and analyze it. In a fundamental sense, that light was as much a part of the star as clay is part of the earth. Light is energy, and it is the energy of a star that makes it shine—that forces it through a radiant life, and enables it to die.

An analysis of the light that comes from the hot gaseous surface of a star makes it possible to say definitely what the star is made of. Each atom gives light of certain definite colors, by which it can be immediately recognized. By passing starlight through a prism, and thus splitting it up into a rainbow band of colors—the spectrum—we determine what kinds of atoms are sending forth the light by which we see a star. We find that the stars are made of silicon and aluminum, oxygen and hydrogen, carbon and iron, and of calcium, magnesium, sodium, nickel, nitrogen, sulphur, and many other substances familiar to us who live on earth.

This discovery is the more surprising because the conditions are very different from those that exist on earth. The coolest stars have atmospheres at a temperature of about three thousand degrees Fahrenheit, and the hottest may be at thirty thousand degrees. Can iron and oxygen survive such fierce heat as this?

The iron atom has a definite structure. Twenty-six electrons whirl around the iron core, and if the atom is in a suitable environment, not too hot, they give out light of all the different colors that are characteristic of the atom of iron under the so-called normal conditions of our earth. But if the atom of iron is heated to sixteen thousand degrees—about the temperature of the surface of Sirius—it can no longer remain as it was, and one of the twenty-six electrons flies off altogether. And because the array of wave lengths given by an atom depends entirely on the number of its electrons, the spectrum is completely transformed by the change, although the atom is still an iron atom, for it still has an iron core. From the spectrum of Sirius we can tell not only that there is iron on the star, but also that the iron atoms have been crippled by the loss of one of their electrons.

The same thing happens to all atoms on a hot star; we can observe atoms of oxygen, nitrogen, sulphur, and silicon, that have lost two or even three electrons. Within the stars—even those that are comparatively cool outside— the atoms may lose twenty, thirty, or forty electrons, each according to its means, so that there is little left but the core. The atoms are still atoms of nitrogen, carbon, sulphur, silicon, and so on, because the central cores still survive—the energy available is not great enough to break down atomic nuclei. Incidentally we may note that the normal state of an atom is the stripped state, with few electrons around the core; the amount of matter in the universe that consists of atoms with their full quota of electrons is very small, though such atoms happen to be relatively common on earth because of our comparatively low temperature.

We cannot take a part of Sirius and analyze it chemically. There is, however, one star of which we can very easily obtain a piece, though too large a piece for convenient analysis. Because the planets were born from the atmosphere of the sun, the earth is a good sample of the building material of the stars. It is difficult—indeed it is impossible—to analyze the earth. At best we can examine only a thin layer at the surface, knowing that the composition must be different lower down. But it can be said fairly definitely that the seven commonest elements of which earth is made—oxygen, iron, silicon, magnesium, aluminum, calcium, and sodium—are also the commonest constituents of the stars. The parallel extends also to the less common elements. The earth seems to be a representative sample of the material of which the stars are built. . . .

In summary, we are fairly certain of the building materials of the stars, and we even have an inkling of its source, and some idea of the mechanism of upkeep and repair. But of the intimate structure of the stars, the details of the plan, we are more ignorant. There are those who advocate the classical style, with an intensely hot, dense interior that behaves as a perfect gas. There are more radical designs that are built around a liquid interior of fantastic substance unknown on earth. There are hints of modernistic structures that have yet to stand the test of time. On a few details, however, the designers are agreed: the stars are intensely hot within, and the atoms of matter at their centers are stripped, by the conditions of heat, pressure and radiation, virtually to the bare nuclei.

With the known terrestrial atoms as a basis, and the laws of physics and chemistry to work upon, it has been possible to present a sketch of the manner in which the stars are built. As frequently happens, our methods of analysis have outstripped our powers of understanding. The structure that we try to visualize is impressive, even though tentative. Perhaps the universal occurrence of the elements found on the earth is the principal feature of reality in the picture. But in addition there is overwhelming evidence in the night sky that these elements are built, according to some definite plan, into the stars.

SOURCES: *Cecilia Payne-Gaposchkin: An Autobiography and Other Recollections*, ed. Katherine Haramundanis, Cambridge: Cambridge UP, 1984; *The Universe of Stars*, ed. Harlow Shapley and Cecilia H. Payne, 2nd edition, Cambridge, Mass.: Harvard Observatory, 1929.

63

You Have Broken Newton's Back

George Bernard Shaw

No one thinks of George Bernard Shaw (1856–1950) as a cosmologist, but on 28 October 1930, at a dinner in London's Savoy Hotel, he was asked to propose a toast for the evening's guest of honor, someone who was indeed a cosmologist: Albert Einstein. The dinner could not have been more delicious than Shaw's speech.

There are great men who are great men among small men, but there are also great men who are great among great men, and that is the sort of great man whom you have among you here tonight. Napoleon and other great men of his type, they were makers of empires, but there is an order of men who get beyond that. They are not makers of empires, but they are makers of universes, and when they have made those universes, their hands are unstained by the blood of any human being on earth. They are very rare. I go back 2,500 years, and how many of them can I count in that period? I can count them on the fingers of my two hands: Pythagoras, Ptolemy, Aristotle, Copernicus, Kepler, Galileo, Newton, Einstein, and I still have two fingers left vacant.

Since the death of Newton, three hundred years have passed, nine generations of men, and those nine generations of men have not enjoyed the privilege which we are enjoying here tonight of standing face to face with one of those eight great men and looking forward to the privilege of hearing his voice; and another three hundred years may very well pass before another generation will enjoy that privilege. And I must—even among those eight men—I must make a distinction. I have called them makers of universes, but some of them were only repairers of universes. Only three of them made universes. Ptolemy made a universe. Newton made a universe which lasted for three hundred years. Einstein made a universe which I suppose you want me to say will never stop, but I don't know how long it will last. . . .

All these great men, what have they been doing? Ptolemy, as I say created a universe, Copernicus proved that Ptolemy was wrong, Kepler proved that Copernicus was wrong, Galileo proved that Aristotle was wrong, and now you are expecting me to say that Newton proved that they were all wrong. But you forget, when science reached Newton, science came up against that incalculable, that illogical, that hopelessly inconsequent and extraordinarily natural phenomenon, an Englishman. That had never happened to it before. As an Englishman, Newton was able to combine mental power so extraordinary that if I were speaking fifty years ago, as I am old enough to have done, I should have said that his was the greatest mind that any man had ever been endowed with, and he contrived to combine the exercise of that wonderful mind with credulity, with superstition, with delusion, which it would not have imposed on a moderately intelligent rabbit. As an Englishman also, he knew his people, he knew his language, he knew his own soul; and knowing that language, he knew that an honest thing was a square thing, an honest bargain was a square deal, an honest man was a square man, who acted on the square—that is to say, the universe that he creates has above everything to be a rectilinear universe.

Now see the dilemma in which this placed Newton. He knew his universe, he knew that it consisted of heavenly bodies all in motion, and he also knew that the one thing that you cannot do to any body in motion whatsoever is to make it move in a straight line. You may fire it out of a cannon with the strongest charge that you can put into it; you may have the cannon contrived to have, as they say, the flattest trajectory. In other words, motion will not go in a straight line. . . . [But] mere fact will never stop an Englishman. Newton invented a straight line. . . .

I advisedly say he invented the force which would make the straight line fit the straight lines of his universe—and bend them—and that was the force of gravitation. And when he had invented this force, he had created a universe which was wonderful and consistent in itself and which was thoroughly British. And when applying his wonderful genius, when he had completed the book of that universe, what sort of book was it? It was a book which told you the station of all the heavenly bodies; it showed the rate at which they were travelling; it gave you the exact hour at which they would arrive at such and such a point to make an eclipse or at which they would strike this earth. . . . In other words, it was not a magical marvelous thing like a Bible. It was a matter-of-fact British thing like a Bradshaw [an English railway timetable].

For three hundred years we believed in that Bradshaw and in the Newtonian universe, as I suppose no system has ever been believed in before. The more educated we were, the more firmly we believed in it. I believed in it. I was brought up to believe in it. Then an amazing thing happened. A young professor got up in the middle of Europe and, without betraying any consciousness of saying anything extraordinary, he addressed himself to our astronomers, and he said, "Excuse me, gentlemen, but if you will principally observe the next eclipse of the sun you will find out what is wrong with the perihelion of Mercury," and all Europe staggered. It said, "Something wrong—something wrong in the Newtonian universe—how can that be?" And we said, "This man is a blasphemer, burn him alive, confute him, madman!" But the astronomers only looked rather foolish and they said, "Oh, let us wait for the eclipse." But we said, "No, this is not a question of an eclipse. This man has said there is something wrong with the perihelion of Mercury. Do you mean to say there is something wrong with the perihelion of Mercury?" And then they said, "Oh, yes, we knew it all along." They said, "Newton knew it." "Then why did you not tell us so before?" Our faith began to shake, and we said, "If this young man says when the eclipse comes and gets away with it, then the next thing that he will be doing, he will be questioning the existence of gravitation." And the young professor smiled, and he said, "No, I mean no harm to gravitation; gravitation is a very useful hypothesis, and after all it gives you fairly healthy results. But personally and for my part I can do without it."

And we said, "What do you mean, do without it? What about the apple?" The young professor said, "What happened to that apple is really a very curious and interesting thing. You see, Newton did not know what happened to the apple. The only real authority upon the subject of what happened to the apple was the apple itself! Now apples are very intelligent. If you watch apples carefully you will learn that they behave much more sensibly than men often do, but unfortunately we do not know their language." And the professor said, "What Newton ought to have done would be to see something fall that could tell the story afterwards, could explain itself. He should have reflected that not only apples fall but men fall. And," he said, "I, instead of sitting about in orchards and watching apples fall, what did I do? I frequented cities in quarters where building operations were going on. I knew, as a man of science, that it was statistically certain that sooner or later I should see a man fall off a scaffolding, and I did. I went to that man in hospital, and after condoling him in the usual fashion, saying how sorry I was for his accident and how he was, I came to business. I said, 'When you came off that scaffolding, did the earth attract you?' The man said, 'Certainly not, *gar nicht*, on the contrary the earth repelled me with such violence that here I am in hospital with most of my bones broken!'" And the professor could only say, "Well, my friend, you have been lucky enough to escape without breaking your own back, but you have broken Newton's back."

That was very clear, and we turned round and we said, "Well, this is all very well, but what about the straight line? If there is no gravitation, why do not the heavenly bodies travel in a straight line right out of the universe?" The professor said, "Why should they? That is not the way the world is made. The world is not a British rectilinear world. It is a curvilinear world, and the heavenly bodies go in curves because that is the natural way for them to go." And at that word the whole Newtonian universe crumbled up and vanished and was succeeded by the Einsteinian universe.

(For more on the story about the man who fell from the scaffolding, see Chapter 60.)

SOURCE: *The Religious Speeches of Bernard Shaw*, ed. Warren Sylvester Smith, University Park, PA: Pennsylvania State UP, 1963.

64

The Realm of the Nebulae

Edwin Hubble

While the conception of the expanding universe is grounded in the theoretical physics of Einstein, it is principally Edwin Hubble (1889–1953) who laid its observational foundation.

To understand how profoundly cosmology changed in the first third of the twentieth century, however, we need to see where it stood at the end of the nineteenth, particularly on the question of the galaxies and their relations to each other. It is worth remembering that "galaxy" originally referred specifically to the Milky Way (we can still see the etymological connection between "lactic" and "galactic"). Although Kant had theorized "island universes" (the term Weltinseln *being popularized by Alexander von Humboldt), by the start of the twentieth century the idea seemed to have lost its currency. In* The System of the Stars *(second edition, 1905), Agnes Clerke asserts:*

The question whether nebulae are external galaxies hardly any longer needs discussion. It has been answered by the progress of research. No competent thinker, with the whole of the available evidence before him, can now, it is safe to say, maintain any single nebula to be a star system of coordinate rank with the Milky Way. A practical certainty has been attained that the entire contents, stellar and nebular, of the sphere belong to one mighty aggregation, and stand in ordered mutual relations within the limits of one all-embracing scheme. All-embracing, that is to say, so far as our capacities of knowledge extend. With the infinite possibilities beyond, science has no concern.

More or less the same view was espoused on into the early 1920s by Harlow Shapley (1885–1972). In the mid-twenties, however, Hubble, using the 100-

inch reflecting telescope at Mt. Wilson Observatory in California, began to publish his findings concerning the distances of the nebulae, findings which turned Clerke's "practical certainty" upside down.

Before turning to Hubble's own account of his conclusions, however, we need to highlight two important points. First, the term "nebula" means "cloud," and there are really two separate categories of astronomical nebulae. In brief, there are the nebulae that truly are clouds—the sort that William Huggins in August 1864 proved were made up of gases (see chapter 52). And then there are those nebulae which appear cloudy but which are in fact composed of stars that the naked eye cannot resolve. It is mainly this latter category that is at issue in the question of "external galaxies."

Second, the early twentieth century saw a breakthrough in techniques for estimating astronomical distances. Most notably, in 1912 an article appeared in the Harvard College Observatory Circular ascribed to E. C. Pickering, the Observatory's director, but actually written by Henrietta Swan Leavitt, entitled "Periods of 25 Variable Stars in the Small Magellanic Cloud." Leavitt's work on a class of stars known as Cepheid variables provided a tool critical for the solution of problems related to cosmic scale. Each Cepheid variable star displays a periodic variation in brightness, and the variations have a specific "rhythm." What Leavitt discovered by studying such "rhythmic" stars within the Small Magellanic Cloud—which can all be assumed to be roughly equidistant from us—is that those stars' periods are proportionate to their luminosity: a longer period (slower "rhythm") means greater luminosity. This principle then allows one to determine the relative distances of stars even when one does not know their actual distances.

To put this into familiar terms, suppose we are sitting at one end of a darkened stadium and have to answer some skill-testing questions about how far away certain flashing light bulbs are from us. We are not told the wattage of these bulbs; but we are told that the higher the wattage of the bulb, the more slowly it will flash. Accordingly, if we see two bulbs, A and B, that have the same brightness (= apparent luminosity) and they are flashing at the same rate, we will know that A and B are equidistant from us. But if we see two other bulbs, C and D, apparently of the same brightness, but C is flashing twice as fast as D, then we will know that D is actually much farther from us than C is, for there's no other way to explain why a higher wattage bulb would appear only as bright as one that is in fact less luminous.

This basic principle discovered by Leavitt, combined with mathematics rather more sophisticated than I have divulged, became a central component in astronomers' increasing ability to estimate distances of the nebulae, and without such estimations the thorniest questions concerning the sizes of and distances between cosmic components may have gone unanswered.

Let us move to Hubble's account of the question of the nebulae as it stood in the late nineteenth century.

The status of the nebulae . . . was undetermined because the distances were wholly unknown. They were definitely beyond the limits of direct measurement, and the scanty, indirect evidence bearing on the problem could be interpreted in various ways. The nebulae might be relatively nearby objects and hence members of the stellar system, or they might be very remote and hence inhabitants of outer space. At this point, the development of nebular research came into immediate contact with the philosophical theory of island universes. The theory represented, in principle, one of the alternative solutions of the problem of nebular distances. The question of distances was frequently put in the form: Are nebulae island universes? . . .

The solution came [in 1924], largely with the help of a great telescope, the 100-inch reflector. . . . Several of the most conspicuous nebulae were found to be far beyond the limits of the galactic system [i.e., the Milky Way]—they were independent, stellar systems in extragalactic space. Further investigations demonstrated that the other, fainter nebulae were similar systems at greater distances, and the theory of island universes was confirmed.

The 100-inch reflector partially resolved a few of the nearest, neighboring nebulae into swarms of stars. Among these stars various types were recognized which were well known among the brighter stars in [our own] galactic system. The intrinsic luminosities (candle powers) were known, accurately in some cases, approximately in others. Therefore, the apparent faintness of the stars in the nebulae indicated the distances of the nebulae.

The most reliable results were furnished by Cepheid variables, but other types of stars furnished estimates of orders of distance, which were consistent with the Cepheids. . . . With the nature of the nebulae known and the scale of nebular distances established, the investigations proceeded. . . .

Investigations of the observable region as a whole have led to two results of major importance. One is the homogeneity of the region—the uniformity of the large-scale distribution of nebulae. The other is the velocity-distance relation.

The small-scale distribution of nebulae is very irregular. Nebulae are found singly, in pairs, in groups of various sizes, and in clusters. The galactic system [the Milky Way] is the chief component of a triple nebula in which the Magellanic Clouds are the other members. The triple system, together with a few additional nebulae, forms a typical small group that is isolated in the general field of nebulae. The members of this local group furnished the first distances, and the Cepheid criterion of distance is still confined to the group.

When large regions of the sky, or large volumes of space, are compared, the irregularities average out and the large-scale distribution is sensibly uniform. The distribution over the sky is derived by comparing the numbers of nebulae brighter than a specified limit of apparent faintness, in sample areas scattered at regular intervals. The true distribution is confused by local obscuration. No nebulae are seen within the Milky Way, and very few along the borders. Moreover, the apparent distribution thins out, slightly but systematically, from the poles to the borders of the Milky Way. The explanation is found in the great clouds of dust and gas which are scattered throughout the stellar system, largely in the galactic plane. These clouds hide the more distant stars and nebulae. Moreover, the sun is embedded in a tenuous medium which behaves like a uniform layer extending more or less indefinitely along the galactic plane. Light from nebulae near the galactic poles is reduced about one fourth by the obscuring layer, but in the lower latitudes, where the light-paths through the medium are longer, the absorption is correspondingly greater. It is only when these various effects of galactic obscuration are evaluated and removed, that the nebular distribution over the sky is revealed as uniform, or isotropic (the same in all directions). . . .

The observable region is not only isotropic but homogeneous as well—it is much the same everywhere and in all directions. The nebulae are scattered at average intervals of the order of two million light-years or perhaps two hundred times the mean diameters. The pattern might be represented by tennis balls fifty feet apart.

The order of the mean density of matter in space can also be roughly estimated if the (unknown) material between the nebulae is ignored. If the nebular material were spread evenly through the observable region, the smoothed-out density would be of the general order of 10^{-29} or 10^{-28} grams per cubic centimeter—about one grain of sand per volume of space equal to the size of the earth. . . .

(More recent estimates reduce this number to about 10^{-30} g/cm²)

THE VELOCITY-DISTANCE RELATION

The foregoing sketch of the observable region has been based almost entirely upon results derived from direct photographs. The region is homogeneous and the general order of the mean density is known. The next—and last—property to be discussed, the velocity-distance relation, emerged from the study of spectrograms.

When a ray of light passes through a glass prism (or other suitable device) the various colors of which the light is composed are spread out in an ordered sequence called a spectrum. The rainbow is, of course, a familiar example. The sequence never varies. The spectrum may be long or short, depending on the apparatus employed, but the order of the colors remains unchanged. Position in the spectrum is measured roughly by colors, and more precisely by wave-lengths, for each color represents light of a particular wave-length. From the short waves of the violet, they steadily lengthen to the long waves of the red.

The spectrum of a light source shows the particular colors or wave-lengths which are radiated, together with their relative abundance (or intensity), and thus gives information concerning the nature and the physical condition of the light source. An incandescent solid radiates all colors, and the spectrum is *continuous* from violet to red (and beyond in either direction). An incandescent gas radiates only a few isolated colors and the pattern, called an *emission* spectrum, is characteristic for any particular gas.

A third type, called an *absorption* spectrum and of special interest for astronomical research, is produced when an incandescent solid (or equivalent source), giving a continuous spectrum, is surrounded by a cooler gas. The gas absorbs from the continuous spectrum just those colors which the gas would radiate if it were itself incandescent. The result is a spectrum with a continuous background interrupted by dark spaces called absorption lines. The pattern of dark absorption lines indicates the particular gas or gases that are responsible for the absorption. . . .

The nebulae in general show absorption spectra similar to the solar spectrum, as would be expected for systems of stars among which the solar type predominated. . . . [But] nebular spectra are peculiar in that the lines are not in the usual positions found in nearby light sources. They are displaced toward the red of their normal position, as indicated by suitable comparison spectra. The displacements, called red-shifts, increase, on the average, with the apparent faintness of the nebula that is being observed. Since apparent faintness measures distance, it follows that red-shifts increase with distance. Detailed investigation shows that the relation is linear.

Small microscopic shifts, either to the red or to the violet, have long been known in the spectra of astronomical bodies other than nebulae. These displacements are confidently interpreted as the results of motion in the line of sight—radial velocities of recession (red-shifts) or of approach (violet-shifts). The same interpretation is frequently applied to the red-shifts in nebular spectra and has led to the term "velocity-distance" relation for the observed relation between red-shifts and apparent faintness. On this assumption, the nebulae are supposed to be rushing away from our region of space, with velocities that increase directly with distance. . . .

A completely satisfactory interpretation of red-shifts is a question of great importance, for the velocity-distance relation is a property of the observable region as a whole. The only other property that is known is the uniform distribution of nebulae. Now the observable region is our sample of the universe. If the sample is fair, its observed characteristics will determine the physical nature of the universe as a whole.

And the sample may be fair. As long as explorations were confined to the stellar system, the possibility did not exist. The system was known to be isolated. Beyond lay a region, unknown, but necessarily different from the star-strewn space within the system. We now observe that region—a vast sphere, through which comparable stellar systems are uniformly distributed. There is no evidence of a thinning-out, no trace of a physical boundary. There is not the slightest suggestion of a supersystem of nebulae isolated in a larger world. Thus, for purposes of speculation, we may apply the principle of uniformity, and suppose that any other equal portion of the universe, selected at random, is much the same as the observable region. We may assume that the realm of the nebulae is the universe.

The very clarity of Hubble's style and its lack of grandiosity can combine to make us miss the power of his conclusions. We have become perhaps blasé about the idea of an expanding universe, but it is worth remembering that even Einstein resisted the pull of his own theories when they pointed in this direction. In the mid-1920s, those who first read the evidence that the universe is homogeneous, isotropic, and expanding were surely astonished. Hubble's 1936 retrospective of these developments, before lapsing back into a flat, "lab-report" tone, hints at the excitement he and his fellow cosmologists must have felt.

As a mere criterion of distance the [velocity-distance] relation is a valuable aid to nebular research. . . . [But it] is not merely a powerful aid to research; it is also a general characteristic of our sample of the universe—one of the very few that are known. Until lately, the explorations of space had been confined to relatively short distances and small volumes—in a cosmic sense, to comparatively microscopic phenomena. Now, in the realm of the nebulae, large-scale, macroscopic phenomena of matter and radiation could be examined. Expectations ran high. There was a feeling that almost anything might happen and, in fact, the velocity-distance relation did emerge as the mists receded. This was of the first importance for, if it could be fully interpreted, the relation would probably contribute an essential clue to the problem of the structure of the universe.

Observations show that details in nebular spectra are displaced toward the red from their normal positions, and that the red-shifts increase with apparent faintness of the nebulae. Apparent faintness is confidently interpreted in terms of distance. Therefore, the observational result can be restated—red-shifts increase with distance. . . .

Thorough investigation of the problem has led to the following conclusions. Several ways are known in which red-shifts might be produced. Of them all, only one will produce large shifts without introducing other effects which should be conspicuous, but which are not observed. This explanation interprets red-shifts as Doppler effects, that is to say, as velocity-shifts, indicating actual motion of recession. It may be stated with some confidence that red-shifts are velocity-shifts or else they represent some hitherto unrecognized principle in physics.

The interpretation as velocity-shifts is generally adopted by theoretical investigators, and the velocity-distance relation is considered as the observational basis for theories of an expanding universe. Such theories are widely current. They represent solutions of the cosmological equation, which follow from the assumption of a nonstatic universe.

Source: Edwin Hubble, *The Realm of the Nebulae*, New Haven: Yale UP, 1936.

65

Driven to Admit Anti-Chance

Arthur Stanley Eddington

The early 1930s were a time of great cosmological ferment. Earlier contributions of Einstein and Hubble were continuing to arouse discussion about the very nature of time and space and about how the new physics should affect our conception of the universe, including its beginning and its end. One of the greatest contributors to these discussions was Sir Arthur Stanley Eddington (1882–1944), who had been the first to learn of and then to promote Einstein's general theory of relativity—indeed also the first to test it, when he led an expedition in 1919 to measure the deflection of light during the solar eclipse of that year. (See Cecilia Payne's recollection of his lecture on relativity, Chapter 62.)

In the following excerpt, in the course of a penetrating philosophical and physical discussion of the nature of time, including entropy as time's melancholy "signpost," Eddington makes his famous remark about finding the idea of "a beginning of the present order of nature repugnant"—which provided a foil for a reply by Georges Lemaître seven weeks later (see next chapter). Eddington begins with a discussion of time's direction.

If, then, we are looking for an end of the world—or, instead of an end, an indefinite continuation for ever and ever—we must start off in one of the two time directions. How shall we decide which of these two directions to take? It is an important question. Imagine yourself in some unfamiliar part of space-time so as not to be biased by conventional landmarks or traditional standards of reference. There ought to be a signpost with one arm marked "To the future" and the other arm marked "To the past." My first business is to find this signpost, for if I make a mistake and go the wrong way I shall lead you to what is no doubt an "end of the world," but it will be that end which is more usually described as the *beginning*.

In ordinary life the signpost is provided by consciousness. Or perhaps it would be truer to say that consciousness does not bother about signposts; but wherever it finds itself it goes off on urgent business in a particular direction, and the physicist meekly accepts its lead and labels the course it takes "To the future." . . .

ENTROPY AND DISORGANIZATION

Leaving aside the guidance of consciousness, we have found it possible to discover a kind of signpost for time in the physical world. The signpost is of rather a curious character, and I would scarcely venture to say that the discovery of the signpost amounts to the same thing as the discovery of an objective "going on of time" in the universe. But at any rate it serves to discriminate past and future, whereas there is no corresponding objective distinction of left and right. The distinction is provided by a certain measurable quantity called entropy. Take an isolated system and measure its entropy S at two instants t_1 and t_2 without employing the intuition of consciousness, which is too disreputable a witness to trust in mathematical physics. The rule is that the instant which corresponds to the greater entropy is the later. . . . This is the famous second law of thermodynamics.

Entropy is a very peculiar conception, quite unlike the conceptions ordinarily employed in the classical scheme of physics. We may most conveniently describe it as the measure of disorganization of a system. Accordingly, our signpost for time resolves itself into the law that disorganization increases from past to future. It is one of the most curious features of the development of physics that the entropy outlook grew up quietly alongside the ordinary analytical outlook for many years. Until recently it always "played second fiddle"; it was convenient for getting practical results, but it did not pretend to convey the most penetrating insight. But now it is making a bid for supremacy, and I think there is little doubt that it will ultimately drive out its rival. . . .

It is possible for the disorganization of a system to become complete. The state then reached is called thermodynamic equilibrium. The entropy can increase no further, and, since the second law of thermodynamics forbids a decrease, it remains constant. Our signpost for time disappears; and so far as that system is concerned, time ceases to go on. That does not mean that time ceases to exist; it exists and extends just as space exists and extends, but there is no longer any one-way property. It is like a one-way street on which there is never any traffic.

Let us return to our signpost. Ahead there is ever-increasing disorganization. Although the sum total of organization is diminishing, certain parts of the universe are exhibiting a more and more highly specialized organization; that is the phenomenon of evolution. But ultimately this must be swallowed up in the advancing tide of chance and chaos, and the whole universe will reach a state of complete disorganization—a uniform featureless mass in thermodynamic equilibrium. This is the end of the world. Time will *extend* on and on, presumably to infinity. But there will be no definable sense in which it can be said to *go* on. Consciousness will obviously have disappeared from the physical world before thermodynamical equilibrium is reached, and . . . there will remain nothing to point out a direction in time.

THE BEGINNING OF TIME

It is more interesting to look in the opposite direction—towards the past. Following time backwards, we find more and more organization in the world. If we are not stopped earlier, we must come to a time when the matter and energy of the world had the maximum possible organization. To go back further is impossible. We have come to an abrupt end of space-time—only we generally call it the "beginning."

I have no "philosophical axe to grind" in this discussion. Philosophically, the notion of a beginning of the present order of nature is repugnant to me. I am simply stating the dilemma to which our present fundamental conception of physical law leads us. I see no way round it; but whether future developments of science will find an escape I cannot predict. The dilemma is this: Surveying our surroundings, we find them to be far from a "fortuitous concourse of atoms." The picture of the world, as drawn in existing physical theories, shows arrangement of the individual elements for which the odds are multillions[1] to 1 against an origin by chance. Some people would like to call this non-random feature of the world purpose or design; but I will call it non-committally anti-chance. We are unwilling to admit in physics that anti-chance plays any part in the reactions between the systems of billions of atoms and quanta that we study; and indeed all our experimental evidence goes to show that these are governed by the laws of chance. Accordingly, we sweep anti-chance out of the laws of physics—out of the differential equations. Naturally, therefore, it reappears in the boundary conditions, for it must be got into the scheme somewhere. By sweeping it far enough away from the sphere of our current physical problems, we fancy we have got rid

---

[1] I use "multillions" as a general term for numbers of order $10^{10^{10}}$ or larger [Eddington's note].

of it. It is only when some of us are so misguided as to try to get back billions of years into the past that we find the sweepings all piled up like a high wall and forming a boundary—a beginning of time—which we cannot climb over.

A way out of the dilemma has been proposed which seems to have found favor with a number of scientific workers. I oppose it because I think it untenable, not because of any desire to retain the present dilemma. . . .

FLUCTUATIONS

The loophole to which I [refer] depends on the occurrence of chance fluctuations. If we have a number of particles moving about at random, they will in the course of time go through every possible configuration, so that even the most orderly, the most non-chance configuration, will occur by chance if only we wait long enough. When the world has reached complete disorganization (thermodynamic equilibrium) there is still infinite time ahead of it, and its elements will thus have opportunity to take up every possible configuration again and again. If we wait long enough, a number of atoms will, just by chance, arrange themselves in systems as they are at present arranged in this room; and, just by chance, the same sound-waves will come from one of these systems of atoms as are at present emerging from my lips; they will strike the ears of other systems of atoms, arranged just by chance to resemble you, and in the same stages of attention or somnolence. This mock Mathematical Association meeting must be repeated many times over—an infinite number of times, in fact. . . . Do not ask me whether I expect you to believe that this will really happen. "Logic is logic. That's all I say."

So, after the world has reached thermodynamical equilibrium the entropy remains steady at its maximum value, except that "once in a blue moon" the absurdly small chance comes off and the entropy drops appreciably below its maximum value. When this fluctuation has died out, there will again be a very long wait for another coincidence giving another fluctuation. It will take multillions of years, but we have all infinity of time before us. There is no limit to the amount of the fluctuation, and if we wait long enough we shall come across a big fluctuation which will take the world as far from thermodynamical equilibrium as it is at the present moment. If we wait for an enormously longer time, during which this huge fluctuation is repeated untold numbers of times, there will occur a still larger fluctuation which will take the world as far from thermodynamical equilibrium as it was one second ago.

The suggestion is that we are now on the downward slope of one of these fluctuations. It has quite a pleasant subtlety. Is it chance that we happen to be running down the slope and not toiling up the slope? Not at all. So far as the physical universe is concerned, we have *defined* the direction of time as the direction from greater to less organization, so that, on whichever side of the mountain we stand, our signpost will point downhill. In fact, on this theory, the going on of time is not a property of time in general, but is a property of the slope of the fluctuation on which we are standing. Again, although the theory postulates a universe involving an extremely improbable coincidence, it provides an infinite time during which the most improbable coincidence might occur. Nevertheless, I feel sure that the argument is fallacious.

If we put a kettle of water on the fire there is a chance that the water will freeze. If mankind goes on putting kettles on the fire until $t= \infty$, the chance will one day come off and the individual concerned will be somewhat surprised to find a lump of ice in his kettle. But it will not happen to *me*. Even if tomorrow the phenomenon occurs before my eyes, I shall not explain it this way. I would much sooner believe in interference by a demon than in a coincidence of that kind coming off; and in doing so I shall be acting as a rational scientist. The reason why I do not at present believe that devils interfere with my cooking arrangements and other business is [that] I have become convinced by experience that nature obeys certain uniformities which we call laws. I am convinced because these laws have been tested over and over again. But it is possible that every single observation from the beginning of science which has been used as a test has just happened to fit in with the law by a chance coincidence. It would be an improbable coincidence, but I think not quite so improbable as the coincidence involved in my kettle of water freezing. So if the event happens and I can think of no other explanation, I shall have to choose between two highly improbable coincidences: (*a*) that there are no laws of nature and that the apparent uniformities so far observed are merely coincidences; (*b*) that the event is entirely in accordance with the accepted laws of nature, but that an improbable coincidence has happened. I choose the former because mathematical calculation indicates that it is the less improbable. I reckon a sufficiently improbable coincidence as something much more disastrous than a violation of the laws of nature; because my whole reason for accepting the laws of nature rests on the assumption that improbable coincidences do not happen—at least, that they do not happen in my experience.

Similarly, if logic predicts that a mock meeting of the Mathematical Association will occur just by a fortuitous arrangement of atoms before $t= \infty$, I reply that I cannot possibly accept that as being the explanation of a meeting of the Mathematical Association in $t= 1931$. We must be a little careful over this, because there is a trap for the unwary. The year 1931 is not an absolutely random date between $t= -\infty$ and $t= +\infty$. We must not argue that because for only $1/x$th of time between $t= -\infty$ and $t= \infty$ a fluctuation as great as the present one is in operation, therefore the chances are x to 1 against such a fluctuation occurring in the year 1931. For the purposes of the present discussion, the important characteristic of the year 1931 is that it belongs to a period during which there exist in the universe beings capable of speculating about the universe and its fluctuations. Now I think it is clear that such creatures could not exist in a universe in thermodynamical equilibrium. A considerable degree of deviation is required to permit of living beings. Therefore it is perfectly fair for supporters of this suggestion to wipe out of account all those multillions of years during which the fluctuations are less than the minimum required to permit of the development and existence of mathematical physicists. That greatly diminishes x, but the odds are still overpowering. The *crude* assertion would be that (unless we admit something which is not chance in the architecture of the universe) it is practically certain that at any assigned date the universe will be almost in the state of maximum disorganization. The *amended* assertion is that (unless we admit something which is not chance in the architecture of the universe) it is practically certain that a universe containing mathematical physicists will at any assigned date be in the state of maximum disorganization which is not inconsistent with the existence of such creatures. I think it is quite clear that neither the original nor the amended version applies. We are thus driven to admit anti-chance; and apparently the best thing we can do with it is to sweep it up into a heap at the beginning of time.

SOURCE: Arthur S. Eddington, "The End of the World: From the Standpoint of Mathematical Physics," *Nature* (Supplement), 127.3203 (March 21, 1931).

66

Did the Expansion Start from the Beginning?

Georges Édouard Lemaître

A further central figure in cosmology in the 1930s was the Belgian cleric and physicist Georges Édouard Lemaître (1894–1966), most famous as an eloquent proponent of an expanding universe. Lemaître first published his Einsteinian description of cosmic expansion in an obscure Belgian journal in 1927, but his work did not become well known until 1931, when Eddington received a copy and arranged for its translation and republication. Lemaître went on to hypothesize a "primal atom" from which the universe took its beginning.

Lemaître, in a short reply in Nature *published only seven weeks after Eddington's article excerpted in the previous chapter, raises an issue that continues to stimulate interest today: the relationship between cosmogony and quantum physics. He does not take on Eddington's whole argument; rather he uses Eddington's distaste vis-à-vis a beginning (and his subsequent wrestling with classical physics' blend of chance and determinism) as an occasion for pointing towards a model which, as his metaphors indicate, Lemaître considers more plausible and more exciting, as well he might.*

Sir Arthur Eddington states that, philosophically, the notion of a beginning of the present order of nature is repugnant to him. I would rather be inclined to think that the present state of quantum theory suggests a beginning of the world very different from the present order of nature. Thermodynamical principles from the point of view of quantum theory may be stated as follows: (1) Energy of constant total amount is distributed in discrete quanta.

(2) The number of distinct quanta is ever increasing. If we go back in the course of time we must find fewer and fewer quanta, until we find all the energy of the universe packed in a few or even in a unique quantum.

Now, in atomic processes, the notions of space and time are no more than statistical notions; they fade out when applied to individual phenomena involving but a small number of quanta. If the world has begun with a single quantum, the notions of space and time would altogether fail to have any meaning at the beginning; they would only begin to have a sensible meaning when the original quantum had been divided into a sufficient number of quanta. If this suggestion is correct, the beginning of the world happened a little before the beginning of space and time. I think that such a beginning of the world is far enough from the present order of nature to be not at all repugnant.

It may be difficult to follow up the idea in detail as we are not yet able to count the quantum packets in every case. For example, it may be that an atomic nucleus must be counted as a unique quantum, the atomic number acting as a kind of quantum number. If the future development of quantum theory happens to turn in that direction, we could conceive the beginning of the universe in the form of a unique atom, the atomic weight of which is the total mass of the universe. This highly unstable atom would divide in smaller and smaller atoms by a kind of super-radioactive process. Some remnant of this process might, according to Sir James Jeans's idea, foster the heat of the stars until our low atomic number atoms allowed life to be possible.

Clearly the initial quantum could not conceal in itself the whole course of evolution; but, according to the principle of indeterminacy, that is not necessary. Our world is now understood to be a world where something really happens; the whole story of the world need not have been written down in the first quantum like a song on the disc of a phonograph. The whole matter of the world must have been present at the beginning, but the story it has to tell may be written step by step.

As contributor to a forum discussion in Nature *later in 1931, Lemaître explains more fully his ideas regarding the need for the hypothesis which, with subsequent modifications, came to be known as the Big Bang. Lemaître's discussion fuses difficulty with lucidity and has become best known for its image of the present universe as "ashes and smoke of bright but very rapid fireworks." It also starts to grapple with the problem of cosmic time-scale: if the universal expansion has been continuous, then in principle one may "fix a limit", as regards the beginning. (Lemaître's ballpark minimum of "a few hundred thousand million years" is about ten times as long as the most re-*

cent estimates.) And not least of Lemaître's interests is the evidence which cosmic rays may provide as fossils of cosmic evolution.

The expansion of the universe is a matter of astronomical facts interpreted by the theory of relativity, with the help of assumptions as to the homogeneity of space, without which any theory seems to be impossible. . . . I shall . . . try to show that the universe must be expanding, or rather that the most necessary processes of evolution are contradictory to the view that space is and always has been static.

It has been pointed out by Sir Arthur Eddington that a static universe is unstable, and he proposed the problem of finding the possible causes of its expansion. He suggested that such a cause might be the formation of condensations. I obtained recently a solution of this problem, and the main results are as follows:

When the expansion is already started, the effect of kinetic energy or pressure of radiation is quite negligible. On the contrary, pressure is the chief factor in the question of instability of a static universe. If the pressure were rigorously zero, the expansion could never appear. But, if the pressure (or kinetic energy) is not zero, any diminution of pressure must start the expansion. For example, a world full of radiation starts expanding as soon as the radiation can transform itself into matter.

When condensations exist or are formed, the problem is complicated by gravitational effects; but it can be shown that the general expansion of the universe depends entirely on the density of kinetic energy or of pressure at the places where the gravitational influences of the condensations cancel one another. I call these places (for brevity) "neutral zones." Condensation in itself has no direct effect whatever on the stability of the universe; but condensations would necessarily induce a rarefaction at the neutral zone and so a diminution of the density of kinetic energy at the neutral zone; and this must induce expansion.

We can conclude that any general process of condensation, occurring in a world where the kinetic energy does not vanish, must induce expansion. Therefore, practically, the expectation of Sir Arthur Eddington is fully confirmed. For example, formation of stars out of a primeval gas starts the expansion; formation of extra-galactic nebulae out of a uniform mass of gas or of stars starts the expansion. I think that these results add much weight to the fact that the actual velocity of expansion fixes a limit to the time scale of the evolution, as we must rule out of our speculations every process which would start a premature expansion of space.

Even if we had no experimental evidence of the expansion of space, considerations of stability would fix a limit to the time-scale of evolution. The reason is that, if the universe has existed for too long a time, any general process of condensation would be contradictory to the actual value of the density of matter. Although this quantity is not known with great accuracy, its value may give some idea of the maximum scale of evolution. I find that any general process of condensation, even of very moderate intensity, cannot have happened earlier than a few hundred thousand million years ago.

As stated by Sir James Jeans, this brings almost complete chaos into the already chaotic problem of stellar evolution. A complete revision of our cosmological hypothesis is necessary, the primary condition being the test of rapidity. We want a "fireworks" theory of evolution. The last two thousand million years are slow evolution: they are ashes and smoke of bright but very rapid fireworks.

Near the end of his contribution to the Nature *forum, Lemaître articulates the question that was to set the agenda for the entire generation of cosmologists following his own.*

If I had to ask a question of the infallible oracle alluded to by Sir James Jeans [in an earlier contribution to the *Nature* forum], I think I should choose this: "Has the universe ever been at rest, or did the expansion start from the beginning?" But, I think, I would ask the oracle not to give the answer, in order that a subsequent generation would not be deprived of the pleasure of searching for and of finding the solution.

SOURCE: Georges Édouard Lemaître, "The Beginning of the World from the Point of View of Quantum Theory," *Nature* 127.3210 (May 9, 1931); "The Evolution of the Universe," *Nature* (Supplement) 128.3234 (October 24, 1931).

67

This Big Bang Idea

Fred Hoyle

Sir Fred Hoyle (b. 1915) was one of the most prominent English cosmologists of his day and one of the most eloquent and imaginative. Together with two other Cambridge astronomers, Hermann Bondi and Thomas Gold, he founded the "steady state" model of the universe, which recognized cosmic expansion but postulated the continuous creation of matter that counterbalanced the expansion, resulting in a steady state. This model was framed in direct opposition to the alternative explanation, namely that at a finite time in the past the universe exploded into being. The latter model, now the standard model among cosmologists, Hoyle was the first to refer to as the "big bang" idea. Thus one of the ironies of cosmological history is that Hoyle's eloquence provided the catchy name for the very model that has eclipsed his own.

Hoyle's ideas and his skill as a communicator can be seen in the following selections from the popular book he published in 1950 based on a series of radio lectures delivered for the BBC.

We must move on to consider the explanations that have been offered for [the] expansion of the universe. First I will consider the older ideas—that is to say, the ideas of the nineteen-twenties and the nineteen-thirties—and then I will go on to offer my own opinion. Broadly speaking, the older ideas fall into two groups. One was that the universe started its life a finite time ago in a single huge explosion, and that the present expansion is a relic of the violence of this explosion. This big bang idea seemed to me to be unsatisfactory even before detailed examination showed that it leads to serious difficulties. For when we look at our own galaxy there is not the smallest sign that such an explosion ever occurred. This might not be such a cogent argument if on such theories our galaxy were much younger than the whole universe. But this is not so. In fact, in some of these theories there is the obvious contradiction that the universe comes out to be younger than our astrophysical estimates of the age of our galaxy. But the really serious difficulty arises when we try to reconcile the idea of an explosion with the requirement that the galaxies have condensed out of diffuse background material. The two concepts of explosion and condensation are obviously contradictory, and it is easy to show, if you postulate an explosion of sufficient violence to explain the expansion of the universe, that condensations looking at all like the galaxies could never have been formed.

It is worth noting that the problem Hoyle refers to here concerning the condensation of otherwise exploding matter is to a large degree what animates interest in the "wrinkles" or "ripples" indicated by study of the cosmic background radiation (see Chapter 76)—which to the satisfaction of many resolve what Hoyle claims here is an obvious contradiction.

In any case, Hoyle's solution is to have recourse to the idea of creation, an idea which, as Hoyle points out, is one that big bang cosmologists themselves cannot readily avoid (although more recently those such as Hawking, Rees, and Smolin have tried to avoid it). Thus what is unique about Hoyle's hypothesis is not the assumption of creation but the assumption of continuous creation.

Although I think there is no doubt that every galaxy we now observe to be receding from us will in about 10,000,000,000 years have passed entirely beyond the limit of vision of an observer in our galaxy, yet I think that such an observer would still be able to see about the same number of galaxies as we do now. By this I mean that new galaxies will have condensed out of the background material at just about the rate necessary to compensate for those that are being lost as a consequence of their passing beyond our observable universe. At first sight it might be thought that this could not go on indefinitely because the material forming the background would ultimately become exhausted. But again I do not believe that this is so, for it seems likely that new material is constantly being created so as to maintain a constant density in the background material. So we have a situation in which the loss of galaxies, through the expansion of the universe, is compensated by the condensation of new galaxies, and this can continue indefinitely. . . .

From time to time people ask where the created material comes from. Well, it does not come from anywhere. Material simply appears—it is created. At one time the various atoms composing the material do not exist and at a later time they do. This may seem a very strange idea and I agree that it is, but in science it does not matter how strange an idea may seem so long as it works—that is to say, so long as the idea can be expressed in a precise form and so long as its consequences are found to be in agreement with observation. In any case, the whole idea of creation is queer. In the older theories all the material in the universe is supposed to have appeared at one instant of time, the whole creation process taking the form of one big bang. For myself I find this idea very much queerer than continuous creation.

Perhaps you may think that the whole question of the creation of the universe could be avoided in some way. But this is not so. To avoid the issue of creation it would be necessary for all the material in the universe to be infinitely old, and this it cannot be. For if this were so, there could be no hydrogen left in the universe. Hydrogen is being steadily converted into helium and the other elements throughout the universe and this conversion is a one-way process—that is to say, hydrogen cannot be produced in any appreciable quantity through the breakdown of the other elements. How comes it then that the universe consists almost entirely of hydrogen? If matter were infinitely old this would be quite impossible. So we see that the universe being what it is, the creation issue simply cannot be dodged.

Moving to issues not quite so final, we see how Hoyle's pictorially precise imagination and sense of analogy allow him to narrate a physical process that in principle is far beyond our actual powers of direct observation:

Now I must introduce you to the idea that this immense disk of gas and stars [i.e., the galaxy] is in motion, that it is turning round in space like a great wheel. How then do the stars move? The main motion of a star consists of an orbit that is roughly a circle with its center at the center of the galaxy. The sun and the planets move together as a group around such an orbit. The speed of this motion is nearly 1,000,000 miles an hour. But in spite of this seemingly tremendous speed it nevertheless takes the sun and its retinue of planets about 200,000,000 years to make a round trip of the galaxy. At this stage I should like you to reflect on how many ways you are now moving through space. You have a speed of about 1,000 miles an hour round the polar axis of the earth. You are rushing with the earth at about 70,000 miles an hour along its pathway round the sun. There are also some slight wobbles due to the gravitational attraction of the moon and the other planets. On top of all this you have the huge speed of nearly 1,000,000 miles an hour due to your motion around the galaxy.

There is an obvious analogy between the motion of the earth around the sun, and the motion of the sun and planets around the galaxy. But the analogy must not be pushed too far. In the solar system the sun contains most of the material, and it lies at the center, whereas there is no specially large blob at the center of the galaxy. Also in the solar system the planets move in nearly circular orbits of very different sizes, whereas in the galaxy many stars have orbits of nearly the same size. . . .

Astronomers are generally agreed that the galaxy started its life as a rotating flat disk of gas with no stars in it. The gas was probably distributed much as it is now. That is to say, the diameter of the disk was about 60,000 light years, but its thickness was only about ten light years. A fairly good idea of the rotation of the disk can be got by thinking of the rotation of a thin wheel. Once again this analogy must not be followed too strictly. There would everywhere be small perturbations in the detailed motions of the various bits of gas, especially near the edge of the disk. To assume a complete absence of such perturbations would be rather like supposing that the flow of water in a whirlpool is entirely smooth, being devoid of ripples and small eddies.

How does a rotating disk of very diffuse gas give birth to compact stars? Such a disk would be what mathematicians call gravitationally unstable. That is to say, the attractive force of gravitation would exaggerate any irregularities that were present in it at the beginning. From this it can be deduced that the gas would have to break up into a large number of separate irregular clouds. This prediction, first made by Jeans, has been confirmed by observation, which shows that the interstellar gas is indeed composed of clouds. The distance across an individual cloud usually lies between ten and a hundred light years. That is to say, the clouds are small compared with the diameter of the disk, but not with its thickness. Once clouds have condensed like this, gravitation again exaggerates all the small initial irregularities that they happen to contain. So further condensation would take place in each cloud. At this point it is only necessary to say "and so on," for by repeating the condensation process a sufficient number of times, we must eventually arrive at the particularly dense sort of condensation that we call a star. To sum up the stages—first a whirling disk of gas, then eddies, clouds, condensations, and finally stars.

Granted, then, that gravitation must lead to the condensation of stars within the rotating galactic disk of gas, let us consider the simplest case of this happening. This is when a star is formed out of a roughly spherical blob of gas. On account of the very diffuse nature of the gas, it is clear that such a blob has to be enormously compressed before a star can be formed out of it. In fact, the blob has to condense to about a millionth of its original diameter. So compared with the gas clouds a star is a body of very small dimensions, and this explains why direct collisions or even close approaches between stars are events of extreme rarity, whereas collisions between the gas clouds themselves are quite common. In spite of the large numbers of the stars there is plenty of room for them to move about without seriously interfering with each other.

Why does stellar condensation ever stop contracting? Perhaps I had better clear up this question before we go any further. As a condensation shrinks, its internal temperature rises, and when this becomes sufficiently high, energy begins to be generated in the interior. This is due to hydrogen being converted into helium by nuclear transmutations. . . . A stage is eventually reached when the energy so generated is adequate to balance the radiation escaping from the surface of the star. Contraction then ceases and the body becomes a normal star like the sun.

SOURCE: Fred Hoyle, *The Nature of the Universe*, Oxford: Basil Blackwell, 1950.

Part Six

BEGINNINGS AND ENDS

68

Incomprehensible Magnitude, Unimaginable Darkness

Werner Gitt

Werner Gitt (b. 1937) is head of data processing at the Physikalisch-Technische Bundesanstalt in Braunschweig, Germany. From his perspective as a specialist in information theory, he has written numerous popular books on religion and science. In the two excerpts included here, he offers simple models presuming minimal mathematical sophistication to convey, as much as one can in earthly terms, a sense of cosmic proportions, in particular of distance, density, and darkness.

The immensity of distances within the observable universe makes it hard for us to form any accurate conception of cosmic proportions. Even distances to the nearest stars defy meaningful measurement by any earthly scale. But for those who enjoy the company of numbers, here are four models that convey some impression of our cosmic neighborhood.

MODEL #1

Let's begin by adopting a model in which everything is miniaturized on a scale of 1:100,000,000,000 (=1:100 billion). Given such a drastic reduction, one centimeter in the model corresponds to one million kilometers in reality. Accordingly, the 1.392 million-kilometer diameter of the sun shrinks to 1.4 centimeters, about the size of a cherry. A small grain of sand—the earth—revolves around the cherry at a distance of a meter and a half. Almost 8 meters from the sun is Jupiter, 1.4 millimeters in diameter; and 59 meters from the sun is Pluto, 0.05 millimeters in size. If we imagine the sun-cherry located in New York City, then the star that is our closest neighbor, Alpha Centauri, is 410 kilometers away in Rochester. We would find the next closest star (Barnard's Star) 560 kilometers away in Toronto. But the Andromeda galaxy, which in reality is 2.3 million light-years away, now shatters our attempt to construe the universe on an earthly scale, for *in our model* it would be one and a half times as far away as *in reality* the earth is from the sun. The farthest perceivable distance in our universe— from here to Quasar Q1208011, 12.4 billion light years away—produces in our model a distance corresponding to 7800 times the actual distance between the earth and the sun. And so our attempt to model actual distances itself collapses under the weight of the incomprehensible magnitude of the universe.

MODEL #2

Another way of indicating the immensity of cosmic expanses is to imagine future space flights. Unmanned space flights to the planets of our solar system are already feasible. Voyager I, launched on September 5, 1977, and Voyager II, a few weeks earlier on August 20, 1977, were projected to traverse the solar system and then to leave it forever. Voyager II reached Jupiter on July 9, 1979; Saturn on November 12, 1981; and Uranus on January 24, 1986. On August 25, 1989, Voyager's cameras, aimed at Neptune, discovered two new Neptunian moons. But news of the discovery was delayed, because, even at the speed of light, the radio signal took four hours and six minutes to reach earth. If we ourselves board a space ship bound for Proxima Centauri, and travel at 100,000 km/h—twice the speed of Voyager II, but still only one ten-thousandth the speed of light—our journey will last 46,000 years, one way! That is how distant—and how humanly unreachable—is even the nearest star.

MODEL #3

Now let's imagine we shrink a human being to the size of a hydrogen atom. How much room do we need for the whole population of the world? Would a living room be large enough? Or a matchbox? A thimble? A pinhead? The answer is astounding: Even a bacterium about two or three thousandths of a centimeter wide could contain the population of a thousand earths. On this scale the earth itself would have a diameter of one millimeter, the moon only

0.3 millimeters, and the sun 11 centimeters. Likewise on this scale the distance from the earth to the sun would be 12 meters, and Saturn (9 millimeters in diameter) would revolve about the sun at a distance of 111 meters. Proxima Centauri would lie at a distance corresponding to the actual mileage between Chicago and San Francisco. Rigel in the constellation Orion would lie 50,000 kilometers beyond the moon.

MODEL #4

We'll conclude our attempts to imagine the magnitude of stellar space with a fourth thought-experiment. The well-known Andromeda galaxy has a diameter of 150,000 light-years. If we had a postcard-size photograph of this galaxy and stuck a pin through it, the pinhole would correspond in actuality to a gigantic breach with an aperture of 600 light-years. If we wanted to travel across this cosmic pinprick in the comfort of a jet airliner traveling 1000 kilometers per hour, then we had better allow plenty of time for the journey—about 650 million years!

In a subsequent chapter, with Job 26:7 as epigraph—"He spreads out the northern [skies] over empty space; he suspends the earth over nothing" (NIV)—Gitt expounds the emptiness and darkness of the universe.

SPACE IS EMPTY

Certain stars comprise such huge amounts of matter that it boggles the imagination. The mass of each component of Plaskett's star (HD47-129, a binary), is more than 55 times that of our sun. This means 18.4 million times the mass of the earth, which weighs in at 5.98×10^{24} kilograms. If we put our entire Milky Way galaxy on the scales, the readout would indicate a total mass 200 billion times that of the sun. In the face of such numbers, the expression "space is empty" may sound almost ridiculous. But it is only when we consider the masses side by side with the magnitude of space itself that we achieve a true sense of the actual proportions.

Consider first the situation in our Milky Way galaxy. On average, one "sun" is located in a volume of 350 cubic light-years. A cubic light-year is a cube each of whose edges is one light-year in length and whose volume is 8.47×10^{38} cubic kilometers or 6×10^{20} times that of the sun. Such a volume entails a median mass-density in the Milky Way of about 7×10^{-23} grams per cubic centimeter. If we take into account that 70% of all visible matter in the universe consists of hydrogen, the simplest and lightest atom, and that the rest except for a tiny fraction is helium, the next-simplest element, then an appropriate comparison is with a single atom of hydrogen of mass 1.66 x 10^{-24} grams. If we then imagine the entire mass of the Milky Way distributed evenly throughout the volume in which its stars are located, then each cubic centimeter of our galaxy contains the equivalent of only forty hydrogen atoms.

By comparison, air under normal conditions contains roughly 27 x 10^{18} molecules per cubic centimeter. On earth, the most extreme vacuum is achieved under special laboratory conditions, for example with very small volumes and helium cooling, and at a pressure of 10^{-12} pascals. Yet such an extreme vacuum still contains 270 molecules per cubic centimeter. This means that the average density of the Milky Way, with its 200 billion stars spread across space, is ten times "thinner" than the most extreme vacuum achievable on earth. Therefore, if the galaxies of our cosmos are mere "island universes" of matter—but islands of such low mass-density—then how much "emptier" appears interstellar space in the context of the universe as a whole! Estimates of these densities vary from author to author but fall in the realm of 10^{-31} to 10^{-34} grams per cubic centimeter [10^{-30} is now a more common estimate—Ed.]. Accordingly, the universe as a whole is on average several million times "emptier" even than the already "empty" vastness of the galaxies. All of which makes the biblical expression "empty space" (Job 26:7) appear quite up to date.

SPACE IS DARK

It is a commonplace that, except for moonlight, the sky at night is dark. On their journey to the moon, the astronauts saw the earth as a blue pearl against a black background. And yet our night sky is not utterly dark. Its slight illumination has essentially two causes: 1. the "airglow" in the outer layers of earth's atmosphere triggered by X-rays from the sun; and 2. the diffuse illumination originating from the Milky Way. Both of these sources of light are results of our special position in the universe: that is, close to the sun and within the Milky Way. Beyond the neighborhood of the stars and outside of the galaxies, space is dark indeed. If our earth had no atmosphere, then even at noon the sun would stand naked against a coal-black sky. Only on account of the special properties of our atmosphere, scattering light at the high end of the spectrum, does our cloudless daytime sky appear blue. . . .

Moreover, only because we are located *within* a galaxy do we experience a starry sky at night. How different it would be if the earth and our solar system were located somewhere far beyond any galaxy. By night, we would discover not a single star in the sky: we would be enshrouded by the unimaginable darkness of intergalactic space.

SOURCE: Translated from Werner Gitt, *Signale aus dem All: Wozu gibt es Sterne?*, 2nd ed., Bielefeld: CLV, 1995, and revised in consultation with Werner Gitt, *Stars and Their Purpose: Signposts in Space*, trans. Jaap Kies, Bielefeld: CLV, 1996.

69

That All-But-Eternal Crimson Twilight

Arthur C. Clarke

Arthur C. Clarke (b.1917) joined the fledgling British Interplanetary Society in 1935, spent World War II as a Royal Air Force radar instructor, and in 1945 wrote the first specific proposal for communications satellites—then bordering on science fiction, now an everyday part of our lives. (Clarke describes the "prehistory of comsats" in an article subtitled "How I Lost a Billion Dollars in My Spare Time"!) Clarke went on to become one of the most famous writers of popular science and science fiction in the last half of the twentieth century, and is probably best known for his collaboration with Stanley Kubrick on the movie 2001: A Space Odyssey, based on one of Clarke's short stories. The following sketch of the last stages of the sun's "biography" illustrates Clarke's characteristic fusion of capacious imagination, acute technical detail, and narrative vigor.

Stars, like individuals, age and change. As we look out into space, we see around us stars at all stages of evolution. There are faint blood-red dwarfs so cool that their surface temperature is a mere 4,000 degrees Fahrenheit; there are searing ghosts blazing at 100,000 degrees, and almost too hot to be seen, for the greater part of their radiation is in the invisible ultraviolet. Obviously, the "daylight" produced by any star depends upon its temperature; today (and for ages past, as for ages to come) our sun is about 10,000 degrees F, and this means that most of its light is concentrated in the yellow band of the spectrum, falling slowly in intensity toward both the longer and the shorter waves.

The yellow "hump" will shift as the sun evolves, and the light of day will change accordingly. It is natural to assume that as the sun grows older, and uses up its hydrogen fuel—which it is now doing at the spanking rate of half a billion tons *a second*—it will become steadily colder and redder.

But the evolution of a star is a highly complex matter, involving chains of interlocking nuclear reactions. According to one theory, the sun is still growing hotter, and will continue to do so for several billion years. Probably life will be able to adapt itself to these changes, unless they occur catastrophically, as would be the case if the sun exploded into a nova. In any event, whatever the vicissitudes of the next five or ten billion years, at long last the sun will settle down to the white dwarf stage.

It will be a tiny thing, not much bigger than the earth, and, therefore, too small to show a disc to the naked eye. At first, it will be hotter than it is today, but because of its minute size it will radiate very little heat to its surviving planets. The daylight of that distant age will be as cold as moonlight, but much bluer, and the temperature of earth will have fallen to 300 degrees below zero. If you think of mercury lamps on a freezing winter night, you have a faint mental picture of high noon in the year 7,000,000,000 A.D.

Yet that does not mean that life—even life as we know it today—will be impossible in the solar system; it will simply have to move in toward the shrunken sun. The construction of artificial planets would be child's play to the intelligences we can expect at this date; indeed, it will be child's play to us in a few hundred years' time.

Around the year 10,000,000,000 the dwarf sun will have cooled back to its present temperature, and hence to the yellow color that we know today. From a body that was sufficiently close to it—say only a million miles away—it would look exactly like our present sun, and would give just as much heat. There would be no way of telling, by eye alone, that it was actually a hundred times smaller, and a hundred times closer.

So matters may continue for another five billion years; but at last the inevitable will happen. Very slowly, the sun will begin to cool, dropping from yellow down to red. Perhaps by the year 15,000,000,000 it will become a red dwarf, with a surface temperature of a mere 4,000 degrees. It will be nearing the end of the evolutionary track, but reports of its death will be greatly exaggerated. For now comes one of the most remarkable, and certainly least appreciated, results of modern astrophysical theories.

When the sun shrinks to a dull red dwarf, it will not be dying. It will just be starting to live—*and everything that has gone before will be merely a fleeting prelude to its real history.*

For a red dwarf, because it is so small and so cool, loses energy at such an incredibly slow rate that it can stay in business for *thousands* of times longer than a normal-sized white or yellow star. We must no longer talk in billions, but of trillions of years if we are to measure its life span. Such figures are, of course, inconceivable (for that matter, who can think of a thousand years?). But we can nevertheless put them into their right perspective if we relate the life of a star to the life of a man.

On this scale, the sun is but a week old. Its flaming youth will continue for another month; then it will settle down to a sedate adult existence which may last at least eighty years.

Life has existed on this planet for two or three days of the week that has passed; the whole of human history lies within the last second, and there are eighty years to come.

In the wonderful closing pages of *The Time Machine*, the young H. G. Wells described the world of the far future, with a blood-red sun hanging over a freezing sea. It is a somber picture that chills the blood, but our reaction to it is wholly irrelevant and misleading. For we are creatures of the dawn, with eyes and senses adapted to the hot light of today's primeval sun. Though we should miss beyond measure the blues and greens and violets which are the fading afterglow of creation, they are all doomed to pass with the brief billion-year infancy of the stars.

But the eyes that will look upon that all-but-eternal crimson twilight will respond to the colors that we cannot see, because evolution will have moved their sensitivity away from the yellow, somewhere out beyond the visible red. The world of rainbow-hued heat they see will be as rich and colorful as ours and as beautiful; for a melody is not lost if it is merely transposed an octave down into the bass.

So now we know that Shelley, who was right in so many things, was wrong when he wrote:

> Life, like a dome of many-colored glass,
> Stains the white radiance of eternity.

For the radiance of eternity is not white: it is infrared.

SOURCE: Arthur C. Clarke, "The Light of Common Day," in *Voices from the Sky: Previews of the Coming Space Age*, New York: Harper & Row, 1965.

70

The Cosmic Oasis

Hans Blumenberg

Hans Blumenberg (1920–1996) was a German philosopher and intellectual historian whose work is marked by what one critic calls "a plenitude of insightful ideas." He ranks with Thomas Kuhn as a preeminent historiographer of Copernicanism; and in the tradition of Kant and Lambert he treats the Copernican revolution as part—and as paradigmatic—of an ongoing process in the history of humankind's consciousness of itself and the world. Toward the end of his mammoth study The Genesis of the Copernican World *(1975), Blumenberg meditates on the significance of "reflexive telescopics" both for the early development of Copernicanism and for the lunar missions of the 1960s and 1970s.*

Copernicus had deduced from the shape of the earth's shadow in eclipses of the moon that the earth is precisely spherical. Thus, without auxiliary optical means, he ushered in the practice of reflexive vision: In a star (in the classical sense) he found confirmation that the earth too is a star. When Galileo aimed his telescope at the moon, he saw a duplicate of the earth: forests, seas, continents, and islands. But at the same time he deduced from the secondary moonlight that the earth "shone" like a star. . . . The earth had become a star, with the consequence that stars now could only be multiple earths. . . .

Exerting himself to look out into space, man did not descry something entirely different and alien; rather, what was held out to him was a cosmic mirror of his own world, of its history and potential. Whether this had only been due to the anthropomorphic narrowness of language and the feebleness of description was something that was bound to emerge, at the latest, with the first steps in astronautics. It was just as much to be expected as it was to be feared that the first people on the moon would describe the landscape, which had never been seen, with the means with which they were familiar. . . . Thus, in fact, in February 1969, the commander of *Apollo 8*, Frank Borman, told the Royal Society in London that the closest thing to a likeness of the surface of the moon would be the Mojave Desert in California.

Blumenberg goes on to discuss the very terrestrial trappings of the first moon landing, including a "fluttering" American flag and transmitted comments about the lunar "weather" on the day of the landing. The very familiarity of the descriptions raised suspicions among some, at the time, that the whole lunar expedition was merely a hoax.

The first footprints in the dust of the moon: A deceiving demon, and a very small caliber one in comparison to Descartes', could easily have produced them for the world as a theatrical illusion. During this decade of astronautics only one single picture could not have been invented, but simply went beyond anything the imagination could have anticipated: the picture of the earth from space. If one tries to relate the centuries of imaginative effort and cosmic curiosity to the event, then the both unexpected and heart-stopping peripety of the gigantic departure from the earth was this one thing, that in the sky above the moon one sees the earth. Kepler had described it in advance, but in this case knowledge was not the important thing. In August 1966, *Lunar Orbiter II* transmitted, from its orbit of the moon, the first picture of earth shining over a lunar landscape. . . . Only mobility and color in the transmitted picture allowed one to grasp the uniqueness of the moment of cosmic reflection that enabled man to experience—above the lifeless desert of something that had once had the unattainable quality of a star—the seemingly living star, the earth. This roundabout view of the earth exonerates Cardinal Francesco Barberini, the nephew of Urban VIII, who (as Castelli writes to Galileo on 6 February 1630) had blamed Copernicanism for degrading the earth by making it a star. . . .

[This] backward view of the earth . . . brought to an end the Copernican trauma of the earth's having the status of a mere point—of the annihilation of its importance by the enormity of the universe. Something that we do not yet fully understand has run its course: The successive increases in the disproportion between the earth and the universe, between man and totality, have lost their significance—without its having been necessary to retract the theoretical effort. The astronautical success was a disproof, *in extremis*, of ancient expectations, and it also destroyed the Enlightenment's myth of reason's being compelled to compare itself with the cosmos. One can also put it this way: Equivalence is established between the microscopic and the telescopic sides of reality—absence of difference, in a sense that no longer has any tinge of Pascal's abysses of the infinities.

A decade of intensive attention to astronautics has produced a surprise that is, in an insidious way, pre-Copernican. The earth has turned out to be a cosmic exception.

Blumenberg summarizes the astronautical decade from roughly the mid-1960s to the mid-1970s, in short, by suggesting that the main thing that we humans found in space was the earth. The accumulation of concrete evidence concerning what lies beyond (be it on the moon or on other planets within the solar system) shattered "the (open or secret) assumption that mankind still has some option other than the earth. What presented itself to human view was nothing but scarred and cratered worlds or stifling hot hells with no indications of potentialities for life." Moreover, while admitting that there can be no "proof of the uninhabitability and uninhabitedness of distant planets," Blumenberg points to the remoteness of the possibility that our race could ever gain any knowledge of life or "reason" elsewhere in the universe.

The billions of solar systems in the universe may imply the probability of living creatures here or there, but the order of magnitude of the distances between them destroys, at the same time, what for metaphysical speculation could still be called the "meaning" of this state of affairs. The slowness of the speed of light makes the final decision as to what the superabundance of suns and the frequency of planets, the emptiness of space and the probability of living creatures add up to for reason. Contrary to its title, relativistic physics set the absolute limit against which all technical finesse becomes powerless and in view of which even our Milky Way already far exceeds the order of magnitude that would allow communicative "simultaneity." This reflection concludes, as the "dreadful cynicism" of nature, what had begun as reason's self-consolation. The insurmountable wall constituted by the speed of light for the first time denies man an existence that has cosmic importance.

Part of the euphoria of the astronautic departure and race is the metaphor of the "mothership earth"—that is, of the mere foothold, in the universe, for centrifugal activities in space, activities that now brace themselves only episodically, and with one foot, on the ground from which they started. Only half of this is a metaphor of intimacy and security; the other half is one of mobility and transiency. The centrifugal impetus of astronautics is like a remnant of the special value assigned to the stellar reality by metaphysics, and of its corresponding degradation of the earth as the dregs of the universe. When gravity seemed to be victorious, it became synonymous with a burden. A sufficient reason why the earth is not the mothership of astronautics is that it is the solidity of its ground to which the spaceships so speedily return. The reflexiveness of Copernican vision is repeated in the movements by which seeing earth was to be followed by walking on it. It is more than a triviality that the experience of returning to the earth could not have been had except by leaving it. The cosmic oasis on which man lives—this miracle of an exception, our own blue planet in the midst of the disappointing celestial desert—is no longer "also a star," but rather the only one that seems to deserve this name.

SOURCE: Hans Blumenberg, *Die Genesis der kopernikanischen Welt* (1975); *The Genesis of the Copernican World*, trans. Robert M. Wallace, Cambridge, Mass.: MIT Press, 1987.

71

The Very Womb of Life

James Lovelock

*James Lovelock (b. 1919) has mounted perhaps the most famous effort in re-
cent decades to view the earth not just as a container for organisms but as an
organism itself. This "Gaia hypothesis" (named, at the suggestion of Lovelock's novelist friend William Golding, after the ancient Greek goddess of the
earth) can be viewed as a fusion of both scientific thinking and mythmaking:
much of it based on warranted fact and observation, but much also based on
anthropomorphic thinking and grand leaps of inference. The beginning of
Lovelock's "genesis account" combines these features of his thought in a way
that is conspicuously cosmological, including a faint echo of earlier views
(see Chapters 37, 38, 46) of the order of the world as akin to that of a watch.*

Little is known about the origin of life on our planet and still less about the
course of its early evolution. But if we review what we know concerning the
earth's beginnings in the context of the universe from which it was formed,
we can at least make intelligent guesses about the environment in which life,
and potentially Gaia, began, and set about ensuring their mutual survival.

We know, from observations of events in our own galaxy, that the stellar
universe resembles a living population, in which at any time may be found
people of all ages from infants to centenarians. As old stars, like old soldiers,
fade away, while others expire more spectacularly in an explosive blaze of
glory, fresh incandescent globes with their satellite moths are taking shape.
When we examine spectroscopically the interstellar dust and gas clouds from
which new suns and planets condense, we find that these contain an abundance of the simple and compound molecules from which the chemical build-

ing blocks of life can be assembled. Indeed, the universe appears to be littered with life's chemicals. Nearly every week there is news from the astronomical front of yet another complex organic substance found far away in space. It seems almost as if our galaxy were a giant warehouse containing the spare parts needed for life.

If we can imagine a planet made of nothing but the component parts of watches, we may reasonably assume that in the fullness of time—perhaps 1,000 million years—gravitational forces and the restless motion of the wind would assemble at least one working watch. Life on earth probably started in a similar manner. The countless number and variety of random encounters between individual molecular components of life may have eventually resulted in a chance association of parts which together could perform a life-like task, such as gathering sunlight and using its energy to contrive some further action which would otherwise have been impossible or forbidden by the laws of physics. (The ancient Greek myth of Prometheus stealing fire from heaven and the biblical story of Adam and Eve tasting the forbidden fruit may have far deeper roots in our ancestral history than we realize.) Later, as more of these primitive assembly-forms appeared, some successfully combined and from their union more complex assemblies emerged with new properties and powers, and united in their turn, the product of fruitful associations being always a more potent assembly of working parts, until eventually there came into being a complex entity with the properties of life itself: the first micro-organism and one capable of using sunlight and the molecules of the environment to produce its own duplicate.

The odds against such a sequence of encounters leading to the first living entity are enormous. On the other hand, the number of random encounters between the component molecules of the earth's primeval substance must have been incalculable. Life was thus an almost utterly improbable event with almost infinite opportunities of happening. So it did. . . .

It seems almost certain that close in time and space to the origin of our solar system, there was a supernova event. A supernova is the explosion of a large star. Astronomers speculate that this fate may overtake a star in the following manner: a star burns, mostly by fusion of its hydrogen and, later, helium atoms, the ashes of its fire in the form of other heavier elements such as silicon and iron accumulate at the center. If this core of dead elements, no longer generating heat and pressure, should much exceed the mass of our own sun, the inexorable force of its own weight will be enough to cause its collapse in a matter of seconds to a body no larger than a few thousand cubic miles in volume, although still as heavy as a star. The birth of this extraordinary object, a neutron star, is a catastrophe of cosmic dimensions. Although the details of this and other similar catastrophic processes are still obscure, it is obvious that we have here, in the death throes of a large star, all the ingredients for a vast nuclear explosion. The stupendous amount of light, heat, and hard radiation produced by a supernova event equals at its peak the total output of all the other stars in the galaxy.

Explosions are seldom one hundred percent efficient. When a star ends as a supernova, the nuclear explosive material, which includes uranium and plutonium together with large amounts of iron and other burnt-out elements, is disturbed around and scattered in space just as in the dust cloud from a hydrogen bomb. Even today, aeons later, there is still enough of the unstable explosive material remaining in the earth's crust to enable the reconstitution on a minute scale of the original event.

Binary, or double, star systems are quite common in our galaxy, and it may be that at one time our sun, that quiet and well-behaved body, had a large companion which rapidly consumed its store of hydrogen and ended as a supernova. Or it may be that the debris of a nearby supernova explosion mingled with the swirl of interstellar dust and gases from which the sun and its planets were condensing. In either case, our solar system must have been formed in close conjunction with a supernova event. There is no other credible explanation of the great quantity of exploding atoms still present on the earth. The most primitive and old-fashioned Geiger counter will indicate that we stand on fall-out from a vast nuclear explosion. Within our bodies, no less than three million atoms rendered unstable in that event still erupt every minute, releasing a tiny fraction of the energy stored from that fierce fire of long ago. . . .

Thus life probably began under conditions of radioactivity far more intense than those which trouble the minds of certain present-day environmentalists. Moreover, there was neither free oxygen nor ozone in the air, so that the surface of the earth would have been exposed to the fierce unfiltered ultra-violet radiation of the sun. The hazards of nuclear and of ultra-violet radiation are much in mind these days and some fear that they may destroy all life on earth. Yet the very womb of life was flooded by the light of these fierce energies.

SOURCE: James E. Lovelock, *Gaia: A New Look at Life on Earth*, Oxford: Oxford UP, 1979.

72

The Urge to Trace the History of the Universe

Steven Weinberg

Besides Stephen Hawking, probably no one in the last quarter of the twentieth century did more to lay the groundwork for a public awareness of Big Bang cosmology than Steven Weinberg (b. 1933), much decorated physicist, Nobel-prize-winner, and author of The First Three Minutes. *Also like Hawking, Weinberg has been most frequently quoted for making claims with patent philosophical and theological implications. He both begins and ends* The First Three Minutes, *moreover, by employing literary categories. The first of these is* narrative, *and he begins his own genesis story (published in 1977) with a gently critical examination of another—one whose lack of completeness and coherence provides a foil for the account to follow.*

The origin of the universe is explained in the *Younger Edda*, a collection of Norse myths compiled around 1220 by the Icelandic magnate Snorri Sturleson. In the beginning, says the *Edda*, there was nothing at all. "Earth was not found, nor Heaven above, a Yawning-gap there was, but grass nowhere." To the north and south of nothing lay regions of frost and fire, Niflheim and Muspelheim. The heat from Muspelheim melted some of the frost from Niflheim, and from the liquid drops there grew a giant, Ymer. What did Ymer eat? It seems there was also a cow, Audhumla. And what did *she* eat? Well, there was also some salt. And so on.

I must not offend religious sensibilities, even Viking religious sensibilities, but I think it is fair to say that this is not a very satisfying picture of the origin of the universe. Even leaving aside all objections to hearsay evidence, the story raises as many problems as it answers, and each answer requires a new complication in the initial conditions.

We are not able merely to smile at the *Edda*, and forswear all cosmological speculation—the urge to trace the history of the universe back to its beginnings is irresistible. From the start of modern science in the sixteenth and seventeenth centuries, physicists and astronomers have returned again and again to the problem of the origin of the universe.

However, an aura of the disreputable always surrounded such research. I remember that during the time that I was a student and then began my own research (on other problems) in the 1950s, the study of the early universe was widely regarded as not the sort of thing to which a respectable scientist would devote his time. Nor was this judgment unreasonable. Throughout most of the history of modern physics and astronomy, there simply has not existed an adequate observational and theoretical foundation on which to build a history of the early universe.

Now, in just the past decade, all this has changed. A theory of the early universe has become so widely accepted that astronomers often call it "the standard model." It is more or less the same as what is sometimes called the "big bang" theory, but supplemented with a much more specific recipe for the contents of the universe.

Weinberg's account is now familiar, having been summarized and redigested many times. However, his own overview is worth excerpting both because it is so authoritative and compact, and because it makes clear the status of the model as a model—highly useful and influential, yet still plagued by more than a hint of mystery about the very *beginning (though perhaps a lot less so than the story about the giant and the cow!).*

In the beginning there was an explosion. Not an explosion like those familiar on earth, starting from a definite center and spreading out to engulf more and more of the circumambient air, but an explosion which occurred simultaneously everywhere, filling all space from the beginning, with every particle of matter rushing apart from every other particle. "All space" in this context may mean either all of an infinite universe, or all of a finite universe which curves back on itself like the surface of a sphere. Neither possibility is easy to comprehend, but this will not get in our way; it matters hardly at all in the early universe whether space is finite or infinite.

At about one-hundredth of a second, the earliest time about which we can speak with any confidence, the temperature of the universe was about a hundred thousand million (10^{11}) degrees Centigrade. This is much hotter than in the center of even the hottest star, so hot, in fact, that none of the components of ordinary matter, molecules, or atoms, or even the nuclei of atoms, could have held together. Instead, the matter rushing apart in this explosion consisted of various types of the so-called elementary particles, which are the subject of modern high-energy nuclear physics.

. . . One type of particle that was present in large numbers is the electron, the negatively charged particle that flows through wires in electric currents and makes up the outer parts of all atoms and molecules in the present universe. Another type of particle that was abundant at early times is the positron, a positively charged particle with precisely the same mass as the electron. In the present universe positrons are found only in high-energy laboratories, in some kinds of radioactivity, and in violent astronomical phenomena like cosmic rays and supernovas, but in the early universe the number of positrons was almost exactly equal to the number of electrons. In addition to electrons and positrons, there were roughly similar numbers of various kinds of neutrinos, ghostly particles with no mass or electric charge whatever. Finally, the universe was filled with light. This does not have to be treated separately from the particles—the quantum theory tells us that light consists of particles of zero mass and zero electrical charge known as photons. . . . Every photon carries a definite amount of energy and momentum depending on the wavelength of the light. To describe the light that filled the early universe, we can say that the number and the average energy of the photons was about the same as for electrons or positrons or neutrinos.

These particles—electrons, positrons, neutrinos, photons—were continually being created out of pure energy, and then after short lives being annihilated again. Their number therefore was not preordained, but fixed instead by a balance between processes of creation and annihilation. From this balance we can infer that the density of this cosmic soup at a temperature of a hundred thousand million degrees was about four thousand million (4×10^9) times that of water. There was also a small contamination of heavier particles, protons and neutrons, which in the present world form the constituents of atomic nuclei. (Protons are positively charged; neutrons are slightly heavier and electrically neutral.) The proportions were roughly one proton and one neutron for every thousand million electrons or positrons or neutrinos or photons. This number—a thousand million photons per nuclear particle—is the crucial quantity that had to be taken from observation in order to work out the standard model of the universe. The discovery of the cosmic radiation background . . . was in effect a measurement of this number.

As the explosion continued the temperature dropped, reaching thirty thousand million (3×10^{10}) degrees Centigrade after about one-tenth of a second; ten thousand million degrees after about one second; and three thousand million degrees after about fourteen seconds. This was cool enough so that the electrons and positrons began to annihilate faster than they could be recreated out of the photons and neutrinos. The energy released in this annihilation of matter temporarily slowed the rate at which the universe cooled, but the temperature continued to drop, finally reaching one thousand million degrees at the end of the first three minutes. It was then cool enough for the protons and neutrons to begin to form into complex nuclei, starting with the nucleus of heavy hydrogen (or deuterium), which consists of one proton and one neutron. The density was still high enough (a little less than that of water) so that these light nuclei were able rapidly to assemble themselves into the most stable light nucleus, that of helium, consisting of two protons and two neutrons.

At the end of the first three minutes the contents of the universe were mostly in the form of light, neutrinos, and antineutrinos. There was still a small amount of nuclear material, now consisting of about 73 percent hydrogen and 27 percent helium, and an equally small number of electrons left over from the era of electron-positron annihilation. This matter continued to rush apart, becoming steadily cooler and less dense. Much later, after a few hundred thousand years, it would become cool enough for electrons to join with nuclei to form atoms of hydrogen and helium. The resulting gas would begin under the influence of gravitation to form clumps, which would ultimately condense to form the galaxies and stars of the present universe. However, the ingredients with which the stars would begin their life would be just those prepared in the first three minutes.

The standard model sketched above is not the most satisfying theory imaginable of the origin of the universe. Just as in the *Younger Edda*, there is an embarrassing vagueness about the very beginning, the first hundredth of a second or so. Also, there is the unwelcome necessity of fixing initial conditions, especially the initial thousand-million-to-one ratio of photons to nuclear particles. We would prefer a greater sense of logical inevitability in the theory.

This "we would prefer" of Weinberg's should not be read as a sign of lack of respect for the standard model. However, it does tacitly acknowledge the relationship between theories on the one hand and human longing for coherence and completeness on the other. Weinberg is perhaps most frequently quoted with respect to the frustration of this longing. The claim that "the more the universe seems comprehensible, the more it also seems pointless" is undoubtedly his most famous single sentence (quoted again in the following chapter). But we should notice that it ends his penultimate *paragraph, not the book overall. The final paragraph of* The First Three Minutes *finds Weinberg again seeking meaning—and consolation—and once more invoking literary categories as part of that process, though this time his appeal is not to narrative but to drama.*

If there is no solace in the fruits of our research, there is at least some consolation in the research itself. Men and women are not content to comfort themselves with tales of gods and giants, or to confine their thoughts to the daily affairs of life; they also build telescopes and satellites and accelerators, and sit at their desks for endless hours working out the meaning of the data they gather. The effort to understand the universe is one of the very few things that lifts human life a little above the level of farce, and gives it some of the grace of tragedy.

SOURCE: Steven Weinberg, *The First Three Minutes: A Modern View of the Origin of the Universe*, New York: Basic Books, 1977.

73

To Transform the Universe
on a Cosmological Scale

John Barrow and Frank Tipler

Although Brandon Carter was the first person to coin the term "anthropic principle," it is hard to imagine anyone providing a more comprehensive, or more audacious, discussion of it than that provided by John Barrow (b.1952) and Frank Tipler (b. 1947) in their seven-hundred-page encyclopedic study The Anthropic Cosmological Principle *(1986). It is also hard to imagine any account of recent cosmology that neglects their work, despite the impossibility of summing it up in brief excerpts. It is both exciting and amazing to see how an apparent truism—that any cosmological theory must take seriously those conditions necessary for the presence of theorists in the actual cosmos being theorized—can generate such powerful implications for enquiries both scientific and philosophical. A large proportion of Barrow and Tipler's work is highly technical, but its value for the general reader is in the explicit connections it forges with discussions across history about the meaningfulness of the world, and with the kinds of questions any human being asks when faced with doctrines of physics such as the second law of thermodynamics. They begin with the issue of where minds or Mind fits into our picture of the physical universe.*

The central problem of science and epistemology is deciding which postulates to take as fundamental. The perennial solution of the great idealistic philosophers has been to regard Mind as logically prior, and even materialistic philosophers consider the innate properties of matter to be such as to al-low—or even require—the existence of intelligence to contemplate it; that is, these properties are necessary or sufficient for life. Thus the existence of Mind is taken as one of the basic postulates of a philosophical system. Physi-cists, on the other hand, are loath to admit any consideration of Mind into their theories. Even quantum mechanics, which supposedly brought the ob-server into physics, makes no use of intellectual properties; a photographic plate would serve equally well as an "observer." But, during the past fifteen years there has grown up amongst cosmologists an interest in a collection of ideas, known as the Anthropic Cosmological Principle, which offer a means of relating mind and observership directly to the phenomena traditionally within the encompass of physical science.

The expulsion of Man from his self-assumed position at the center of Na-ture owes much to the Copernican principle that we do not occupy a privi-leged position in the universe. This Copernican assumption would be regarded as axiomatic at the outset of most scientific investigations. How-ever, like most generalizations it must be used with care. Although we do not regard our position in the universe to be central or special in every way, this does not mean that it cannot be special in *any* way. This possibility led Bran-don Carter to limit the Copernican dogma by an "Anthropic Principle" to the effect that "our location in the universe is necessarily privileged to the ex-tent of being compatible with our existence as observers." The basic features of the universe, including such properties as its shape, size, age and laws of change, must be *observed* to be of a type that allows the evolution of ob-servers, for if intelligent life did not evolve in an otherwise possible universe, it is obvious that no one would be asking the reason for the observed shape, size, age and so forth of the universe. At first sight such an observation might appear true but trivial. However, it has far-reaching implications for physics. It is a restatement of the fact that any observed properties of the universe that may initially appear astonishingly improbable can only be seen in their true perspective after we have accounted for the fact that certain properties of the universe are necessary prerequisites for the evolution and existence of any observers at all. The measured values of many cosmological and physical quantities that define our universe are circumscribed by the necessity that we observe from a site where conditions are appropriate for the occurrence of biological evolution and at a cosmic epoch exceeding the astrophysical and biological timescales required for the development of life-supporting envi-ronments and biochemistry.

What we have been describing is just a grandiose example of a type of in-trinsic bias that scientists term a "selection effect." For example . . . if a rat-catcher tells you that all rats are more than six inches long because he has never caught any that are shorter, you should check the size of his traps before drawing any far-reaching conclusions about the length of rats. Even though you are most likely to see an elephant in a zoo, that does not mean that all elephants are in zoos, or even that most elephants are in zoos. . . .

The fact that modern astronomical observations reveal the visible universe to be close to fifteen billion light years in extent has provoked many vague generalizations about its structure, significance and ultimate purpose. Many a philosopher has argued against the ultimate importance of life in the universe by pointing out how little life there appears to be compared with the enormity of space and the multitude of distant galaxies. But the Big Bang cosmological picture shows this up as too simplistic a judgement. Hubble's classic discovery that the universe is in a dynamic state of expansion revealed that its size is inextricably bound up with its age. The universe is fifteen billion light years in size because it is fifteen billion years old. Although a universe the size of a single galaxy would contain enough matter to make more than one hundred billion stars the size of our sun, it would have been expanding for less than a single year.

We have learned that the complex phenomenon we call "life" is built upon chemical elements more complex than hydrogen and helium gases. Most biochemists believe that carbon, on which our own organic chemistry is founded, is the only possible basis for the *spontaneous* generation of life. In order to create the building blocks of life—carbon, nitrogen, oxygen and phosphorus—the simple elements of hydrogen and helium which were synthesized in the primordial inferno of the Big Bang must be cooked at a more moderate temperature and for a much longer time than is available in the early universe. The furnaces that are available are the interiors of stars. There, hydrogen and helium are burnt into the heavier life-supporting elements by exothermic nuclear reactions. When stars die, the resulting explosions which we see as supernovae, can disperse these elements through space and they become incorporated into planets and, ultimately, into ourselves. This stellar alchemy takes over ten billion years to complete. Hence, for there to be enough time to construct the constituents of living beings, the universe must be at least ten billion years old and therefore, as a consequence of its expansion, at least ten billion light years in extent. We should not be surprised to observe that the universe is so large. No astronomer could exist in one that was significantly smaller. The Universe needs to be as big as it is in order to evolve just a single carbon-based life form.

We should emphasize that this selection of a particular size for the universe actually does *not* depend on accepting most biochemists' belief that only carbon can form the basis of spontaneously generated life. Even if their belief is false, the fact remains that *we are a carbon-based intelligent life-form which spontaneously evolved on an earthlike planet around a star of G2 spectral type, and any observation we make is necessarily self-selected by this absolutely fundamental fact.* In particular, a life-form which evolved spontaneously in such an environment must necessarily see the universe to be at least several billion years old and hence see it to be at least several billion light years across. This remains true even if non-carbon life-forms abound in the cosmos. Non-carbon life-forms are not necessarily restricted to seeing a minimum size to the universe, but *we* are. Human bodies are measuring instruments whose self-selection properties *must* be taken into account, just as astronomers *must* take into account the self-selection properties of optical telescopes. Such telescopes tell us about radiation in the visible band of the electromagnetic spectrum, but it would be completely illegitimate to conclude from purely optical observations that all of the electromagnetic energy in the universe is in the visible band. Only when one is aware of the self-selection of optical telescopes is it possible to consider the possibility that non-visible radiation exists. Similarly, it is essential to be aware of the self-selection which results from our being *Homo sapiens* when trying to draw conclusions about the nature of the universe. In a sense, the Weak Anthropic Principle may be regarded as the culmination of the Copernican Principle, because the former shows how to separate those features of the universe whose appearance depends on anthropocentric selection, from those features which are genuinely determined by the action of physical laws.

In fact, the Copernican Revolution was initiated by the application of the Weak Anthropic Principle. The outstanding problem of ancient astronomy was explaining the motion of the planets, particularly their retrograde motion. Ptolemy and his followers explained the retrograde motion by invoking an epicycle, the ancient astronomical version of a new physical law. Copernicus showed that the epicycle was unnecessary; the retrograde motion was due to an anthropocentric selection effect: we were observing the planetary motions from the vantage point of the moving earth.

The Weak Anthropic Principle (which Barrow and Tipler abbreviate as "WAP") is their relatively uncontroversial first step towards stronger, and much more controversial, versions of the principle. Put simply, it is the starting point for a teleological cosmology, one that is concerned with the telos, the purpose, the whole point of the universe.

The teleological import of selection effects is well introduced by philosopher Richard Swinburne's "firing squad" analogy:

On a certain occasion the firing squad aim their rifles at the prisoner to be executed. There are twelve expert marksmen in the firing squad, and they fire twelve rounds each. However, on this occasion all 144 shots miss. The prisoner laughs and comments that the event is not something requiring any explanation because if the marksmen had not missed, he would not be here to observe them having done so. But of course the prisoner's comment is absurd; the marksmen all having missed is indeed something requiring explanation; and so too is what goes with it—the prisoner being alive to observe it. And the explanation will be either that it was an accident . . . or that it was planned. (John Leslie, ed., *Physical Cosmology and Philosophy*, p. 165.)

Here is a strong family resemblance to the older, cosmological argument of William Paley for the existence of a "watchmaker" God. It is the "argument from design" with a new twist. And a large part of the twist, for Barrow and Tipler (though not for Swinburne), is that the design requires no divine designer.

In short, in Barrow and Tipler's argument, the WAP leads on to the SAP (Strong Anthropic Principle), which leads to the FAP (Final Anthropic Principle), which are defined as follows:

WAP: "Features of the universe which appear to us astonishingly improbable, *a priori*, can only be judged in their correct perspective when due allowance has been made for the fact that certain properties of the universe are necessary if it is to contain carbonaceous astronomers like ourselves."

SAP: "The universe must have those properties which allow life to develop within it at some stage in its history."

FAP: "Intelligent information-processing must come into existence in the universe, and, once it comes into existence, it will never die out." [One waggish reviewer—Martin Gardner—suggested this version be called the Completely Ridiculous Anthropic Principle.]

Clearly if a grand new scheme of teleology is to be well founded, its founders must mount a defense against contrary evidence pointing to universal dysteleology.

Modern science presents a critical problem for teleological arguments. The very notion of teleology, that there is some goal to which the universe is heading, strongly suggests a steady improvement as this goal is approached. Although progress was not strictly allowed by the Newtonian physics of the day, the defenders of the teleological argument before the nineteenth century generally held this optimistic view. Meliorism even survived Darwin's destruction of traditional teleology. Darwin himself felt that his theory of evolution justified such an optimistic view. As he wrote in the closing pages of the first edition of *On the Origin of Species*:

> As all the living forms of life are the lineal descendants of those which lived long before the Silurian epoch, we may feel certain that the ordinary succession by generation has never once been broken, and that no cataclysm has desolated the whole world. Hence we may look with some confidence to a secure future of equally inappreciable length. And as natural selection works solely by and for the good of each being, all corporeal and mental endowments will tend to progress towards perfection.

Darwin wrote these words in 1859, just slightly after the formulation of the Second Law of Thermodynamics, but before its dysteleological implications became generally known. The great German physicist Hermann von Helmholtz was the first to point out, in an article published in 1854, that the Second Law suggested the universe was using up all its available energy, and thus within a finite time all future changes must cease; the universe and all living things therein must die when the universe reaches this final state of maximum entropy. This is the famous "Heat Death" of the universe. It strongly denies the universe is progressing toward some goal; but rather is using up the store of available energy which existed in the beginning. The universe is actually moving from a higher state to a lower state. The universe, in other words, is not teleological, but *dysteleological*!

. . . This Heat Death concept had a profoundly negative effect on the optimism of the late nineteenth and early twentieth centuries. The popular books on cosmology written in the 1930s by the British astronomers Jeans and Eddington were particularly important in making the general public aware of the Heat Death. The new attitude this produced concerning the relationship between Man and the Cosmos was epitomized in 1903 in a famous passage of Bertrand Russell's:

> . . . the world which science presents for our belief is even more purposeless, more void of meaning, [than a world in which God is malevolent]. Amid such a world, if anywhere, our ideals henceforward must find a home. That man is the product of causes which had no prevision of the end they were achieving; that his origin, his growth, his hopes and fears, his loves and his beliefs, are but the outcome of accidental collocations

of atoms; that no fire, no heroism, no intensity of thought and feeling, can preserve an individual life beyond the grave; that all the labors of the ages, all the devotion, all the inspiration, all the noonday brightness of human genius, are destined to extinction in the vast death of the solar system, and the whole temple of Man's achievement must inevitably be buried beneath the debris of a universe in ruins—all these things, if not quite beyond dispute, are yet so nearly certain that no philosophy which rejects them can hope to stand. Only within the scaffolding of these truths, only on the firm foundation of unyielding despair, can the soul's habitation henceforth be safely built.

The dysteleology of the long-term evolution of the universe did not worry Russell. He suggested it meant we should take a short-term view of life:

I am told that that sort of view is depressing, and people will sometimes tell you that if they believed that, they would not be able to go on living. Do not believe it; it is all nonsense. Nobody really worries much about what is going to happen millions of years hence. . . .

But some people were unable to take a short-term view. For example, by the end of his life, Charles Darwin's own optimism had been severely shaken by the prospect of the Heat Death, which he learned about in the course of the late nineteenth-century debates on the age of the earth. As Darwin recorded in his *Autobiography*:

[consider] . . . the view now held by most physicists, namely, that the sun with all the planets will in time grow too cold for life, unless indeed some great body dashes into the sun and thus gives it fresh life—believing as I do that man in the distant future will be a far more perfect creature than he now is, it is an intolerable thought that he and all other sentient beings are doomed to complete annihilation after such long-continued slow progress.

It is against the background of just such pessimism and dysteleology, and without recourse to any religious solution transcending materialistic physics, that Barrow and Tipler lay out their immanent eschatology. It is a vision in which machines—our machines and our software—inherit not only the earth but the entire cosmos.

When we investigate the relationship between intelligent life and the cosmos, one fact stands out at the present time: there is no evidence whatsoever of intelligent life having any significant effect upon the universe in the large. As we have discussed at length [earlier], the evidence is very strong that intelligent life is restricted to a single planet, which is but one of nine circling a star which itself is only one of about 10^{11} stars in the galaxy and our galaxy is but one of some 10^{12} galaxies in the visible universe. Indeed, one of the seeming implications of science as it has developed over the past few centuries is that mankind is an insignificant accident lost in the immensity of the cosmos. The evolution of the human species was an extremely fortuitous accident, one which is unlikely to have occurred elsewhere in the visible universe.

It has appeared to most philosophers and scientists over the past century that mankind is forever doomed to insignificance. Both our species and all our works would disappear eventually, leaving the universe devoid of mind once more. This world view was perhaps most eloquently stated by Bertrand Russell in the passage we quoted, but the same sentiment has recently been expressed by the Nobel-prize-winning physicist Steven Weinberg in his popular book on cosmology, *The First Three Minutes*:

> It is almost irresistible for humans to believe that we have some special relation to the universe, that human life is not just a more-or-less farcical outcome of a chain of accidents reaching back to the first three minutes [of the universe's existence], but that we were somehow built in from the beginning. . . . It is very hard to realize that [the entire earth] is just a tiny part of an overwhelmingly hostile universe. It is even harder to realize that this present universe has evolved from an unspeakably unfamiliar early condition, and faces a future extinction of endless cold or intolerable heat. The more the universe seems comprehensible, the more it also seems pointless.

These ideas neglect to consider one extremely important possibility: Although mankind—and hence life itself—is at present confined to one insignificant, doomed planet, this confinement may not be perpetual. Bertrand Russell wrote his gloomy lines at the turn of the century, and at that time space travel was viewed as an impossibility by almost all scientists. But we have landed men on the moon. We *know* space travel is possible. We argued [earlier] that even interstellar travel is possible. Thus once space travel begins, there are, in principle, no further physical barriers to prevent *Homo sapiens* (or our descendants) from eventually expanding to colonize a substantial portion, if not all, of the visible cosmos. Once this has occurred, it becomes quite reasonable to speculate that the operations of all these intelligent beings could begin to affect the large scale evolution of the universe. If this is true, it would be in *this* era—in the far future near the Final State of the Universe—that the true significance of life and intelligence would manifest itself. Present-day life would then have cosmic significance because of what future life may someday accomplish. . . .

A species capable of rapid technological innovation has existed in the universe for only about 40,000 years. This species has just begun to take the first, faltering steps to leave its place of origin. In the time to come, it and its descendant species could conceivably change structural features of the universe.

To say that intelligent life has some global cosmological significance is to say that intelligent life will someday begin to transform and continue to transform the universe on a cosmological scale. . . . As our discussion of dysteleology . . . and Weinberg's remarks make abundantly clear, until recently scientists did not believe the physical laws could ever permit intelligent life to act on a cosmological scale. In part this belief is based on the notion that intelligent life means *human life*. Weinberg points out that the ultimate future of the universe involves great cold or great heat, and that human life—the species *Homo sapiens*—cannot survive in either environment. We must agree with him. The ultimate state of the universe appears to involve one of these environments, and thus *Homo sapiens* must eventually become extinct. This is the inevitable fate of any living species. As Darwin expressed it in the concluding pages of the *Origin of Species*:

> Judging from the past, we may safely infer that not one living species will transmit its unaltered likeness to a distant futurity.

But though our species is doomed, our civilization and indeed the values we care about may not be. We emphasized [earlier] that from the behavioral point of view intelligent *machines* can be regarded as people. These machines may be our ultimate heirs, our ultimate descendants, because under certain circumstances they could survive forever the extreme conditions near the Final State. Our civilization may be continued indefinitely by them, and the values of humankind may thus be transmitted to an arbitrarily distant futurity.

Barrow and Tipler's concluding paragraphs concerning what they call the Omega Point are perhaps as logically comprehensive an apocalypse as has ever been written:

Finally, the time is reached when life has encompassed the entire universe and regulated all matter contained therein. Life begins to manipulate the dynamical evolution of the universe as a whole, forcing the horizons to disappear, first in one direction, and then another. The information stored continues to increase. . . .

From our [earlier] discussion . . . we see that if life evolves in all of the many universes in a quantum cosmology, and if life continues to exist in all of these universes, then *all* of these universes, which include *all* possible histories among them, will approach the Omega Point. At the instant the Omega Point is reached, life will have gained control of *all* matter and forces not only in a single universe, but in all universes whose existence is logically possible; life will have spread into *all* spatial regions in all universes which could logically exist, and will have stored an infinite amount of information including *all* bits of knowledge which it is logically possible to know. And this is the end.

SOURCE: John D. Barrow and Frank J. Tipler, *The Anthropic Cosmological Principle*, Oxford: Oxford UP, 1986.

74

The No Boundary Condition

Stephen Hawking

*Stephen Hawking (b. 1942) is the most widely recognized, and instantly rec-
ognizable, icon of cosmology at the end of the twentieth century and the beginning of the twenty-first. He would no doubt be recorded as a great
scientist and thinker even in the absence of electronic media, though it is
these that have captured and purveyed his image. This holder of Isaac Newton's Cambridge professorship, this victim of Lou Gehrig's disease, sitting
crumpled in a wheelchair and speaking by means of an artificial apparatus
about the very nature of space and time—with brilliance, clarity, and a bland
American accent!—somehow captures the glory, the chutzpah, the incon-
gruity, the pathos, and the heroism that together characterize humankind's
attempts to imagine the universe.*

*Beyond the electronic media, Hawking's writing conveys a voice that is ac-
cessible, familiar, and at once profound and playful. He is said to have remarked that each additional equation (beyond the one that he does cite,
E=MC²) would cut sales of* A Brief History of Time *in half, and that each successive mention of God would double them. Indeed, in spite of clichés about
science and theology being non-intersecting circles, parts of Hawking's discussion do have a quite theological flavor. In his famous proposal that the
universe be conceived as possessing no initial singularity—truly no starting
point, no beginning—his physical theory's theological implications, and perhaps "atheological" motivations, can scarcely be ignored.*

In the classical theory of gravity, which is based on real space-time, there are
only two possible ways the universe can behave: either it has existed for an
infinite time, or else it had a beginning at a singularity at some finite time in

the past. In the quantum theory of gravity, on the other hand, a third possibility arises. Because one is using Euclidean space-times, in which the time direction is on the same footing as direction in space, it is possible for space-time to be finite in extent and yet to have no singularities that formed a boundary or edge. Space-time would be like the surface of the earth, only with two more dimensions. The surface of the earth is finite in extent but it doesn't have a boundary or edge: if you sail off into the sunset, you don't fall off the edge or run into a singularity. . . .

If Euclidean space-time stretches back to infinite imaginary time, or else starts at a singularity in imaginary time, we have the same problem as in the classical theory of specifying the initial state of the universe: God may know how the universe began, but we cannot give any particular reason for thinking it began one way or another. On the other hand, the quantum theory of gravity has opened up a new possibility, in which there would be no boundary to space-time and so there would be no need to specify the behavior at the boundary. There would be no singularities at which the laws of science broke down and no edge of space-time at which one would have to appeal to God or some new law to set the boundary conditions of space-time. One could say: "The boundary condition of the universe is that it has no boundary." The universe would be completely self-contained and not affected by anything outside itself. It would neither be created nor destroyed. It would just BE. . . .

I'd like to emphasize that this idea that time and space should be finite without boundary is just a *proposal*: it cannot be deduced from some other principle. Like any other scientific theory, it may initially be put forward for aesthetic or metaphysical reasons, but the real test is whether it makes predictions that agree with observation. This, however, is difficult to determine in the case of quantum gravity, for two reasons. First, . . . we are not yet sure exactly which theory successfully combines general relativity and quantum mechanics, though we know quite a lot about the form such a theory must have. Second, any model that described the whole universe in detail would be much too complicated mathematically for us to be able to calculate exact predictions. One therefore has to make simplifying assumptions and approximations—and even then, the problem of extracting predictions remains a formidable one. . . .

Under the no boundary proposal one learns that the chance of the universe being found to be following most of the possible histories is negligible, but there is a particular family of histories that are much more probable than the others. These histories may be pictured as being like the surface of the earth, with the distance from the North Pole representing imaginary time and the size of a circle of constant distance from the North Pole representing the spatial size of the universe. The universe starts at the North Pole as a single point. As one moves south, the circles of latitude at constant distance from the North Pole get bigger, corresponding to the universe expanding with imaginary time. The universe would reach a maximum size at the equator and would contract with increasing imaginary time to a single point at the South Pole. Even though the universe would have zero size at the North and South poles, these points would not be singularities, any more than the North and South Poles on earth are singular. The laws of science will hold at them, just as they do at the North and South Poles on earth.

The history of the universe in real time, however, would look very different. At about ten or twenty thousand million years ago, it would have a minimum size, which was equal to the maximum radius of the history in imaginary time. At later real times, the universe would expand like the chaotic inflationary model proposed by Linde (but one would not now have to assume that the universe was created somehow in the right sort of state). The universe would expand to a very large size and eventually it would collapse again into what looks like a singularity in real time. Thus, in a sense, we are still all doomed, even if we keep away from black holes. Only if we could picture the universe in terms of imaginary time would there be no singularities.

If the universe really is in such a quantum state, there would be no singularities in the history of the universe in imaginary time. It might seem therefore that my more recent work had completely undone the results of my earlier work on singularities. But . . . the real importance of the singularity theorems was that they showed that the gravitational field must become so strong that quantum gravitational effects could not be ignored. This in turn led to the idea that the universe could be finite in imaginary time but without boundaries or singularities. When one goes back to the real time in which we live, however, there will still appear to be singularities. The poor astronaut who falls into a black hole will still come to a sticky end; only if he lived in imaginary time would he encounter no singularities.

This might suggest that the so-called imaginary time is really the real time, and that what we call real time is just a figment of our imaginations. In real time, the universe has a beginning and an end at singularities that form a boundary to space-time and at which the laws of science break down. But in imaginary time, there are no singularities or boundaries. So maybe what we call imaginary time is really more basic, and what we call real is just an idea that we invent to help us describe what we think the universe is like. But . . . a scientific theory is just a mathematical model we make to describe our observations: it exists only in our minds. So it is meaningless to ask: Which is real, "real" or "imaginary" time? It is simply a matter of which is the more useful description. . . .

The proposed no boundary condition leads to the prediction that it is extremely probable that the present rate of expansion of the universe is almost the same in each direction. This is consistent with the observations of the microwave background radiation, which show that it has almost exactly the same intensity in any direction. If the universe were expanding faster in some directions than in others, the intensity of the radiation in those directions would be reduced by an additional red shift.

Further predictions of the no boundary condition are currently being worked out. A particularly interesting problem is the size of the small departures from uniform density in the early universe that caused the formation first of the galaxies, then of stars, and finally of us. The uncertainty principle implies that the early universe cannot have been completely uniform because there must have been some uncertainties or fluctuations in the positions and velocities of the particles. Using the no boundary condition, we find that the universe must in fact have started off with just the minimum possible nonuniformity allowed by the uncertainty principle. The universe would have then undergone a period of rapid expansion, as in the inflationary models. During this period, the initial nonuniformities would have been amplified until they were big enough to explain the origin of the structures we observe around us. In an expanding universe in which the density of matter varied slightly from place to place, gravity would have caused the denser regions to slow down their expansion and start contracting. This would lead to the formation of galaxies, stars, and eventually even insignificant creatures like ourselves. Thus all the complicated structures that we see in the universe might be explained by the no boundary condition for the universe together with the uncertainty principle of quantum mechanics.

The idea that space and time may form a closed surface without boundary also has profound implications for the role of God in the affairs of the universe. With the success of scientific theories in describing events, most people have come to believe that God allows the universe to evolve according to a set of laws and does not intervene in the universe to break these laws. However, the laws do not tell us what the universe should have looked like when it started—it would still be up to God to wind up the clockwork and choose how to start it off. So long as the universe had a beginning, we could suppose it had a creator. But if the universe is really completely self-contained, having no boundary or edge, it would have neither beginning nor end: it would simply be. What place, then, for a creator?

SOURCE: Stephen W. Hawking, *A Brief History of Time: From the Big Bang to Black Holes*, New York: Bantam, 1988.

75

Prisons of Light

Kitty Ferguson

To judge from her writing and her curriculum vitae, Kitty Ferguson (b. 1941) does things with flair. She was born in Texas, became a professional musician in New York, then in her 40s retired from music and began writing about an- other life-long interest: physics and cosmology. While in Cambridge, Eng- land, she met Stephen Hawking and, with his cooperation, wrote his biography. In a subsequent book, Prisons of Light, *she again takes upon her- self the task of biographer, only this time her subject is the life of black holes.*

Stars spend most of their adult years converting hydrogen to helium, and that's what keeps them alive. For all those millennia there is a balance—a tug-of-war, if you will—between two closely matched opposing teams. On the one hand there is gravity, the force that continues to draw the atoms and particles closer together. Left unopposed, gravity would cause the star to col- lapse in on itself. We who depend on a star for our continued existence are fortunate that gravity does not go unopposed. Paradoxically, gravity brings about nuclear fusion and thus assures that there IS an opposing team. The heat released in the nuclear reactions (as hydrogen becomes helium) creates enough pressure in the gas to balance the gravitational attraction, and the star doesn't collapse. Think of the way the pressure in a balloon holds the walls of the balloon apart. The rubber walls pull toward each other, trying to come together—which they will do quite readily if we let the gas out of the balloon. The pressure in a star doesn't allow it to collapse—which it would do quite readily if the nuclear reactions stopped occurring and gravity were allowed to have its way. The heat released in nuclear reactions also makes up for the heat lost as the star radiates light into space.

Picture this tug-of-war, then. Pressure from the heat released in nuclear reactions is one team, gravity the other. Occasionally gravity appears to be winning. When the accumulation of helium ash snuffs out the central furnace of the star, the core begins to collapse, but the collapse itself soon produces higher temperatures and the star reignites, now fusing helium into carbon atoms and swelling into a red giant. . . . In some stars the snuffing out and reignition occur several times, with stars fusing atoms into heavier and heavier elements.

The competition eventually ends with gravity the victor. When most of the hydrogen atoms have fused into helium atoms and some of those into carbon atoms and, in more massive stars, into heavier elements, the star finally exhausts its usable fuel. Nuclear reactions happen less and less often as the fuel runs out, but the star continues to radiate light into space even when it is no longer able to make up for the heat lost in this radiation. There is less and less heat and pressure to counteract gravitational attraction. After patiently holding its own for millions or billions of years, gravity wins the tug-of-war. The cooling star begins to shrink and collapse in on itself. . . .

Three possible fates now await a collapsing star.

1. It may settle down and spend its old age as a "white dwarf" . . . with a radius of a few thousand miles, not too much smaller than the earth but with a density of hundreds of tons per cubic inch. We observe many white dwarfs in our own part of the Milky Way galaxy. There is one orbiting Sirius (the Dog Star), the brightest star in our night sky. [And our] sun will probably end up as a white dwarf.
2. It may not settle as a white dwarf but instead crunch down until its circumference is approximately a mere 100 kilometers, ending up as a "neutron star" with an incredible density of millions of tons per cubic inch. "Pulsars" . . . are neutron stars.
3. It may continue to collapse and form a "black hole."

Which will it be? What causes some stars to retire as white dwarfs, while others collapse to neutron stars and still others are doomed to crunch all the way down to black holes?

After all the fuel for nuclear reactions has been used up, gravity meets a fresh opponent—the "exclusion principle." All particles of ordinary matter—the particles that make up atoms, which in turn make up such things as this book, you and me, stars, gases—obey this principle. The exclusion principle insures that particles keep their distance and there will be empty space in atoms. Without it we would have no stars, or people, or any other of the familiar objects in our universe. We would have something more like dense soup.

The exclusion principle requires that no more than two electrons can occupy the same region of space at the same time. Electrons—moving in clouds around the nuclei of atoms—are paired together in cells (or "orbitals"). Electrons protest their confinement in these cells by moving erratically, shaking, flying around, kicking forcefully against adjacent electrons. This motion is called "degenerate motion," and the pressure it produces is "electron degeneracy pressure." This pressure keeps electrons from being pulled into the nucleus of the atom.

As a star collapses, the clouds of electrons around the nuclei of the atoms in the star get squashed until the electrons are confined in cells many times smaller than they usually could move around in. In this situation an electron begins to behave in part like a wave. . . . It stands to reason that the length of the wave cannot be larger than the cell the electron is in. Shorter wavelengths mean higher energy. Higher energy implies more rapid motion. It follows that the denser the matter in the star becomes, the smaller the cells will be, the smaller the cells, the shorter the electrons' wavelengths must be; the shorter the wavelengths, the higher the electrons' energy; the higher the energy, the faster the electrons' motion; the faster this motion, the larger the electron degeneracy pressure it produces. It is this pressure that will continue to support the star against gravity.

This scheme sounds as though it would continue to work in a highly satisfactory manner. As the pull of gravity increased, so would the pressure opposing it. There is, however, a hitch. Our universe has a speed limit which for all practical purposes seems to be unbreakable. That speed limit is the speed of light, approximately 300,000 kilometers (or 186,000 miles) per second. Degenerate electrons can't move faster than the speed of light. But even short of that, when matter is so dense that degenerate electrons move at *near* the speed of light, matter has serious difficulty supporting itself against the squeeze of gravity. Can it succeed?

In the late 1920s Subrahmanyan Chandrasekhar, a young Indian physicist then at the University of Cambridge, calculated that if a star's mass is less than 1.4 times the mass of our sun, gravity will not be able to overpower this exclusion principle repulsion among the electrons. The star shrinks and becomes a white dwarf, but, because of the exclusion principle, it shrinks no further. That star will not become a black hole. We now call this mass of 1.4 solar masses the "Chandrasekhar limit." Only in a star whose mass is *more* than the Chandrasekhar limit will gravity overcome the exclusion principle among the electrons and be the victor in this second competition.

Let's suppose now that we are dealing with a slightly more massive star, more than 1.4 solar masses. Is every star with a mass over the Chandrasekhar limit destined to be a black hole?

The Russian scientist Lev Davidovich Landau . . . called attention to another possible final state for a star. When gravity has overpowered the exclusion principle repulsion among the electrons, as the star squashes down, electrons are squeezed into the atomic nuclei and combine with protons in the nuclei to form additional neutrons. After a while the core of the star is almost entirely made of neutrons. Neutrons must obey the exclusion principle as surely as electrons must. The resistance to squeezing, partly due to degeneracy pressure and partly due to another force called the strong nuclear force, is stronger than it was previously among the electrons, presenting gravity with an even more formidable opponent. . . . Nevertheless, the current consensus is that the maximum allowed mass for a "neutron star" lies between 1.5 and 3 solar masses and is most likely to be about 3 solar masses. A neutron star is smaller and denser than a white dwarf.

However, suppose we have a star a little more massive than the maximum allowed mass for neutron stars. Is *that* star destined to become a black hole? Not necessarily. A star may lose a considerable amount of mass in a late-in-life explosion. However, if it doesn't lose enough to bring it below the maximum allowed mass for neutron stars, about 3 solar masses, gravity *will* overpower both the exclusion principle repulsion among the neutrons and the strong nuclear force, and the star will continue to collapse. When it has reached a size not much smaller than it would have been had it remained a neutron star, the star will form a black hole.

Common sense suggests that there is something suspicious about this description. Why should it be the *more* massive stars that end up smallest, as black holes, and *less* massive stars that end up larger, as white dwarfs or neutron stars? Shouldn't it be the other way around? That question gives us a good excuse for a closer look at gravity.

Physicist John A. Wheeler . . . asks us to think of gravity as a universal democratic system, with every particle in the universe casting a vote that can affect every other particle in the universe. Though no particle of matter has much gravitational influence by itself, when particles join forces and vote as a bloc (the earth, for instance, or a star), that bloc can wield enormous influence. The combined votes of all the particles in this great hulk of a planet under our feet obviously constitute a significant amount of gravitational clout. . . .

Getting back to the stars whose life stories we have been tracing—it should be clear that saying one star is more massive than another means that there are more particles of matter in it. The more particles of matter, the more gravitational attraction. Hence, the more massive the star, the more powerful the gravity team is in the tug-of-war, the more powerful the potential "squeeze," and the more likely it is that the squeeze will be able to overpower the exclusion principle.

Gravity was the force that gave birth to the star in the cloud of gas. It was the force that kept the star in balance for all those years, not allowing it to fly apart but also contributing to the processes that prevented its collapse. Now, after millions or even billions of years have passed, gravity is the force that claims the star as its victim, squeezing all its enormous mass to something the size of the sun . . . then to the size of the earth . . . to the size of the moon . . . to the size of London . . . St. James's Park . . . the lake in the park . . . a duck on the lake . . . a tennis ball . . . a marble . . . the head of a pin . . . the point of a pin . . . a microbe . . . For a star ending up more massive than about 3 solar masses, we know of no power capable of halting this catastrophic collapse. The star continues to crunch down even after it has become a black hole. It's best to think of a black hole *not* as a star, but as what happens to spacetime around a star that goes on collapsing to near infinite density.

In a subsequent chapter Ferguson leaves the biographical approach behind and adopts an astronaut's-eye "view" of a black hole and the experience of "tidal effects" (an aspect of gravity also discussed very early in the twentieth century by G. H. Darwin; see Chapter 55).

Although it is the earth's gravity that keeps a NASA space shuttle orbiting the earth rather than haring off into outer space, astronauts aboard the shuttle don't feel any pull of the earth's gravity once they are in orbit. We have probably all seen pictures of them floating around the cabin of the shuttle. Their weightlessness is not caused by the spacecraft being too far away to be affected by the earth's gravity. It's due to the fact that the downward acceleration of the spacecraft exactly cancels the pull of that gravity. The astronauts, with their spacecraft, are in free fall, as surely as they would be if the shuttle were plummeting directly toward the earth. In free fall you don't feel any gravity. Remember that an orbiting spacecraft *is* falling, but its orbital motion causes it to fall *around* the earth, continually missing the earth, overshooting it again and again. That, roughly speaking, is what happens in any free fall orbit. A handy catch-phrase to keep in mind is "Falling free or orbiting 'round, gravity cannot be found." . . .

Similarly, although it is the moon's and the earth's gravities that keep each of them in free fall orbit around the other (or, to be more precise, around their common center of mass), we can't think of them or anything "on board" either of them (such as the ocean) as directly experiencing the pull of the other's gravity, any more than astronauts in free fall orbit feel the earth's gravity. If this seems unlikely, recall that although it is the sun's gravity that causes the earth to orbit rather than fly off into outer space beyond the solar system, we don't weigh any less when the sun is directly overhead. . . .

Nevertheless, the ocean tides do occur, and we have already said that it's gravity that causes this tidal effect. What is this "effect"? We find that the side of the earth closest to the moon (approximately 12,000 km nearer to the moon than the opposite side of the earth) bulges out toward the moon, and the side of the earth furthest from the moon bulges out away from the moon. The earth is stretched slightly, first in one direction and then in another, depending upon where the moon is in its orbit. The stretch doesn't occur only where there are great bodies of water, but those areas of the earth stretch more readily than other areas, and the effect there is noticeable. Whatever part of the ocean is closest to the moon bulges out in the direction of the moon, making a high tide. Whatever part of the ocean is furthest from the moon bulges out in the direction away from the moon, making another high tide. Halfway in between we have low tides. We are right to think gravity is the culprit behind all this, and it isn't entirely accurate in our catch-phrase to say that "gravity cannot be found." In order to understand how the deed is done, you need to know something about "tidal gravity."

We do well to understand Ferguson's explanation of tidal effect against the backdrop of the discussion of negative (or "noncontractile") spacetime curvature in Chapter 60. Imagine that, while in a space capsule in free fall towards the earth, we conduct an experiment with four marbles (also in free fall, of course) arranged in a square, like a miniature vertical baseball diamond, with "second base" and "home plate" in a straight line towards the center of the earth and "first" and "third" equidistant from the earth. Two things will happen simultaneously to change the shape of the "diamond." First, because the two marbles we are calling "first" and "third" are both falling toward the center of the earth, their trajectories are not parallel but convergent. So "first" and "third" are actually getting closer together as they fall. But "second" and "home" are getting farther apart, because an object's rate of acceleration increases the closer it is to the earth, and the "home plate" marble is in fact closer, so its velocity of fall is increasing more rapidly than that of "second base." Thus our "diamond" is both narrowing and lengthening. And the same effect takes place within the structure of any single object that is not absolutely rigid. If we repeat our experiment with a perfectly spherical rubber ball, it will behave like a soccer ball that wants to become a rugby ball. Ferguson uses a similar example of a gel lozenge that undergoes the same narrowing and elongation as a result of gravitational forces.

This vertical stretch and lateral squeeze are called "tidal gravitational forces" or "tidal gravity." Looking at the same stretch and squeeze through Einstein's eyes, curvature of spacetime and tidal gravity are the same thing.

The earth and the moon, in free fall orbit, are not falling directly toward one another as the lozenge is falling toward the earth in our example. But, just as surely as if they were, they are in free fall, and nothing on board the moon or earth can feel the direct gravity of the other. However, as happened with the gel lozenge, the earth, the moon, and things on board them can be affected by the *difference* in the strength of the gravitational pull from place to place and the *difference* that makes in the way they "fall." The upshot is that the side of the earth nearest the moon bulges out moonward, making a high tide. The opposite side gets "left behind" a little, making a second high tide. In between there are low tides. If the entire earth's surface were as flexible as the gel lozenge, we would not have to go to the shore to experience the tides. . . . The moon undergoes a similar stretching as an indirect effect of the earth's gravity. However, there are no bodies of water on the moon to make the effect obvious, and the moon is no gel lozenge.

Ferguson makes the distinction between direct gravity and tidal gravity by pointing out that although the sun's direct gravity is 180 times stronger at the earth's surface than the moon's is, lunar tides are much stronger than solar tides. To understand why this is so we may again recall the discussion in Chapter 60. If we remember the analogy of the cannon ball on the trampoline, we can see that (by analogy with the sun) a large cannon ball on the trampoline might create a greater stretching, and a greater slope, to the fabric of the trampoline than a small one would. But if our rolling marble is a long way from the large cannon ball, the change in its rate of acceleration might be less than the change in its rate of acceleration would be when it is very close to a smaller cannon ball (as the earth and moon are relatively close together). We recall, moreover, that relativists describe this change in rate of acceleration as "degree of negative spacetime curvature." Thus we can recognize that in the vicinity of a black hole, this change—this extreme negative curvature—will be "noncontractile" with a vengeance.

Though tides are known to cause problems for us living on the earth, most of us don't often have reason to consider tidal effects a major threat to our well-being. Travelling near a black hole, we are advised to treat them with far greater respect. From the above [discussion] we can guess that it isn't the enormous strength of gravity that will present a hazard in this case, even *within* the black hole, but rather the extremely rapid *change* in the strength of gravity as one approaches or moves away from the event horizon. In our example, the "highest" and "lowest" of the falling objects fell at slightly different rates. Near a black hole these same two objects would fall at *vastly* different rates. If they are two parts of a lozenge, they will likewise fall at vastly different rates and the lozenge won't last long as a lozenge. If they are two parts of our ship—or of you or me—we are in trouble.

Gravitational attraction or spacetime curvature increases rapidly as one approaches the event horizon and falls off just as rapidly as one moves away from the hole. The change may be over short, even microscopic, distances. The effect on one side of the ship, the side facing the black hole, could differ drastically from the effect on the side away from the black hole. The closer the spacecraft approaches the black hole, the greater the difference in these effects on various parts of its structure will become. At some point the ship will not merely be stretched. The ship and its passengers will be ripped asunder.

SOURCE: Kitty Ferguson, *Prisons of Light—Black Holes*, Cambridge: Cambridge UP, 1996.

76

A Very Lumpy Universe

George Smoot

George Smoot (b. 1945) was a member of the team that designed and conducted experiments using the now-famous Cosmic Background Explorer (COBE) satellite. These experiments looked for the "fingerprint" of the Big Bang and sought evidence that would explain why the universe is not "smooth" but "lumpy." For if the matter in the universe were distributed evenly, given an actual average cosmic density of roughly 10^{-30} grams per cubic centimeter, then no galaxies, stars, or planets would have come into being. Put another way, if the Big Bang theory is true, then the actual present lumpiness of the universe presupposes variations in the density of matter at a very early stage of cosmic history. By laboriously examining the "static" left over from the Big Bang, Smoot and his team were able to detect and to map these variations, or "wrinkles." The COBE team's announcement of their findings at the American Physical Society meeting on 23 April, 1992 aroused great interest not only in scientific circles but also in the press. In Wrinkles in Time, *the book recounting the project from its beginnings, Smoot, with co-author Keay Davidson, summarizes the event and its importance.*

For me, the moment marked the culmination of an eighteen-year search, and for cosmology a major milestone on the long journey to understanding the nature of the universe. Very simply, the discovery of the wrinkles salvaged big bang theory at a time when detractors were attacking in increasing numbers. The result indicated that gravity could indeed have shaped today's universe from the tiny quantum fluctuations that occurred in the first fraction of a second after creation. When Stephen Hawking later commented that we had made "the most important discovery of the century, if not of all time," he may have been exaggerating, but it was still momentous. Before the discovery, our understanding of the origin and history of the universe rested on four major observations: first, the darkness of the night sky; second, the composition of the elements, with the great preponderance of hydrogen and helium over the heavier elements; third, the expansion of the universe; fourth, the existence of the cosmic background radiation, the afterglow of a fiery creation.

The discovery of the wrinkles that were present in the fabric of time at three hundred thousand years after creation becomes the fifth pillar in this intellectual edifice and gives us a way of understanding how structures of all sizes, from galaxies to superclusters, could have formed as the universe evolved during the past 15 billion years.

The evolution of the universe is effectively the change in distribution of matter through time—moving from a virtual homogeneity in the early universe to a very lumpy universe today, with matter condensed as galaxies, clusters, superclusters, and even larger structures. We can view that evolution as a series of phase transitions, in which matter passes from one state to another under the influence of decreasing temperature (or energy). We are all familiar with the fact that steam, on cooling, condenses: This is a phase transition from a gaseous to a liquid state. Reduce the temperature further, and eventually the water freezes, making a phase transition from liquid to the solid state. In the same way, matter has gone through a series of phase transitions since the first instant of the big bang.

At a ten-millionth of a trillionth of a trillionth of a trillionth (10^{-42}) of a second after the big bang—the earliest moment about which we can sensibly talk, and then only with some suspension of disbelief—all the universe we can observe today was the tiniest fraction of the size of a proton. Space and time had only just begun. (Remember, the universe did not expand *into* existing space after the big bang; its expansion created space-time as it went.) The temperature at this point was a hundred million trillion trillion (10^{32}) degrees, and the three forces of nature—electromagnetism and the strong and weak nuclear forces—were fused as one. Matter was undifferentiated from energy, and particles did not yet exist.

By a ten-billionth of a trillionth of a trillionth of a second (10^{-34}) inflation had expanded the universe (at an accelerating rate) a million trillion trillion (10^{30}) times, and the temperature had fallen to below a billion billion billion (10^{27}) degrees. The strong nuclear force had separated, and matter underwent its first phase transition, existing now as quarks (the building blocks of protons and neutrons), electrons, and other fundamental particles.

The next phase transition occurred at a ten-thousandth of a second, when quarks began to bind together to form protons and neutrons (and antiprotons and antineutrons). Annihilations of particles of matter and antimatter began, eventually leaving a slight residue of matter. All the forces of nature were now separate.

The temperature had fallen sufficiently after about a minute to allow protons and neutrons to stick together when they collided, forming the nuclei of hydrogen and helium, the stuff of stars. This soup of matter and radiation, which initially was the density of water, continued expanding and cooling for another three hundred thousand years, but it was too energetic for electrons to stick to the hydrogen and helium nuclei to form atoms. The energetic photons existed in a frenzy of interactions with the particles in the soup. The photons could travel only a very short distance between interactions. The universe was essentially opaque.

When the temperature fell to about 3,000 degrees, at three hundred thousand years, a crucial further phase transition occurred. The photons were no longer energetic enough to dislodge electrons from around hydrogen and helium nuclei and so atoms of hydrogen and helium formed and stayed together. The photons no longer interacted with the electrons and were free to escape and travel great distances. With this decoupling of matter and radiation, the universe became transparent, and radiation streamed in all directions—to course through time as the cosmic background radiation we experience still. The radiation released at that instant gives us a snapshot of the distribution of matter within the universe at three hundred thousand years of age. Had all matter been distributed evenly, the fabric of space would have been smooth, and the interaction of photons with particles would have been homogeneous, resulting in a completely uniform cosmic background radiation. Our discovery of the wrinkles reveals that matter was not uniformly distributed, that it was already structured, thus forming the seeds out of which today's complex universe has grown.

Those regions of the universe with higher concentration of matter exerted more gravitational attraction and therefore curved space positively; less dense areas had less gravitational attraction, resulting in less curvature of space. When radiation and matter decoupled three hundred thousand years later, the suddenly released flux of cosmic background photons bore the imprint of these distortions of space, showing the wrinkles we see in our maps: Radiation traveling from the denser areas looks cooler than the average background; that from less dense, warmer.

Matter in the universe . . . is of two kinds—dark matter and visible matter—and their role in gravitational formation of structure is different. Dark matter, which by its nature is unaffected by radiation but responsive to gravity, would have started forming structure much earlier than visible matter, which is buffeted by the energetic flux of photons. Molded by the contours of space that originated as quantum fluctuations in the inflationary universe, dark matter could have begun to aggregate, under the influence of gravity, as early as ten thousand years after the big bang. At three hundred thousand years, the decoupling of matter and radiation freed ordinary visible matter to be attracted to the structures formed by dark matter. As the visible matter aggregates, stars and galaxies form. A nice image here is the way cobwebs, often unseen in ordinary light, become strikingly visible when dew that settles on their strands during the night is lit by the morning sun. The gossamer network of galaxies we see in the night sky is the shimmering dew on a cosmic cobweb, as visible matter outlines the shape of structures of invisible dark matter, to which it has been drawn by gravitational attraction.

Because of the limitations of the differential microwave radiometer aboard COBE, the resolution of our maps is relatively poor. The smallest objects we can see as wrinkles are enormous, leading to structures as large as or larger than the Great Wall, a vast, concentrated sheet of galaxies stretching many hundreds of millions of light-years across. When we can achieve greater resolution, I expect us to be able to discern structures the size of galaxies. Despite current limitations, however, the message of our results—the message that engendered so much relief among cosmologists that April day—was clear. Fred Hoyle once claimed that big bang theory failed because it could not account for the early formation of galaxies [see the first part of Chapter 67]. The COBE results prove him wrong. The existence of the wrinkles in time as we see them tell us that big bang theory, incorporating the effect of gravity, can explain not only the early formation of galaxies but also the aggregation within 15 billion years of the massive structures we know to be present in today's universe. This is a triumph for theory and observation.

SOURCE: George Smoot and Keay Davidson, *Wrinkles in Time*, New York: William Morrow, 1993.

77

A Cosmic Archipelago

Martin Rees

Among the toughest philosophical conundrums we face in considering the origin and nature of the cosmos is this: Our universe displays an array of characteristics that are mind-numbingly improbable. According to Arthur Eddington in the early 1930s (see Chapter 65), "The picture of the world, as drawn in existing physical theories, shows arrangement of the individual elements for which the odds are multillions [of order $10^{10^{10}}$] to 1 against an origin by chance." In the chapter after this one, Lee Smolin estimates the chances simply of there being a universe containing stars at one in 10^{229}. Are we left then to conclude "it just so happened" that the universe is "fine tuned" exactly as it is? Or must we assert that in some providential sense the universe was designed? Many cosmologists writing late in the twentieth century undertook to avoid the latter conclusion by means of attempts to model the plausibility the former.

One such is Britain's Astronomer Royal, Sir Martin Rees (b. 1942), who has proposed the hypothesis of a "multiverse" or an "ensemble of universes." The core idea is most simply rendered by a sartorial analogy that Rees points to in a number of his publications. Suppose you come home wearing a new suit that fits you exactly, even though you happen to have quite an unusual build. Your family naturally assumes the suit was made to measure. But you explain to them that, no, with a bit of luck, and a thousand suits in the shop to choose from, you found one that fit just right. Similarly, the hypothesis of the multiverse—of a huge, perhaps infinite ensemble of universes—allows us to behold what looks like a tailor-made universe without concluding that it was purposely made to measure. On this view our universe will still appear special; but, given those racks and racks of other universes, one that fits like ours will not justify our being very surprised.

In a book coauthored with John Gribbin in the 1980s (Cosmic Coincidences), Rees finds another image of the multiverse in the opening words of a short story by Arthur C. Clarke published forty years earlier: "Many and strange are the universes that drift like bubbles in the foam upon the River of Time." In Before the Beginning *(1997), Rees reaches still farther back and cites David Hume's conjecture in the eighteenth century concerning a series of trial-and-error universes that "might have been botched and bungled throughout an eternity ere this system was struck out; much labor lost, many fruitless trials made, and a slow but continual improvement carried out during infinite ages in the art of world-making." Thus not only are there racks and racks of universes to choose from, but the quality of the "tailoring" may have improved over "time."*

Rees sees this general approach to the "specialness" of our universe as ushering in a cosmological sea change of Copernican proportions.

As our universe cooled, its specific mix of energy and radiation, even perhaps the number of dimensions in its space, may have arisen as "accidentally" as the patterns in the ice when a lake freezes. The physical laws were themselves "laid down" in the big bang.

Our universe, and the laws governing it, had to be (in a well-defined sense) rather special to allow our emergence. Stars had to form; the nuclear furnaces that keep them shining had to transmute pristine hydrogen into carbon, oxygen, and iron atoms; a stable environment and vast spans of space and time were prerequisites for the complexities of life on earth.

The apparent fine-tuning on which our existence depends could be a coincidence. I once thought so. But that view now seems too narrow. What's conventionally called "the universe" could be just one member of an ensemble. Countless others may exist in which the laws are different. The universe in which we've emerged belongs to the unusual subset that permits complexity and consciousness to develop. Once we accept this, various apparently special features of our universe—those that some theologians once adduced as evidence for Providence or design—occasion no surprise. . . .

This new concept is, potentially, as drastic an enlargement of our cosmic perspective as the shift from pre-Copernican ideas to the realization that the earth is orbiting a typical star on the edge of the Milky Way, itself just one galaxy among countless others. . . .

Our entire universe may be just one element-one atom, as it were—in an infinite ensemble: a cosmic archipelago. Each universe starts with its own big bang, acquires a distinctive imprint (and its individual physical laws) as it cools, and traces out its own cosmic cycle. The big bang that triggered our entire universe is, in this grander perspective, an infinitesimal part of an elaborate structure that extends far beyond the range of any telescopes.

Some cosmologists speculate that new "embryo" universes can form within existing ones. Implosion to a colossal density (around, for instance, a small black hole) could trigger the expansion of a new spatial domain inaccessible to us. Universes could even be "manufactured"—the experimental challenge is far beyond present human resources, but may become feasible, especially if we recall that our universe has most of its course still to run. No information could be exchanged with a daughter universe, but it could bear the imprint of its parentage. Our own universe might be the (planned or unplanned) outcome of such an event in some preceding cosmos. The traditional theological "argument from design" then reasserts itself in novel guise.

Most naturally created universes would be stillborn in the sense that they could not offer an environment propitious for complex evolution: they would have too short a time span, the wrong number of dimensions, allow no chemistry, or be otherwise maladjusted. But our universe may not be the most complex: others in the ensemble may have richer structure, beyond anything we can imagine.

SOURCE: Martin Rees, *Before the Beginning: Our Universe and Others*, Reading, Mass.: Helix/Perseus Books, 1997.

78

Cosmological
Natural Selection

Lee Smolin

Undoubtedly one of the most innovative, stimulating, and infuriating cosmo-logical studies of the 1990s is The Life of the Cosmos, *by Lee Smolin (b. 1955). Among living cosmologists Smolin has few equals in his knowledge of and respect for intellectual history, and in the degree to which he integrates this knowledge with his dazzling but amiable theorizing. For the common reader perhaps the most endearing dimension of Smolin's work is the way in which he harmonizes tough scientific and philosophical exploration with an ingenuous sense of amazement at the wonders of the universe, as in his dis-cussion of what he calls "The Miracle of Stars."*

Although many different kinds of elementary particles have been discovered, almost all the matter in the universe is made of four kinds: protons, neutrons, electrons, and neutrinos. These interact via four basic forces: gravity, electro-magnetism, and the strong and weak nuclear forces. Each of these forces is characterized by a few numbers. Each has a *range*, which tells us the distance over which the force can be felt. Then, for each kind of particle and each force there is a number which tells us the strength by which that particle par-ticipates in interactions governed by that force. These are called the coupling constants. One of these is the electrical charge, which tells how strongly a particle may interact, or be attracted by, other charged particles. The para-meters of the standard model consist primarily of the masses of the particles and these numbers that characterize the four forces.

In order to understand why the existence of stars is so improbable, it helps to know some basic facts about the four different interactions. We may start with gravity, which is the only universal interaction. Every particle, every form of energy, feels its pull. Its range is infinite, which means that although the gravitational force between two bodies falls off with distance, it is never zero, no matter how far apart the two bodies may be. Gravity has another distinguishing feature, which is that it is always attractive. Any two particles in the universe attract each other through the gravitational interaction.

The strength by which any particle is affected by gravity is proportional to its mass. The actual force between two bodies is given by multiplying the two masses together, and then multiplying the result times a universal constant. This constant is called Newton's gravitational constant; it is one of the parameters of the standard model. The most important thing to know about it is that it is a fantastically small number. Its actual value depends on the units we use, as is the case with many physical constants. For elementary particle physics it is natural to take units in which mass is measured by the proton mass. In these units you or I have a mass of about 10^{28}, for that is how many protons and neutrons it takes to make a human body. By contrast, in these units the gravitational constant is about 10^{-38}. This tiny number measures the strength of the gravitational force between two protons.

The incredible smallness of the gravitational constant is one of the mysteries associated with the parameters of particle physics. Suppose we had a theory that explained the basic forces in the universe. That theory would have to produce, out of some calculation, this ridiculous number, 10^{-38}. How is it that nature is so constructed that one of the key quantities that govern how it works at the fundamental level is so close to zero, but still not zero? This question is one of the most important unsolved mysteries in all of physics.

It may seem strange that a force as weak as gravity plays such an important role on earth and in all the phenomena of astronomy and cosmology. The reason is that, in most circumstances, none of the other forces can act over large distances. For example, in the case of the electrical force, one almost always finds equal numbers of protons and electrons bound together, so that the total charge is zero. This is the reason that most objects, while being composed of enormous numbers of charges, do not attract each other electrically.

Gravity is the only force that is always attractive, which means that it is the only force whose effects must always add, rather than cancel, when one considers aggregates of matter. Thus, when one comes to bodies composed of enormous numbers of particles, such as planets or stars, the tiny gravitational attractions of each of the particles add up and dominate the situation.

The incredible weakness of the gravitational constant turns out to be necessary for the existence of stars. Roughly speaking, this is because the weaker gravity is, the more protons must be piled on top of each other before the pressure in the center is strong enough that the nuclear reactions ignite. As a result, the number of atoms necessary to make a star turns out to grow as the gravitational constant decreases. Stars are so huge exactly because the gravitational constant is so tiny.

It is fortunate for us that stars are so enormous, because this allows them to burn for billions of years. The more fuel a star contains, the longer it can produce energy through nuclear fusion. As a result, a typical star lives for a long time, about ten billion years.

Were the gravitational force somewhat stronger than it actually is, stars would still exist, but they would be much smaller, and they would burn out very much faster. The effect is quite dramatic. If the gravitational force were stronger by only a factor of ten, the lifetime of a typical star would decrease from about ten billion years to the order of ten million years. If its strength were increased by still another factor of ten, making the gravitational force between two protons still an effect of order of one part in 10^{36}, the lifetime of a star would shrink to ten thousand years.

But the existence of stars requires not only that the gravitational force be incredibly weak. Stars burn through nuclear reactions that fuse protons and neutrons into a succession of more and more massive nuclei. For these processes to take place, protons and neutrons must be able to stick together, creating a large number of different kinds of atomic nuclei. For this to happen, it turns out that the actual values of the masses of the elementary particles must be chosen very delicately. Other parameters, such as those that determine the strengths of the different forces, must also be carefully tuned.

Let us think of the three most familiar particles: the proton, neutron, and electron. The neutron, it turns out, has almost the same mass as the proton; it is in fact just slightly heavier, by about two parts in a thousand. In contrast, the electron is much lighter than either; it is about eighteen hundred times lighter than the proton.

In the masses of these three particles there are many mysteries. Why are the neutron and proton so close in mass? Why is the electron so much lighter than the other two particles? But what is most mysterious is that the two small numbers in this problem, the electron mass and the tiny amount by which a neutron is just slightly more massive than a proton, are comparable to each other. The neutron outweighs the proton by only about three electron masses.

We are so used to the idea that protons and neutrons stick together to make hundreds of different stable nuclei, that it is difficult to think of this as an unusual circumstance. But in fact it is. Were the electron's mass not about the same size as the amount that the neutron outweighs the proton, and were each of these not much smaller than the proton's mass, it would be impossible for nuclei to stick together to form stable nuclei. These are then facts of great importance for the world as we know it, for without the many different stable nuclei, there would be no nuclear or atomic physics, no stars, and no chemistry. Such a world would be dramatically uninteresting. . . .

While we are discussing physical constants that must be finely tuned for the universe to contain stars, we may consider another kind of question. Why is the universe big enough that there is room for stars? Why is it not much smaller, perhaps even smaller than an atom? And why does the universe live for billions of years, which is long enough for stars to form? Why should it not instead live just a few seconds? These may seem silly questions, but they are not, because the fact that the universe can become very big and very old depends on a particular parameter of the standard model being extremely tiny. This parameter is called the cosmological constant.

The cosmological constant can be understood as measuring a certain intrinsic density of mass or energy, associated with empty space. That a volume of empty space might itself have mass is a possibility allowed by Einstein's general theory of relativity. If this were sizable, it would be felt by matter, and this would affect the evolution of the universe as a whole. For example, were there enough of it, the whole universe would quickly pull together and collapse gravitationally, as a dead star collapses to a black hole. In order that this not happen, the mass associated with the cosmological constant must be much smaller than any of the masses we have so far mentioned. In units of the proton mass, it can be no larger than about 10^{-40}. If this were not the case, the universe would not live long enough to produce stars. . . .

Perhaps the reader is still not convinced that there is something incredible to be understood here. Let me then go on. We have only discussed gravity; there are three more interactions to consider. These forces are described by still additional parameters. The story for many of these is the same.

We may consider next the force which is most evident in our lives, . . . electromagnetism and light.

The importance of electromagnetism for our modern picture of nature cannot be overstated, as almost all of the phenomena of everyday life which are not due to gravity are manifestations of it. For example, all chemistry is an aspect of electromagnetism. This is because chemical reactions involve rearrangements of electrons in their orbits around atomic nuclei, and it is the electrical force that holds the electrons in those orbits. Light is also an aspect of electromagnetism, for it is a wave traveling through the fields that convey the electric and magnetic forces.

Electromagnetism differs in two important respects from gravity. The first is that electrical force between two fundamental particles is much stronger than their gravitational attraction. The strength of the electrical interaction is measured by a number, which was called alpha by the physicists of the last century, because it is a number of the first importance to science. Alpha, which is essentially a measure of the strength of the electric force between two protons or electrons, has a value of approximately 1/137. Physicists have been wondering about why alpha has this value, without resolution, for the whole of the twentieth century.

The second way in which electricity differs from gravity is that its effect is not only attractive: two electrical charges may attract or repel each other, depending on whether they are alike or unlike.

As we did for gravity, we may ask how important the existence of a force with these properties is for the existence of stars. Light does, indeed, do something essential for stars. For it must be possible for the energy produced in stars to be carried away to great distances. Otherwise, stars could not radiate, and being unable to get rid of the energy they produce, they would simply explode. Light is precisely the medium by which the energy produced in stars is conveyed to the rest of the universe.

However, the existence of electrical forces makes another problem for stars. Like charges repel, and the nucleus of most atoms contains a number of protons, all of like charge, which are packed closely together. What keeps the nuclei from being blown apart by the repulsion of all the protons in them?

There is no way either electricity or gravity could save the situation. What is needed if nuclei are to exist is another force with certain properties. It must act attractively among protons and neutrons, in order to hold the atomic nuclei together. It must be strong enough to counteract the repulsions of all the protons. But it cannot be too strong, otherwise it would be too difficult to break the nuclei apart, and chain reactions of nuclear reactions could not take place inside of stars.

This force must also be short-ranged, otherwise there would be danger of its pulling all the protons and neutrons in the world together into one big nucleus. For the same reason, it cannot act on electrons, otherwise it would pull them into the nuclei, making molecules and chemistry impossible.

It turns out that there is a force with exactly these required properties. It is called the strong nuclear force, and it acts, as it should, only over a range which is more or less equal to the size of an atomic nucleus.

Remarkably, the existence of more than a hundred kinds of stable nuclei is due to the fact that the strength of the attractive nuclear force balances quite well the electrical repulsion of the protons. To see this, it is necessary only to ask how much we have to increase the strength of the electrical force, or decrease the strength of the nuclear force, before no nuclei are stable. The answer is not much. If the strong interaction were only 50% weaker, the electrical repulsion is no longer overcome, and most nuclei become unstable. Going a bit further, perhaps to 25%, all nuclei fall apart. The same effect can also be achieved by holding the strong interaction unchanged and increasing the strength of the electrical repulsions by no more than a factor of about ten.

Thus we see that the simple existence of many species of nuclei, and hence the possibility of a world with the complexity of ours, with many different types of molecules each with distinct chemical properties, is ultimately the result of a rather delicate balance between two of the basic interactions, the electromagnetic and strong nuclear force.

There is, finally, one more basic interaction, which is called the weak nuclear interaction. It is called a nuclear interaction because the scale over which it can act is also about the size of the atomic nucleus. But it is much weaker than the strong nuclear force. It is too weak to play any role binding things together, but it does play an important role in transforming particles into each other. It is this weak interaction that governs the basic nuclear reaction on which the physics of stars is based, by means of which an electron and a proton are transformed into a neutron and a neutrino.

The reader to whom these things are new might pause and ponder the characteristics of these four basic forces, for it is they that give our world its basic shape. With their different properties, they work together to allow a world that is both complex and harmonious. Eliminate any one, or change its range or strength, and the universe around us will evaporate instantly and a vastly different world will come into being.

Would any of these other worlds contain stars? How many could contain life? The answer to both of these questions, as we have seen, is not many.

Physicists are constantly talking about how simple nature is. Indeed, the laws of nature are very simple, and as we come to understand them better they are getting simpler. But, in fact, nature is not simple. To see this, all we need to do is to compare our actual universe to an imagined one that really is simple. Imagine, for example, a homogeneous gas of neutrons, filling the universe at some constant temperature and density. That would be simple. Compared to that possibility, our universe is extraordinarily complex and varied!

Now, what is really interesting about this situation is that while the laws of nature are simple, there is a clear sense in which we can say that these laws are also characterized by a lot of variety. There are only four fundamental forces, but they differ dramatically in their ranges and interaction strengths. Most things in the world are made of only four stable particles: protons, neutrons, electrons, and neutrinos; but they have a very large range of masses, and each interacts with a different mix of the four forces.

The simple observation we have made here is that the variety we see in the universe around us is to a great extent a consequence of this variety in the fundamental forces and particles. That is to say, the mystery of why there is such variety in the laws of physics is essentially tied to the question of why the laws of physics allow such a variety of structures in the universe.

If we are to genuinely understand our universe, these relations between the structures on large scales and the elementary particles must be understood as being something other than coincidence. We must understand how it came to be that the parameters that govern the elementary particles and their interactions are tuned and balanced in such a way that a universe of such variety and complexity arises.

Of course, it is always possible that this is just coincidence. Perhaps before going further we should ask just how probable is it that a universe created by randomly choosing the parameters will contain stars. Given what we have already said, it is simple to estimate this probability. . . . The answer, in round numbers, comes to about one chance in 10^{229}.

To illustrate how truly ridiculous this number is, we might note that the part of the universe we can see from earth contains about 10^{22} stars which together contain about 10^{80} protons and neutrons. These numbers are gigantic, but they are infinitesimal compared to 10^{229}. In my opinion, a probability this tiny is not something we can let go unexplained. Luck will certainly not do here; we need some rational explanation of how something this unlikely turned out to be the case.

I know of three directions in which we might search for the reason why the parameters are tuned to such unlikely values. The first is towards some version of the *anthropic principle*. One may say that one believes that there is a god who created the world in this way, so there would arise rational creatures who would love him. We may even imagine that he prefers our love of him to be a rational choice made after we understand how unlikely our own existence is. While there is little I can say against religious faith, one must recognize that this is mysticism, in the sense that it makes the answers to scientific questions dependent on a faith about something outside the domain of rationality.

A different form of the anthropic principle begins with the hypothesis that there are a very large number of universes. In each the parameters are chosen randomly. If there are at least 10^{229} of them it becomes probable that at least one of them will by chance contain stars. The problem with this is that it makes it possible to explain almost anything, for among the universes one can find most of the other equally unlikely possibilities. To argue this way is not to reason; it is simply to give up looking for a rational explanation. Had this kind of reasoning been applied to biology, the principle of natural selection would never have been found.

A second approach to explaining the parameters is the hypothesis that there is only a single unique mathematically consistent theory of the whole universe. If that theory were found, we would simply have no choice but to accept it as the explanation. But imagine what sense we could then make of our existence in the world. It strains credulity to imagine that mathematical consistency could be the sole reason for the parameters to have the extraordinarily unlikely values that result in a world with stars and life. If in the end mathematics alone wins us our one chance in 10^{229} we would have little choice but to become mystics. This would be an even purer mysticism than the anthropic principle because then even God would have had no choice in the creation of the world.

The only other possibility is much more mundane than these. It is that the parameters may actually change in time, according to some unknown physical processes. The values they take may then be the result of real physical processes that happened sometime in our past. This would take us outside the boundaries of the platonist philosophy, but it seems nevertheless to be our best hope for a completely rational understanding of the universe, one that doesn't rely on faith or mysticism.

Smolin does not define faith or mysticism with the rigor he applies in other areas. Nevertheless, in his effort to avoid them and to find a purely naturalistic (he would say rational*) explanation for this astonishingly unlikely universe, he turns to a possibility that may get around the need to assume that the universe came from nothing. This possibility is that black holes are "locations" for new big bangs. But first, in his informative way, he explains another amazing cosmic conundrum related to the early universe.*

Imagine for a moment that you could see the cosmic black body radiation. You look up at the sky and see a flash of light from a photon that has traveled around ten billion years, from the time of decoupling [the early phase transition at which the early stuff of the universe ceased to be opaque plasma] to your eye. Now, turn your head a few degrees to the right, and wait again for a photon from the black body radiation to come to your eye. Coming to us from different directions, and traveling for such a long time, these two photons come from regions of the universe that were very far apart when they were created. Even taking into account the fact that the universe has expanded a great deal (about a thousand fold) while they were traveling, it is still true that they were very far apart when they began their journeys.

What is remarkable is that, even if they started out from regions of the universe that were very far apart, all of the photons coming from the time the universe was opaque tell the same story. To an accuracy of about one part in a hundred thousand, the temperature at that time seems to have been the same all over the universe.

This is one of the big mysteries of modern cosmology. How is it possible that regions of the universe that were very far apart at that time had, nevertheless, almost precisely the same temperature? Questions like this are usually not hard to answer. We know that a glass of hot water left in a room will eventually cool to the temperature of the room. As a result of the tendency of things to come to equilibrium, when things are in contact for long enough they tend to come to the same temperature. The simplest possibility is then that all the different regions of the universe had been in contact with each other before the moment of decoupling.

Unfortunately, if we believe in the story of cosmology given by general relativity, this cannot have been the case. As the universe is supposed to have been only about a million years old at the time of decoupling, and as nothing can travel faster than light, only regions that were then less than a million light years apart could have had any contact with each other. The problem is that, according to the theory, the universe at this time was much bigger than a million light years across. This means that when we detect the cosmic background radiation coming from two different points in the sky more than a few degrees apart, we are seeing light that originated from regions that up till that time could not have had any kind of contact with each other.

To emphasize how strange this is, let us suppose that the signal of the cosmic background radiation was modulated like a radio broadcast. And let us suppose that, from every corner of the universe, the tune played by the cosmic background radiation was rock 'n' roll, and not only rock 'n' roll, but the same Cosmic Top Ten: The Beatles, Madonna, Bruce Springsteen, Gianna Nannini, etc. How could we account for this? It would be no problem if the different regions had been able to listen to each other, for the appeal of good music (or, if the reader prefers, the economics of cultural penetration) is almost as absolute as the laws of thermodynamics. Indeed, we are not surprised to hear the same music in every restaurant and bar on this planet. But what if the different regions could never have been in contact with each other? It would be as if Hernan Cortes, arriving in the court of Montezuma, heard around him only the songs he had learned in the taverns of Seville. We would then have to believe in a miracle of a thousand simultaneous births of rock 'n' roll, a thousand simultaneous Memphises and Detroits, each totally unaware of the others.

This may seem ridiculous, but it is not much more ridiculous than what is actually seen: many regions which, if we believe the standard theory, could never have been in contact with each other, but in which the temperatures are the same, to fantastic precision.

The reader may be confused by this. Isn't the idea of the "Big Bang" that the whole universe expanded from a point? This is the popular conception, but it is not actually what general relativity says. It is true that if we trace back the history of any particle, we find an initial singularity at which the density of matter becomes infinite. However, what is not true is that all the particles in the universe meet at their first, singular moments. They do not. Instead, they all seem to spring into existence, simultaneously but separately, at the same instant. Just after the first instant of time, the universe already has a finite spatial extent. One million years later, the universe is much larger than one million light years across, leading to the problem we have been discussing.

Of course, it is always possible that all the different regions of the universe were created, separately, with exactly the same conditions. The different regions had the same temperature a million years later, because they were created with the same temperature. This may seem to resolve the question, but it only leaves a different mystery: Why were all the regions created with exactly the same conditions? This does not solve the problem; it only makes it worse by forcing us to imagine that whatever created the universe did it in a way that duplicated the same conditions in an enormous number of separate regions.

Indeed, as long as we believe that the world was born a finite time ago, we have the problem of explaining what the conditions were at the moment of creation. Whether the temperatures were the same everywhere, or whether the pattern of hot spots spelled out "Made in Heaven," we would have the same problem of explaining what the conditions were at the moment of creation.

One escape from this dilemma would be if general relativity were wrong about the early history of the universe. We have already noted that this is quite possible, given that general relativity does not take into account the effects of quantum physics. There are indeed at least two ways that quantum effects might win the universe enough time. The first is called the hypothesis of *cosmological inflation*. The idea is that as the universe expands and cools, it makes a transition between different phases. . . . This transition may have occurred very early in the history of the universe only a fraction of a second after its creation. According to the hypothesis, before the transition the universe was in a phase in which it expanded much more rapidly than it does in its present state. This is called a period of inflation, to contrast it with the present period in which the expansion is much slower. During inflation the universe may double in size every 10^{-35} of a second or so. Because of this, regions of the universe that are now billions of light years apart were initially very, very close to each other. As a result, it becomes possible for all the regions of the universe we can see to have been in contact with each other in the time since its beginning.

The hypothesis of cosmological inflation turns out to have one basic problem, which is that it requires several careful tunings of the parameters of particle physics. This is necessary not to make inflation happen, but instead to make sure that it stops. It is as if the Federal Reserve Board were trying to tune the interest rates now in order to prevent rapid inflation, not only before the next election, but for the next ten billion years. Perhaps they might then be talking about changes in interest rates of a millionth of a percent, which is at least as finely as the parameters in the theory of inflation must be chosen so that the period of rapid expansion lasts only for a very limited time.

Of course, this is only one more problem in which some parameter must be chosen very delicately if the universe is to be as we find it. As it is far from the only one, this cannot be held against the hypothesis of cosmological inflation. If there were a mechanism to tune the proton mass or the cosmological constant to incredibly tiny numbers, it could possibly do the same for the parameters that determine how long cosmological inflation lasts. What is certain is that if . . . quantum mechanics does not get rid of the singularity in our past, then inflation seems necessary to explain why the whole universe seems to have been at the same temperature at the moment of first transparency, only a million years after the first moment of time.

But there is a second possibility, which is that quantum effects might completely eradicate the singularity. In this case there would be no moment of creation. Time would instead stretch indefinitely far into the past. Regardless of inflation, there would have been enough time for all the regions of the universe to come into contact. This would not mean that cosmological inflation is wrong, for there are other reasons one might want to consider it. But in this case we have to ask what happened in the world before the "Big Bang." That term would no longer refer to a moment of creation, but only to some dramatic event that led to the expansion of our region of the universe. In this situation it becomes possible to ask if there were processes which acted before the "Big Bang" to choose the parameters of elementary particle physics.

This last sentence gives the clue to the rest of Smolin's thesis, his radical idea that processes—ones analogous to those of Darwinian evolution—account *for the emergence of "species" of universes fit for survival and complexity.*

From another angle, however, we may also see Smolin's project as something like a conflation of the Big Bang and Steady State theories, with a multiplicity of big bangs—a continuous production of universes—akin to Hoyle's "continuous creation" of matter to replenish the cosmos. And we can also see a similar motivation operating in the case of both Hoyle and Smolin: a desire to avoid at any cost the notion of a radical beginning.

Smolin continues his discussion, pursuing the issue of boundaries, including the crucial role he sees for black holes as locations for the advent of new universes. In this he follows a familiar pattern in the history of cosmology, namely the "pluralizing" of a concept that initially did not admit of a plural: from earth to "earths"; from sun to "suns"; from the galaxy to "galaxies"; and now, in Smolin (as in Rees and Guth), from the universe to "universes."

If we assume that Einstein's general theory of relativity gives a correct description of what happens to a collapsing star, then it is quite certain that what lies inside of each black hole is a singularity. This is in fact exactly what Roger Penrose proved when he found the first of the theorems about singularities.

There is an important difference from the case of the cosmological singularity, which is that in a black hole the singularity lies in the future rather than in the past. According to general relativity every bit of the collapsed star and every particle that falls afterwards into the black hole will end up at a last moment of time, at which the density of matter and strength of the gravitational field become infinite.

However, we do not trust general relativity to give us the whole story about what happens inside a black hole, for the same reason we don't trust it in the cosmological case. As the star is squeezed towards infinite density, it must pass a point at which it has been squeezed so small that effects coming from quantum mechanics are at least as important as the gravitational force squeezing the star. Whether there is a real singularity is then a question that only a theory of quantum gravity can answer.

Many people who work on quantum gravity have faith that the quantum theory will rescue us from the singularities. If so, it may be that time does not come to an end inside of each black hole. At present, despite several very interesting arguments that have recently been invented, the question of what happens inside of a black hole when quantum effects are taken into account remains unresolved.

If time ends, then there is literally nothing more to say. But what if it doesn't? Suppose that the singularity is avoided, and time goes on forever inside of a black hole. What then happens to the star that collapsed to form the black hole? As it is forever beyond the horizon, we can never see what is going on there. But if time does not end, then there is something there, happening. The question is, What?

This is very like the question about what happened "before the Big Bang" in the event that quantum effects allow time to extend indefinitely into the past. There is indeed a very appealing answer to both of these questions, which is that each answers the other. A collapsing star forms a black hole, within which it is compressed to a very dense state. The universe began in a similarly very dense state from which it expands. Is it possible that these are one and the same dense state? That is, is it possible that what is beyond the horizon of a black hole is the beginning of another universe?

This could happen if the collapsing star exploded once it reached a very dense state, but after the black hole horizon had formed around it. If we look from outside of the horizon of the black hole we will never see the explosion, for it lies beyond the range of what we can see. The outside of the black hole is the same, whether or not such an explosion happens inside of it. But suppose we do go inside, and somehow survive the compression down to extremely high density. At a certain point there is an explosion, which has the effect of reversing the collapse of the matter from the star, leading to an expansion. If we survived this also, we would seem to be in a region of the universe in which everything was moving away from each other. It would indeed resemble the early stages of our expanding universe.

This expanding region may then develop much like our own universe. It may first of all go through a period of inflation and become very big. If conditions develop suitably, galaxies and stars may form, so that in time this new "universe" may become a copy of our world. Long after this, intelligent beings may evolve who, looking back, might be tempted to believe that they lived in a universe that was born in an infinitely dense singularity, before which there was no time. But in reality they would be living in a new region of space and time created by an explosion following the collapse of a star to a black hole in our part of the universe.

The idea that a singularity in the future would be avoided by such an explosion is very old; it goes back to the 1930s, long before the idea of a black hole was invented. At this time cosmologists worried about the fate of a universe that neared its final moment of time after expanding and then recontracting. Several cosmologists speculated that we live in what they called a "Phoenix universe," which repeatedly expands and collapses, exploding again each time it becomes sufficiently dense. Such a cosmic explosion was called a "bounce," as the repeated expansions and contractions of the universe are analogous to a bouncing ball.

What we are doing is applying this bounce hypothesis, not to the universe as a whole, but to every black hole in it. If this is true, then we live not in a single universe, which is eternally passing through the same recurring cycle of collapse and rebirth. We live instead in a continually growing community of "universes," each one of which is born form an explosion following the collapse of a star to a black hole.

Recall that we wanted there to be boundaries in the past of our visible universe, when processes might have happened that could somehow choose the laws of physics, or at least select the values of their parameters. What we have learned . . . is that almost inevitably, the existence of such boundaries follows given only the simplest ideas about light and gravity and the basic fact that we live in an expanding universe. Furthermore, we have learned that if we accept the hypothesis that quantum effects eliminate the singularity at the beginning of the universe, and eliminate as well the singularities inside of black holes, we have the possibility that what lies beyond the boundaries is much vaster than our own visible universe. So with a few simple and reasonable hypotheses we get inaccessible regions with as much time as we'd like for processes to have occurred to form the laws of physics as we see them around us.

Smolin leaves his reader, the nonscientist reader at least, worrying that his cosmology is in some sense deeply circular. If we need an evolutionary process of "cosmological natural selection," with a multiplicity of new universes big-banging themselves into existence within black holes, in order to come up with the improbably complex sort of universe which we observe— in particular one that has beaten the 10^{229}-to-one odds against there being stars; and if black holes are themselves collapsed stars—then doesn't Smolin's argument presuppose the existence of the very things his thesis sets out to account for?

The other major uneasiness his reader is left with is this: Part of Smolin's critique of other physical theories of the universe, particularly those that invoke some timeless fundamental theory, centers on his charge that they involve "nostalgia for the absolute." Yet, in its own way, Smolin's thought may itself seem tinged with a kind of nostalgia. In the end, his startling if touching metaphor for the universe is not the Lake District or Mont Blanc, but his own home town, New York City. He leaves us, in short, with what may appear a cosmology less of the Big Bang than of the Big Apple.

For reasons that I thought were quite irrelevant to its content I was drawn to finish this book here, in the greatest city of the planet, my first home. A few weeks ago I took a walk around, looking for a metaphor with which to end this book, a metaphor of the universe constructed, not by a clockmaker standing outside of it but by its elements in a process of evolution, of perhaps negotiation. All of a sudden I realized what I am doing here; for, in its endless diversity and variety, what I love about the city is exactly the way it mirrors the image of the cosmos I have been struggling to bring into focus. The city is the model; it has been all around me, all the time.

Thus the metaphor of the universe we are trying now to imagine, which I would like to set against the picture of the universe as a clock, is an image of the universe as a city, as an endless negotiation, an endless construction of the new out of the old. No one made the city; there is no city-maker, as there is a clock-maker. If a city can make itself, without a maker, why can the same not be true of the universe?

Further, a city is a place where novelty may emerge without violence, where we might imagine a continual process of improvement without revolution, and in which we need respect nothing higher than ourselves, but are continually confronted with each other as the makers of our shared world. We all made it or no one did; we are of it, and to be of it and to be one of its makers is the same thing.

So there never was a God, no pilot who made the world by imposing order on chaos and who remains outside, watching and proscribing. And Nietzsche now also is dead. The eternal return, the eternal heat death, are no longer threats; they will never come, nor will heaven. The world will always be here, and it will always be different, more varied, more interesting, more alive, but still always the world in all its complexity and incompleteness. There is nothing behind it, no absolute or platonic world to transcend it. All there is of Nature is what is around us. All there is of Being is relations among real, sensible things. All we have of natural law is a world that has made itself.

SOURCE: Lee Smolin, *The Life of the Cosmos*, New York: Oxford UP, 1997.

79

The Ultimate Free Lunch

Alan Guth

Typical of the combination of mind-bending insight and occasionally awe-some banality one encounters in modern cosmology is the work of Alan Guth (b. 1947), who is credited with the theory of inflation, *a crucial innovation within the standard model. Inflation must be distinguished carefully from* expansion. *From the time of Hubble (see Chapter 64), the universe has been theorized to be expanding; but inflation, sometimes called "a big bang within the Big Bang," is a mechanism, a physical process, that marks a brief epoch in the development of the universe sometime during its first second of existence. While no anthology excerpt can do justice to inflationary theory, its basic idea is so important, and Guth such a skilled expositor, that at least a taste of his theorizing is not to be missed.*

In attempting to explain how and why this universe appeared in the first place, Guth takes his cue from Edward Tryon, who in a 1973 Nature *article entitled "Is the Universe a Vacuum Fluctuation?" proposed "that our universe is simply one of those things which happen from time to time." In other words, his theory fundamentally challenges the long-standing truism, cited from Lucretius, that "nothing can be created from nothing."*

If inflation is correct, then the inflationary mechanism is responsible for the creation of essentially all the matter and energy in the universe. The theory also implies that the observed universe is only a minute fraction of the entire universe, and it strongly suggests that there are perhaps an infinite number of other universes that are completely disconnected from our own.

Most important of all, the question of the origin of the matter in the universe is no longer thought to be beyond the range of science. After two thousand years of scientific research, it now seems likely that Lucretius was wrong. Conceivably, *everything* can be created from nothing. And "everything" might include a lot more than what we can see. In the context of inflationary cosmology, it is fair to say that the universe is the ultimate free lunch.

The core premise of inflation is a period of strong gravitational repulsion *in the early universe that causes it to double and redouble in size many times within a small interval of time. Although an adequate exposition of ideas such as that of a false vacuum would require more space than we have here, the following is a relatively compact description of Guth's vision of the inflationary process.*

The unusual notion of a material with a constant energy density [even while expanding] has led to the bizarre notion of a negative pressure. The false vacuum actually creates a suction. In using the word "suction," however, we should remember that the suction created in drinking through a soda straw is really just a pressure below that of the surrounding air, while the false vacuum creates a suction even when no pressure is applied from the outside.

Applying these ideas to a supercooled phase transition in the early universe, one might naively guess that the suction of the false vacuum would dramatically slow the expansion of the universe, maybe even reversing it. The truth, however, is exactly the opposite!

The pressure does not slow the expansion of the universe, because a pressure results in a force only if the pressure is nonuniform. For example, if a glass bottle is evacuated and sealed, then the air pressure on the outside will cause the bottle to implode if the walls are not thick enough. However, if the bottle is unsealed and air is allowed to enter, then the air pressure on the inside will very quickly match the pressure on the outside, and the walls will feel no force at all. The false vacuum in a supercooled universe would fill space uniformly, so the forces created by the negative pressure would cancel, like the air pressure inside and outside the open bottle.

Nevertheless, the negative pressure of the false vacuum leads to very peculiar *gravitational* effects. According to Newtonian physics, gravitational fields are produced only by masses. In general relativity, masses are described by their equivalent energy, and the theory implies that any form of energy creates a gravitational field. In addition, general relativity implies that a pressure can create a gravitational field. Under normal circumstances, this contribution is negligibly small: For air at room temperature, the gravitational field caused by the pressure is less than a hundred billionth of the gravitational field caused by the mass density (which is itself very small). In the early universe, however, the pressures were so high that the resulting gravitational fields were important. According to general relativity, a positive pressure creates an attractive gravitational field, as one might guess. A negative pressure, however, creates a repulsive gravitational field. For the false vacuum, the repulsive component to the gravitational field is three times as strong as the attractive component. *The false vacuum actually leads to a strong gravitational repulsion.* . . .

A short calculation shows that the gravitational repulsion causes the universe to expand exponentially. That is, the expansion is described by a *doubling time*, which for typical grand unified theory numbers is about 10^{-37} seconds. In this brief interval of time, all distances in the universe are stretched to double their original size. In two doubling times, the universe would double again, bringing it to four times its original size. After three doubling times it would be eight times its original size, and so on. As has been known since ancient times, such an exponential progression leads rapidly to stupendous numbers.

The exponential sequence was the focus, for example, of the Indian legend of King Shirham, eloquently recounted by George Gamow in *One, Two, Three . . . Infinity*. The king wanted to reward his grand vizier Sissa Ben Dahir for inventing the game of chess, so he asked the vizier to suggest an appropriate gift. Sissa Ben responded with a surprising proposition. He asked for one grain of wheat for the first square on his chessboard, two grains for the second square, four grains for the third square, and so on, until all sixty-four squares would be covered.

The king, it appears, had never studied exponentials, so he happily agreed to this seemingly modest proposal. King Shirham's mathematical education was rapidly advanced, however, as his servants brought in bags and then cartloads of wheat in an attempt to comply with Sissa Ben's request. It soon became obvious . . . that all the wheat in India would not fulfill the king's promise to his vizier. . . .

Cosmological theories tend to go a bit further than Sissa Ben Dahir, typically invoking 100 or more doublings, rather than 64. After 100 doubling times—which is only about 10^{-35} seconds—the universe would be 10^{30} times its original size! For comparison, in standard cosmology the universe would grow during this time interval by only a paltry factor of 10.

There are two difficulties associated with the standard model that the inflationary hypothesis resolves. One is the "horizon problem."

The cosmic background radiation [see Chapter 76] shows us, among other things, that the universe at 300,000 years was incredibly uniform, since the temperature of the radiation is found to be the same in all directions to an accuracy of about one part in 100,000. It is natural to ask, therefore, whether we can understand how this extreme uniformity was established.

The general tendency of objects to come to a uniform temperature is well understood, and is often called by physicists the "zeroth law of thermodynamics." If a hot cup of coffee is placed on a table, it will gradually cool to room temperature. The speed with which heat energy can be moved from one place to another, however, certainly cannot exceed the speed of light, and so the transfer of heat in the early universe is limited by the horizon distance. At 300,000 years after the big bang, the horizon distance was about 900,000 light-years.

Therefore, if one accepts the standard model without the inflationary component, and accordingly extrapolates backwards in time on the presumption of a constant rate of expansion from the beginning, then one has no way of accounting for the homogeneity of temperature exhibited by the cosmic background radiation at t+300,000 years. The inflationary hypothesis, however, allows one to conceive of how the particles of the early universe could have been close enough to each other to exert mutual thermal influence, like the coffee and the ambient air in Guth's example.

The other, perhaps even more daunting problem solved by inflation is the "flatness problem." In the standard model, the force of the Big Bang in combination with the mass and density of the universe has to be just right to produce the optimal rate of expansion. Too much force and the universe will fly apart at such a rate that, among other things, no stars or galaxies could ever form. Too little force, and gravitation causes a recollapse of the universe, again before the cosmic features we so admire have any chance of emerging. Moreover, this balance is critical: *Any deviation from it will magnify itself. Guth gives the analogy of a pencil balancing on its point. In principle, this is possible. But we don't often see it, because any slight imbalance created either by an imprecise positioning of the pencil in the first place or by an external influence such as vibration or air movement will not only adjust the pencil away from its vertical position, but also start a reaction in which the pencil accelerates away from its vertical position.*

The flatness problem . . . concerns the quantity that astronomers call omega, the ratio between the actual mass density of the universe and the critical density. (The critical density, calculated from the expansion rate, is the density that would put the universe just on the borderline between eternal expansion and eventual collapse.) The problem is caused by the instability of the situation in which omega equals one, which is like a pencil balanced on its point.

If omega is exactly equal to one, it will remain exactly one forever. But if omega differed from one by a small amount in the early universe, then the deviation would grow with time, and today omega would be very far from one. Today omega is known to lie between 0.1 and 2, implying that at one second after the big bang omega must have been between 0.999999999999999 and 1.000000000000001. Yet the standard big bang theory offers no explanation of why omega began close to one.

With inflation, however, the flatness problem disappears. The effect of gravity is reversed during the period of inflation, so all the equations describing the evolution of the universe are changed. Instead of omega being driven away from one, as it is during the rest of the history of the universe, during the period of inflation omega is driven toward one. In fact, it is driven toward one with incredible swiftness. In 100 doubling times, the difference between omega and 1 decreases by a factor of 10^{60}. With inflation, it is no longer necessary to postulate that the universe began with a value of omega incredibly close to one. Before inflation, omega could have been 1,000 or 1,000,000, or 0.001 or 0.000001, or even some number further from one. As long as the exponential expansion continues for long enough, the value of omega will be driven to one with exquisite accuracy.

To understand why inflation drives omega toward one, we can begin by recalling why this is called the flatness problem. According to general relativity, the mass density of the universe not only slows the cosmic expansion, but it also causes the universe to curve. If we assume that Einstein's cosmological constant is zero, then any mass density higher than the critical density causes the space to curve back on itself, forming a spatially closed universe. . . . In such a universe, the sum of the angles in a triangle is more than 180°. If the mass density is less than the critical density, then the space is curved in the opposite sense: the sum of the angles in a triangle would be less than 180°. On the borderline between these two cases, when the mass density is equal to the critical density, the space is not curved at all. In this case the space is flat, meaning that ordinary Euclidean geometry is valid, and the sum of the angles in a triangle is exactly 180°. Thus, if we accept the relation between omega and geometry implied by general relativity, then we need only understand why inflation drives the universe toward a state of geometric flatness.

Once the question has been restated in terms of geometric flatness, the answer is as obvious as blowing up a balloon. The more we inflate the balloon, the flatter the surface becomes. . . . To say it another way, inflation makes the universe look flat for the same reason that the surface of the earth appears flat, even though we know that the earth is really round. Since the earth is very large and we view only a small part of it at any one time, the curvature is completely imperceptible.

The standard cosmological evolution would resume at the end of inflation, so any deviation from flatness would begin to grow. The universe, however, would be so nearly flat at the end of inflation that it would remain essentially flat until the present day. Thus, the inflationary theory leads to an important prediction that is in principle testable: The present value of omega should be very precisely equal to one.

SOURCE: Alan H. Guth, *The Inflationary Universe: The Quest for a New Theory of Cosmic Origins*, Reading, Mass.: Addison-Wesley/Helix Books, 1997.

80

Was There a Big Bang?

David Berlinski

David Berlinski (b. 1942) has had a wide-ranging teaching and writing career in the United States and Europe. A novelist as well as a historian of science and mathematics, he is known for his colorful prose—some would call it purple, though "rhapsodic and florid" are the adjectives he prefers—and for his iconoclasm. His "the emperor has no clothes" approach to the orthodoxies of modern science is perhaps best epitomized by his claim that "the image of the fundamental laws of physics zestfully wrestling with the void to bring the universe into being is one that suggests very little improvement over the accounts given by the ancient Norse in which the world is revealed to be balanced on the back of a gigantic ox."

Looking at a few shards of pottery on the desert floor, the archeologist is capable of conjuring up the hanging gardens of the past, the smell of myrrh and honey in the air. His is an act of intellectual reconstruction, one made poignant by the fact that the civilization from which the artifacts spring lies forever beyond the reach of anything but remembrance and the imagination. Cosmology on the grand scale is another form of archeology; the history of the cosmos reveals itself in layers, like the strata of an ancient city.

The world of human artifacts makes sense against the assumption of a continuous human culture. The universe is something else: an old, eerie place with no continuous culture available to enable us to make sense of what we see. It is the hypothesis that the universe is *expanding* that has given cosmologists a unique degree of confidence as they climb down the cliffs of time.

A universe that is expanding is a universe with a clear path into the past. If things are now far apart, they must at one point have been close together;

and if things were once close together, they must at one point have been *hotter* than they are now, the contraction of space acting to compress its constituents like a vise, and so increase their energy. The retreat into the past ends at an initial singularity, a state in which material particles are at no distance from each other and the temperature, density, and curvature of the universe are infinite.

Berlinski summarizes the first phases of cosmic expansion from the first 10^{-43} seconds down to roughly the 400,000 year mark as the universe fills with cosmic background radiation, or CBR (for such a summary, see Smoot's in Chapter 76).

The separation of light and matter allows the galaxies to form, gravity binding the drifting dust in space. At last, the universe fills with matter, the stars settling into the sky, the far-flung suns radiating energy, the galaxies spreading themselves throughout the heavens. On the earth that has been newly made, living things shamble out of the warm oceans, the cosmic archeologist himself finally clambering over the lip of time to survey the scene and take notes on all that has occurred.

Such is the standard version of hot Big Bang cosmology—"hot" in contrast to scenarios in which the universe is cold, and "Big Bang" in contrast to various steady-state cosmologies in which nothing ever begins and nothing ever quite ends. It may seem that this archeological scenario leaves unanswered the question of *how* the show started and merely describes the consequences of some Great Cause that it cannot specify and does not comprehend. But really the question of how the show started answers itself: before the Big Bang there was nothing. *Darkness was upon the face of the deep.*

Notwithstanding the investment made by the scientific community and the general public in contemporary cosmology, a suspicion lingers that matters do not sum up as they should. Cosmologists write as if they are quite certain of the Big Bang, yet, within the last decade, they have found it necessary to augment the standard view by means of various new theories. These schemes are meant to solve problems that cosmologists were never at pains to acknowledge, so that today they are somewhat in the position of a physician reporting both that his patient has not been ill and that he has been successfully revived.

The details are instructive. It is often said, for example, that the physicists Arno Penzias and Robert Wilson observed the remnants of the Big Bang when in 1962 they detected, by means of a hum in their equipment, a signal in the night sky they could not explain. This is not quite correct. What Penzias and Wilson *observed* was simply the same irritating and ineradicable noise that has been a feature of every electrical appliance I have ever owned. What theoreticians like Robert Dicke *inferred* was something else: a connection between that cackle and the cosmic background radiation released into the universe after the era of recombination.

The cosmic hum is real enough, and so, too, is the fact that the universe is bathed in background radiation. But the era of recombination is a shimmer by the doors of theory, something known indirectly and only by means of its effects. And therein lies a puzzle. Although Big Bang cosmology does predict that the universe should be bathed in a milky film of radiation, it makes no predictions about the uniformity of its temperature. Yet, looking at the sky in every direction, cosmologists have discovered that the CBR has the same temperature, to an accuracy of one part in 100,000.

Why should this be so? CBR filled the universe some 400,000 years after the Big Bang; if its temperature thereafter is utterly and entirely the same, some physical agency must have brought this about. But by the time of recombination, the Big Bang had blown up the universe to a diameter of 90,000,000 light years. A physical signal—a light beam, say—sent hustling into the cosmos at Time Zero would, a mere 400,000 years later, be hustling still; by far the greater part of the universe would be untouched by its radiance, and so uninfluenced by the news that it carried. Since, by Einstein's theory of special relativity, nothing can travel faster than light itself, it follows that no physical agency would have had time enough to establish the homogeneity of the CBR, which appears in Big Bang cosmology as an arbitrary feature of the early universe, something that must be assumed and is not explained.

Berlinski here summarizes the hypothesis of early cosmic inflation—what George Smoot calls "the most influential concept in modern cosmology . . . [a] big bang within the big bang" (see previous chapter)—and charges it with arbitrariness. He then shifts his attack to another cardinal premise of Big Bang theory, the phenomenon of galactic redshift.

Streaming in from space, light reaches the earth like a river rich in information, the stars in the sky having inscribed strange and secret messages on its undulations. The universe is very large, light has always whispered; the nearest galaxy to our own—Andromeda—is more than two million light years away. But the universe has also seemed relatively static, and this, too, light suggests, the stars appearing where they have always appeared, the familiar dogs and bears and girdled archers of the constellations making their appointed rounds in the sky each night.

More than anything else, it is this impression that Big Bang cosmology rejects. The cool gray universe, current dogma holds, is a place of extraordinary violence, the galaxies receding from one another, the skin of creation stretching at every spot in space, the whole colossal structure blasting apart with terrible force. And this message is inscribed in light as well.

In one of its incarnations, light represents an undulation of the electromagnetic field; its source is the excitable atom itself, with electrons bouncing from one orbit to another and releasing energy as a result. Each atom has a spectral signature, a distinctive electromagnetic frequency. The light that streams in from space thus reveals something about the composition of the galaxies from which it was sent.

In the 1920s, the characteristic signature of hydrogen was detected in various far-flung galaxies. And then an odd discovery was made. Examining a very small sample of twenty or so galaxies, the American astronomer V. M. Slipher observed that the frequency of the hydrogen they sent into space was shifted to the red portion of the spectrum. It was an extraordinary observation, achieved by means of primitive equipment. Using a far more sophisticated telescope, Edwin Hubble made the same discovery in the late 1920s after Slipher had (foolishly) turned his attention elsewhere.

The galactic redshift, Hubble realized, was an exceptionally vivid cosmic clue, a bit of evidence from far away and long ago, and like all clues its value lay in the questions it prompted. Why should galactic light be shifted to the red and not the blue portions of the spectrum? Why, for that matter, should it be shifted at all?

An invigorating stab in the dark now followed. The pitch of a siren is altered as a police car disappears down the street, the sound waves carrying the noise stretched by the speed of the car itself. This is the familiar Doppler effect. Something similar, Hubble conjectured, might explain the redshift of the galaxies, with the distortions in their spectral signature arising as a reflection of their recessional velocity as they disappeared into the depths.

Observations and inferences resolved themselves into a quantitative relationship. The redshift of a galaxy, cosmologists affirm, and so its recessional velocity, is proportional to its distance and inversely proportional to its apparent brightness or flux. The relationship is known as Hubble's law, even though Hubble himself regarded the facts at his disposal with skepticism.

Hubble's law anchors Big Bang cosmology to the real world. Many astronomers have persuaded themselves that the law represents an observation, almost as if, peering through his telescope, Hubble had noticed the galaxies zooming off into the far distance. This is nonsense. Hubble's law consolidates a number of very plausible intellectual steps. The light streaming in from space is relieved of its secrets by means of ordinary and familiar facts, but even after the facts are admitted into evidence, the relationship among the redshift of the galaxies, their recessional velocity, and their distance represents a complicated inference, an intellectual leap.

The Big Bang rests on the hypothesis that the universe is expanding, and in the end the plausibility of its claims will depend on whether the universe *is* expanding. Astronomers can indeed point to places in the sky where the redshift of the galaxies appears to be a linear function of their distance. But in astrophysics, as in evolutionary biology, it is failure rather than success that is of significance. The astrophysical literature contains interesting and disturbing evidence that the linear relationship at the heart of Hubble's law by no means describes the facts fully.

At the end of World War II, astronomers discovered places in the sky where charged particles moving in a magnetic field sent out strong signals in the radio portion of the spectrum. Twenty years later, Alan Sandage and Thomas Mathews identified the source of such signals with optically discernible points in space. These are the quasars—*quasi stellar radio sources*.

Quasars have played a singular role in astrophysics. In the mid-1960s, Maarten Schmidt discovered that their spectral lines were shifted massively to the red. If Hubble's law were correct, quasars should be impossibly far away, hurtling themselves into oblivion at the far edge of space and time. But for more than a decade, the American astronomer Halton Arp has drawn the attention of the astronomical community to places in the sky where the expected relationship between redshift and distance simply fails. Embarrassingly enough, many quasars seem bound to nearby galaxies. The results are in plain sight: there on the photographic plate is the smudged record of a galaxy, and there next to it is a quasar, the points of light lined up and looking for all the world as if they were equally luminous.

These observations do not comport with standard Big Bang cosmology. If quasars have very large redshifts, they must (according to Hubble's law) be very far away; if they seem nearby, then either they must be fantastically luminous or their redshift has not been derived from their velocity. The tight tidy series of inferences that has gone into Big Bang cosmology, like leverage in commodity trading, works beautifully in reverse, physicists like speculators finding their expectations canceled by the very processes they had hoped to exploit.

Acknowledging the difficulty, some theoreticians have proposed that quasars have been caught in the process of evolution. Others have scrupled at Arp's statistics. Still others have claimed that his samples are too small, although they have claimed this for every sample presented and will no doubt continue to claim this when the samples number in the billions. But whatever the excuses, a great many cosmologists recognize that quasars mark a point where the otherwise silky surface of cosmological evidence encounters a snag.

Within any scientific discipline, bad news must come in battalions before it is taken seriously. Cosmologists can point to any number of cases in which disconcerting evidence has resolved itself in their favor; a decision to regard the quasars with a watchful indifference is not necessarily irrational. The galaxies are another matter. They are central to Hubble's law; it is within the context of galactic observation that the crucial observational evidence for the Big Bang must be found or forged.

The battalions now begin to fill. The American mathematician I. E. Segal and his associates have studied the evidence for galactic recessional velocity over the course of twenty years, with results that are sharply at odds with predictions of Big Bang cosmology. Segal is a distinguished, indeed a great mathematician, one of the creators of modern function theory and a member of the National Academy of Sciences. He has incurred the indignation of the astrophysical community by suggesting broadly that their standards of statistical rigor would shame a sociologist. Big Bang cosmology, he writes,

> owes its acceptance as a physical principle primarily to the uncritical and premature representation of [the redshift-distance relationship] as an empirical fact.... Observed discrepancies ... have been resolved by a pyramid of exculpatory assumptions, which are inherently incapable of noncircular substantiation.

These are strong words of remonstration, but they are not implausible. Having constructed an elaborate scientific orthodoxy, cosmologists have acquired a vested interest in its defense. The astrophysicists J. G. Hoessell, J. E. Gunn, and T .X. Thuan, for example, report with satisfaction that within the structures described by G. O. Abell's *Catalog of Bright Cluster Galaxies* (1958), prediction and observation cohere perfectly to support Hubble's law. Abell's catalog is a standard astronomical resource, used by cosmologists everywhere—but it is useless as evidence for Hubble's law. "In determining whether a cluster meets selection criterion," Abell affirms, "it was assumed that their redshifts were proportional to their distance." If this is what Abell *assumed*, there is little point in asking what conclusions he *derived*.

The fact that the evidence in favor of Hubble's law may be biased does not mean that it is untrue; bias may suggest nothing more than a methodological flaw. But Segal is persuaded that when the evidence is soberly considered, it *does* contravene accepted doctrine, statistical sloppiness functioning, as it so often does, simply to conceal the facts.

A statistical inference is compelling only if the samples upon which it rests are objectively compelling. Objectivity, in turn, requires that the process of sampling be both reasonably complete and unbiased. Segal and his colleagues have taken pains to study samples that within the limits of observation are both. Their most recent study contains a detailed parallel analysis of Hubble's law across four wave bands, one that essentially surveys all stellar objects within each band. The analysis is based on new data drawn from the G. de Vaucoleurs survey of bright cluster galaxies, which includes more than 10,000 galaxies. Hubble's own analysis, it is worthwhile to recall, was limited to twenty galaxies.

The results of their analysis are disturbing. The linear relationship that Hubble saw, Segal and his collaborators cannot see and have not found. Rather, the relationship between redshift and flux or apparent brightness that they have studied in a large number of complete samples satisfies a quadratic law, the redshift varying as the square of apparent brightness. "By normal standards of scientific due process," Segal writes, "the results of [Big Bang] cosmology are illusory."

Cosmologists have dismissed Segal's claims with a great snort of indignation. But the discrepancy from Big Bang cosmology that they reveal is hardly trivial. Like evolutionary biologists, cosmologists are often persuaded that they are in command of a structure intellectually powerful enough to accommodate gross discrepancies in the data. This is a dangerous and deluded attitude. Hubble's law embodies a general hypothesis of Big Bang cosmology—namely, that the universe is expanding—and while the law cannot be established by observation, observation *can* establish that it may be false. A statistically responsible body of contravening evidence has revealed something more than an incidental defect. Indifference to its implications amounts to a decision to place Big Bang cosmology beyond rational inquiry.

Berlinski's critique moves to problems (among others) such as those that arise when mathematical conceptions like "infinite density" are predicated of physical reality, concluding that "the sharp, clean, bracing light that the Big Bang was to have thrown on the very origins of space and time lapses when it is most needed."

Like so many haunting human stories, the scientific story of the Big Bang is circular in the progression of its ideas and circular thus in its deepest nature. Cosmologists have routinely assumed that the universe is expanding because they have been persuaded of FL [Friedmann-Lemaître] cosmology; and they have been persuaded of FL cosmology because they have routinely assumed that the universe is expanding. The pattern would be intellectually convenient if it were intellectually compelling.

If the evidence in favor of Big Bang cosmology is more suspect than generally imagined, its defects are far stronger than generally credited. Whatever else it may be, the universe is a bright, noisy, energetic place. There are monstrously large galaxies in the skies and countlessly many suns burning with fierce thermonuclear fires. Black holes are said to loiter here and there, sucking in matter and light and releasing it slowly in the form of radiation. Whence the energy for the show, the place where accounts are settled? The principles of nineteenth-century physics require that, in one way or another, accounts *must* be settled. Energy is neither created nor destroyed.

Hot Big Bang cosmology appears to be in violation of the first law of thermodynamics. The global energy needed to run the universe has come from nowhere, and to nowhere it apparently goes as the universe loses energy by cooling itself.

This contravention of thermodynamics expresses, in physical form, a general philosophical anxiety. Having brought space and time into existence, along with everything else, the Big Bang itself remains outside any causal scheme. The creation of the universe remains unexplained by any force, field, power, potency, influence, or instrumentality known to physics—or to man. The whole vast imposing structure organizes itself from absolutely nothing.

This is not simply difficult to grasp. It is incomprehensible.

Physicists, no less than anyone else, are uneasy with the idea that the universe simply popped into existence, with space and time "suddenly switching themselves on." The image of a light switch comes from Paul Davies, who uses it to express a miracle without quite recognizing that it embodies a contradiction. A universe that has *suddenly* switched itself on has accomplished something within time; and yet the Big Bang is supposed to have brought space and time into existence.

Having entered a dark logical defile, physicists often find it difficult to withdraw. Thus, Alan Guth writes in pleased astonishment that the universe really did arise from "essentially . . . nothing at all": as it happens, a false vacuum patch "10^{-26} centimeters in diameter" and "10^{-32} solar masses." It would appear, then, that "essentially nothing" has both spatial extension and mass. While these facts may strike Guth as inconspicuous, others may suspect that nothingness, like death, is not a matter that admits of degrees.

The attempt to discover some primordial stuff that can be described both as nothing and as something recalls the Maori contemplating the manifold mysteries of *po*. This apparently gives Stephen Hawking pause. "To ask what happened before the universe began," he has written, "is like asking for a point on the Earth at 91 degrees north latitude." We are on the inside of the great sphere of space and time, and while we can see to the boundaries, there is nothing beyond to see if only because there is nothing beyond. "Instead of talking about the universe being created, and maybe coming to an end," Hawking writes, "one should just say: the universe is."

Now this is a conclusion to which mystics have always given their assent; but having concluded that the universe just "is," cosmologists, one might think, would wish to know why it is. The question that Hawking wishes to evade disappears as a question in physics only to reappear as a question in philosophy; we find ourselves traveling in all the old familiar circles.

Standing at the gate of modern time, Isaac Newton forged the curious social pact by which rational men and women have lived ever since. The description of the physical world would be vouchsafed to a particular institution, that of mathematical physics; and it was to the physicists and not the priests, soothsayers, poets, politicians, novelists, generals, mystics, artists, astrologers, warlocks, wizards, or enchanters that society would look for judgments about the nature of the physical world. If knowledge is power, the physicists have, by this arrangement, been given an enormous privilege. But a social arrangement is among other things a contract: something is given, but something is expected as well. In exchange for their privilege, the physicists were to provide an account of the physical world at once penetrating, general, persuasive, and true.

Until recently, the great physicists have been scrupulous about honoring the terms of their contract. They have attempted with dignity to respect the distinction between what is known and what is not. Even quantum electrodynamics, the most successful theory ever devised, was described honestly by its founder, Richard Feynman, as resting on a number of unwholesome mathematical tricks.

This scrupulousness has lately been compromised. The result has been the calculated or careless erasure of the line separating disciplined physical inquiry from speculative metaphysics. Contemporary cosmologists feel free to say anything that pops into their heads. Unhappy examples are everywhere: absurd schemes to model time on the basis of the complex numbers, as in Stephen Hawking's *A Brief History of Time*; bizarre and ugly contraptions for cosmic inflation; universes multiplying beyond the reach of observation; white holes, black holes, worm holes, and naked singularities; theories of every stripe and variety, all of them uncorrected by any criticism beyond the trivial.

The physicists carry on endlessly because they can. Just recently, for example, Lee Smolin, a cosmologist at [Pennsylvania State University], has offered a Darwinian interpretation of cosmology, a theory of "cosmological natural selection." On Smolin's view, the Big Bang happened within a black hole; new universes are bubbling up all the time, each emerging from its own black hole and each provided with its own set of physical laws, so that the very concept of a law of nature is shown to be a part of the mutability of things.

There is, needless to say, no evidence whatsoever in favor of this preposterous theory. The universes that are bubbling up are unobservable. So, too, are the universes that have bubbled up and those that will bubble up in the future. Smolin's theories cannot be confirmed by experience. Or by anything else. What law of nature could reveal that the laws of nature are contingent? Yet the fact that when Smolin's theory is self-applied it self-destructs has not prevented physicists like Alan Guth, Roger Penrose, and Martin Rees from circumspectly applauding the effort nonetheless.

A scientific crisis has historically been the excuse to which scientists have appealed for the exculpation of damaged doctrines. Smolin is no exception. "We are living," he writes, "through a period of scientific crisis." Ordinary men and women may well scruple at the idea that cosmology is in crisis because cosmologists, deep down, have run out of interesting things to say, but in his general suspicions Smolin is no doubt correct. What we are discovering is that many areas of the universe are apparently protected from our scrutiny, like sensitive files sealed from view by powerful encryption codes. However painful, the discovery should hardly be unexpected. Beyond every act of understanding, there is an abyss.

Like Darwin's theory of evolution, Big Bang cosmology has undergone that curious social process in which a scientific theory is promoted to a secular myth. The two theories serve as points of certainty in an intellectual culture that is otherwise disposed to give the benefit of the doubt to doubt itself. It is within the mirror of these myths that we have come to see ourselves. But if the promotion of theory into myth satisfies one human agenda, it violates another. Myths are quite typically false, and science is concerned with truth. Human beings, it would seem, may make scientific theories or they may make myths, but with respect to the same aspects of experience, they cannot quite do both.

SOURCE: Abridged from David Berlinski, "Was There a Big Bang?" *Commentary*, 105.2 (February 1998).

81

What We Cannot See and Yet Know Must Be There

Vera Rubin

Vera Rubin (b. 1928) has since 1965 been a staff member in the Department of Terrestrial Magnetism at the Carnegie Institution in Washington, D.C., and has published widely on issues related to galactic structure. In 1993, the President of the United States awarded her the National Medal of Science. In the following learned and imaginative article, published in Scientific American Presents *in 1998, Rubin explores one of the enduring mysteries of the cosmos: What is dark matter?[1]*

Imagine, for a moment, that one night you awaken abruptly from a dream. Coming to consciousness, blinking your eyes against the blackness, you find that, inexplicably, you are standing alone in a vast, pitch-black cavern. Befuddled by this predicament, you wonder: Where am I? What is this space? What are its dimensions?

Groping in the darkness, you stumble upon a book of damp matches. You strike one; it quickly flares, then fizzles out. Again, you try; again, a flash and fizzle. But in that moment, you realize that you can glimpse a bit of your surroundings. The next match strike lets you sense faint walls far away. Another flare reveals a strange shadow, suggesting the presence of a big object. Yet

---

another suggests you are moving—or, instead, the room is moving relative to you. With each momentary flare, a bit more is learned.

In some sense, this situation recalls our puzzling predicament on Earth. Today, as we have done for centuries, we gaze into the night sky from our planetary platform and wonder where we are in this cavernous cosmos. Flecks of light provide some clues about great objects in space. And what we do discern about their motions and apparent shadows tells us that there is much more that we cannot yet see.

From every photon we collect from the universe's farthest reaches, we struggle to extract information. Astronomy is the study of light that reaches Earth from the heavens. Our task is not only to collect as much light as possible—from ground- and space-based telescopes—but also to use what we can see in the heavens to understand better what we cannot see and yet know must be there.

Based on 50 years of accumulated observations of the motions of galaxies and the expansion of the universe, most astronomers believe that as much as 90 percent of the stuff constituting the universe may be objects or particles that cannot be seen. In other words, most of the universe's matter does not radiate—it provides no glow that we can detect in the electromagnetic spectrum. First posited some 60 years ago by astronomer Fritz Zwicky, this so-called missing matter was believed to reside within clusters of galaxies. Nowadays we prefer to call the missing mass "dark matter," for it is the light, not the matter, that is missing.

Astronomers and physicists offer a variety of explanations for this dark matter. On the one hand, it could merely be ordinary material, such as ultrafaint stars, large or small black holes, cold gas, or dust scattered around the universe—all of which emit or reflect too little radiation for our instruments to detect. It could even be a category of dark objects called MACHOs (MAssive Compact Halo Objects) that lurk invisibly in the halos surrounding galaxies and galactic clusters. On the other hand, dark matter could consist of exotic, unfamiliar particles that we have not figured out how to observe. Physicists theorize about the existence of these particles, although experiments have not yet confirmed their presence. A third possibility is that our understanding of gravity needs a major revision—but most physicists do not consider that option seriously.

In some sense, our ignorance about dark matter's properties has become inextricably tangled up with other outstanding issues in cosmology—such as how much mass the universe contains, how galaxies formed and whether or not the universe will expand forever. So important is this dark matter to our understanding of the size, shape and ultimate fate of the universe that the search for it will very likely dominate astronomy for the next few decades.

OBSERVING THE INVISIBLE

Understanding something you cannot see is difficult—but not impossible. Not surprisingly, astronomers currently study dark matter by its effects on the bright matter that we do observe. For instance, when we watch a nearby star wobbling predictably, we infer from calculations that a "dark planet" orbits around it. Applying similar principles to spiral galaxies, we infer dark matter's presence because it accounts for the otherwise inexplicable motions of stars within those galaxies.

When we observe the orbits of stars and clouds of gas as they circle the centers of spiral galaxies, we find that they move too quickly. These unexpectedly high velocities signal the gravitational tug exerted by something more than that galaxy's visible matter. From detailed velocity measurements, we conclude that large amounts of invisible matter exert the gravitational force that is holding these stars and gas clouds in high-speed orbits. We deduce that dark matter is spread out around the galaxy, reaching beyond the visible galactic edge and bulging above and below the otherwise flattened, luminous galactic disk. As a rough approximation, try to envision a typical spiral galaxy, such as our Milky Way, as a relatively flat, glowing disk embedded in a spherical halo of invisible material—almost like an extremely diffuse cloud.

Looking at a single galaxy, astronomers see within the galaxy's radius (a distance of about 50,000 light-years) only about one tenth of the total gravitating mass needed to account for how fast individual stars are rotating around the galactic hub.

In trying to discover the amount and distribution of dark matter in a cluster of galaxies, x-ray astronomers have found that galaxies within clusters float immersed in highly diffuse clouds of 100-million-degree gas—gas that is rich in energy yet difficult to detect. Observers have learned to use the x-ray-emitting gas's temperature and extent in much the same way that optical astronomers use the velocities of stars in a single galaxy. In both cases, the data provide clues to the nature and location of the unseen matter.

In a cluster of galaxies, the extent of the x-ray-emitting region and temperature of the gas enable us to estimate the amount of gravitating mass within the cluster's radius, which measures almost 100 million light-years. In a typical case, when we add together the luminous matter and the x-ray-emitting hot gas, we are able to sense roughly 20 to 30 percent of the cluster's total gravitating mass. The remainder, which is dark matter, remains undetected by present instruments.

Subtler ways to detect invisible matter have recently emerged. One clever method involves spotting rings or arcs around clusters of galaxies. These

"Einstein rings" arise from an effect known as gravitational lensing, which occurs when gravity from a massive object bends light passing by. For instance, when a cluster of galaxies blocks our view of another galaxy behind it, the cluster's gravity warps the more distant galaxy's light, creating rings or arcs, depending on the geometry involved. Interestingly, the nearer cluster acts as nature's telescope, bending light into our detectors—light that would otherwise have traveled elsewhere in the universe. Someday we may exploit these natural telescopes to view the universe's most distant objects.

Using computers models, we can calculate the mass of the intervening cluster, estimating the amount of invisible matter that must be present to produce the observed geometric deflection. Such calculations confirm that clusters contain far more mass than the luminous matter suggests.

Even compact dark objects in our own galaxy can gravitationally lens light. When a foreground object eclipses a background star, the light from the background star is distorted into a tiny ring, whose brightness far exceeds the star's usual brightness. Consequently, we observe an increase, then a decrease, in the background star's brightness. Careful analysis of the light's variations can tease out the mass of the dark foreground lensing object.

WHERE IS DARK MATTER?

Several teams search nightly for nearby lensing events, caused by invisible MACHOs in our own Milky Way's halo. The search for them covers millions of stars in the Magellanic Clouds and the Andromeda galaxy. Ultimately, the search will limit the amount of dark matter present in our galaxy's halo.

Given the strong evidence that spiral and elliptical galaxies lie embedded in large dark-matter halos, astronomers now wonder about the location, amount and distribution of the invisible material.

To answer those questions, researchers compare and contrast observations from specific nearby galaxies. For instance, we learn from the motions of the Magellanic Clouds, two satellite galaxies gloriously visible in the Southern Hemisphere, that they orbit within the Milky Way galaxy's halo and that the halo continues beyond the clouds, spanning a distance of almost 300,000 light-years. In fact, motions of our galaxy's most distant satellite objects suggest that its halo may extend twice as far—to 600,000 light-years.

Because our nearest neighboring spiral galaxy, Andromeda, lies a mere two million light-years away, we now realize that our galaxy's halo may indeed span a significant fraction of the distance to Andromeda and its halo. We have also determined that clusters of galaxies lie embedded in even larger systems of dark matter. At the farthest distances for which we can deduce the masses of galaxies, dark matter appears to dwarf luminous matter by a factor of at least 10, possibly as much as 100.

Overall, we believe dark matter associates loosely with bright matter, because the two often appear together. Yet, admittedly, this conclusion may stem from biased observations, because bright matter typically enables us to find dark matter.

By meticulously studying the shapes and motions of galaxies over decades, astronomers have realized that individual galaxies are actively evolving, largely because of the mutual gravitational pull of galactic neighbors. Within individual galaxies, stars remain enormously far apart relative to their diameters, thus little affecting one another gravitationally. For example, the separation between the sun and its nearest neighbor, Proxima Centauri, is so great that 30 million suns could fit between the two. In contrast, galaxies lie close together, relative to their diameters—nearly all have neighbors within a few diameters. So galaxies do alter one another gravitationally, with dark matter's added gravity a major contributor to these interactions.

As we watch many galaxies—some growing, shrinking, transforming or colliding—we realize that these galactic motions would be inexplicable without taking dark matter into account. Right in our own galactic neighborhood, for instance, such interactions are under way. The Magellanic Clouds, our second nearest neighboring galaxies, pass through our galaxy's plane every billion years. As they do, they mark their paths with tidal tails of gas and, possibly, stars. Indeed, on every passage, they lose energy and spiral inward. In less than 10 billion years, they will fragment and merge into the Milky Way.

Recently astronomers identified a still nearer neighboring galaxy, the Sagittarius dwarf, which lies on the far side of the Milky Way, close to its outer edge. (Viewed from Earth, this dwarf galaxy appears in the constellation Sagittarius.) As it turns out, gravity from our galaxy is pulling apart this dwarf galaxy, which will cease to exist as a separate entity after several orbits. Our galaxy itself may be made up of dozens of such previous acquisitions.

Similarly, the nearby galaxy M31 and the Milky Way are now hurtling toward each other at the brisk clip of 130 kilometers (81 miles) per second. As eager spectators, we must watch this encounter for a few decades to know if M31 will strike our galaxy or merely slide by. If they do collide, we will lose: the Milky Way will merge into the more massive M31. Computer models predict that in about four billion years the galactic pair will become one spheroidal galaxy. Of course, by then our sun will have burned out—so others in the universe will have to enjoy the pyrotechnics.

In many ways, our galaxy, like all large galaxies, behaves as no gentle neighbor. It gobbles up nearby companions and grinds them into building blocks for its own growth. Just as Earth's continents slide beneath our feet, so, too, does our galaxy evolve around us. By studying the spinning, twisting and turning motions and structures of many galaxies as they hurtle through space, astronomers can figure out the gravitational forces required to sustain their motions—and the amount of invisible matter they must contain.

How much dark matter does the universe contain? The destiny of the universe hinges on one still unknown parameter: the total mass of the universe. If we live in a high-density, or "closed," universe, then mutual gravitational attraction will ultimately halt the universe's expansion, causing it to contract—culminating in a big crunch, followed perhaps by reexpansion. If, on the other hand, we live in a low-density, or "open," universe, then the universe will expand forever.

Observations thus far suggest that the universe—or, at least, the region we can observe—is open, forever expanding. When we add up all the luminous matter we can detect, plus all the dark matter that we infer from observations, the total still comes to only a fraction—perhaps 20 percent—of the density needed to stop the universe from expanding forever.

I would be content to end the story there, except that cosmologists often dream of, and model, a universe with "critical" density—meaning one that is finely balanced between high and low density. In such a universe, the density is just right. There is enough matter to slow the universe's continuous expansion, so that it eventually coasts nearly to a halt. Yet this model does not describe the universe we actually measure. As an observer, I recognize that more matter may someday be detected, but this does not present sufficient reason for me to adopt a cosmological model that observations do not yet require.

Another complicating factor to take into account is that totally dark systems may exist—that is, there may be agglomerations of dark matter into which luminous matter has never penetrated. At present, we simply do not know if such totally dark systems exist because we have no observational data either to confirm or to deny their presence.

WHAT IS DARK MATTER?

Whatever dark matter turns out to be, we know for certain that the universe contains large amounts of it. For every gram of glowing material we can detect, there may be tens of grams of dark matter out there. Currently the astronomical jury is still out as to exactly what constitutes dark matter. In fact, one could say we are still at an early stage of exploration. Many candidates exist to account for the invisible mass, some relatively ordinary, others rather exotic.

Nevertheless, there is a framework in which we must work. Nucleosynthesis, which seeks to explain the origin of elements after the big bang, sets a limit to the number of baryons—particles of ordinary, run-of-the-mill matter—that can exist in the universe. The limit arises out of the Standard Model of the universe, which has one free parameter—the ratio of the number of baryons to the number of photons.

From the temperature of the cosmic microwave background—which has been measured—the number of photons is now known. Therefore, to determine the number of baryons, we must observe stars and galaxies to learn the cosmic abundance of light nuclei, the only elements formed immediately after the big bang.

Without exceeding the limits of nucleosynthesis, we can construct an acceptable model of a low-density, open universe. In that model, we take approximately equal amounts of baryons and exotic matter (nonbaryonic particles), but in quantities that add up to only 20 percent of the matter needed to close the universe. This model matches all our actual observations. On the other hand, a slightly different model of an open universe in which all matter is baryonic would also satisfy observations. Unfortunately, this alternative model contains too many baryons, violating the limits of nucleosynthesis. Thus, any acceptable low-density universe has mysterious properties: most of the universe's baryons would remain invisible, their nature unknown, and in most models much of the universe's matter is exotic.

EXOTIC PARTICLES

Theorists have posited a virtual smorgasbord of objects to account for dark matter, although many of them have fallen prey to observational constraints. As leading possible candidates for baryonic dark matter, there are black holes (large and small), brown dwarfs (stars too cold and faint to radiate), sun-size MACHOs, cold gas, dark galaxies and dark clusters, to name only a few.

The range of particles that could constitute nonbaryonic dark matter is limited only slightly by theorists' imaginations. The particles include photinos, neutrinos, gravitinos, axioms and magnetic monopoles, among many others. Of these, researchers have detected only neutrinos—and whether neutrinos have any mass remains unknown. Experiments are under way to detect other exotic particles. If they exist, and if one has a mass in the correct range, then that particle might pervade the universe and constitute dark matter. But these are very large "ifs."

To a great extent, the details of the evolution of galaxies and clusters depend on properties of dark matter. Without knowing those properties, it is difficult to explain how galaxies evolved into the structures observed today. As knowledge of the early universe deepens, I remain optimistic that we will soon know much more about both galaxy formation and dark matter.

What we fail to see with our eyes, or detectors, we can occasionally see with our minds, aided by computer graphics. Computers now play a key role in the search for dark matter. Historically, astronomers have focused on observations; now the field has evolved into an experimental science. Today's astronomical experimenters sit neither at lab benches nor at telescopes but at computer terminals. They scrutinize cosmic simulations in which tens of thousands of points, representing stars, gas and dark matter, interact gravitationally over a galaxy's lifetime. A cosmologist can tweak a simulation by adjusting the parameters of dark matter and then watch what happens as virtual galaxies evolve in isolation or in a more realistic, crowded universe.

Computer models can thus predict galactic behavior. For instance, when two galaxies suffer a close encounter, violently merging or passing briefly in the night, they sometimes spin off long tidal tails. Yet from the models, we now know these tails appear only when the dark matter of each galaxy's halo is three to 10 times greater than its luminous matter. Heavier halos produce stubbier tails. This realization through modeling has helped observational astronomers to interpret what they see and to understand more about the dark matter they cannot see. For the first time in the history of cosmology, computer simulations actually guide observations.

New tools, no less than new ways of thinking, give us insight into the structure of the heavens. Less than 400 years ago Galileo put a small lens at one end of a cardboard tube and a big brain at the other end. In so doing, he learned that the faint stripe across the sky, called the Milky Way, in fact comprised billions of single stars and stellar clusters. Suddenly, a human being understood what a galaxy is. Perhaps in the twenty-first century, another—as yet unborn—big brain will put her eye to a clever new instrument and definitively answer, What is dark matter?

SOURCE: Vera Rubin, "Dark Matter in the Cosmos," *Scientific American Presents*, 9.1 (spring 1998).

82

Their Extravagant Smallness
Freeman Dyson and Brian Greene

Although the main emphasis of this book is writings about things macro-scopic—big items like the moon, the solar system, and the universe itself—it is only right to acknowledge an intimate connection between the macro-scopic and the microscopic. In fact, one of the most intriguing theories em-ployed in efforts to understand the universe posits things that are radically submicroscopic, things called superstrings. While no thorough introduction is possible here, Freeman Dyson (b. 1923) and Brian Greene (b. 1962) give us a worthwhile preliminary glimpse into the cosmological relevance of su-perstring theory. One of Dyson's Gifford Lectures in 1985 entitled "Butter-flies and Superstrings" reveals, among other things, the importance to science (and scientists) of sheer beauty.

I will not explain what butterflies and superstrings are. To explain butterflies is unnecessary because everyone has seen them. To explain superstrings is im-possible because nobody has seen them. But please do not think I am trying to mystify you. Superstrings and butterflies are examples illustrating two dif-ferent aspects of the universe and two different notions of beauty. Super-strings come at the beginning and butterflies at the end because they are extreme examples. Butterflies are at the extreme of concreteness, superstrings at the extreme of abstraction. They mark the extreme limits of the territory over which science claims jurisdiction. Both are, in their different ways, beautiful. Both are, from a scientific point of view, poorly understood. Scien-tifically speaking, a butterfly is at least as mysterious as a superstring. When something ceases to be mysterious it ceases to be of absorbing concern to sci-entists. Almost all the things scientists think and dream about are mysteri-ous. . . .

If I were to follow the example of Euclid and try to give a definition of a superstring, it would be something like this: "A superstring is a wiggly curve which moves in a ten-dimensional space-time of peculiar symmetry." Like Euclid's definition of a point ["that which has no parts, or which has no magnitude"], this tells us very little. It gives us only a misleading picture of a wiggly curve thrashing about in a dark room all by itself. In fact a superstring is never all by itself, just as a Euclidean point is never all by itself. The superstring theory has the ten-dimensional space-time filled with a seething mass of superstrings. The objects with which the theory deals are not individual superstrings but symmetry-groups of states in which superstrings may be distributed. The symmetry-groups are supposed to be observable. If the theory is successful, the symmetry-groups derived from the mathematics of superstrings will be found to be in correspondence with the symmetry-groups of fields and particles seen in the laboratory. The correspondence is still far from being established in detail. . . . [But] to have found a theory of the universe which is not mathematically self-contradictory is already a considerable achievement.

The name "superstring" grew out of a historical analogy. The greatest achievement of Einstein was his 1915 theory of gravity. Sixty years later, a new version of Einstein's theory was discovered which brought gravity into closer contact with the rest of physics. The new version of gravity was called "supergravity." About the same time that supergravity was invented, another model of particle interactions was proposed. The new particle model was called "String Theory," because it represented particles by one-dimensional curves or strings. Finally, the same mathematical trick which turned gravity into supergravity turned strings into superstrings. That is how superstrings acquired their name. The name, like the superstring itself, is a mathematical abstraction.

Superstrings have one striking characteristic which is easy to express in words. They are small. They are extravagantly small. Their extravagant smallness is one of the main reasons why we can never hope to observe them directly. To give a quantitative idea of their smallness, let me compare them with other things that are not so small. Imagine, if you can, four things that have very different sizes. First, the entire visible universe. Second, the planet earth. Third, the nucleus of an atom. Fourth, a superstring. The step in size from each of these things to the next is roughly the same. The earth is smaller than the visible universe by about twenty powers of ten. An atomic nucleus is smaller than the earth by twenty powers of ten. And a superstring is smaller than a nucleus by twenty powers of ten. That gives you a rough measure of how far we have to go in the domain of the small before we reach superstrings. . . .

What philosophical conclusions should we draw from the abstract style of the superstring theory? We might conclude, as Sir James Jeans concluded long ago, that the Great Architect of the Universe now begins to appear as a Pure Mathematician, and that if we work hard enough at mathematics we shall be able to read His mind. Or we might conclude that our pursuit of abstractions is leading us far away from those parts of the creation which are most interesting from a human point of view. It is too early yet to come to conclusions. We should at least wait until the experts have decided whether the superstring theory has anything to do with the universe we are living in.

Dyson does return, at the end of his lecture, to the butterfly, both as a beautiful and wonderful thing in itself and, just as important, as a living icon of transcendence: "The Monarch butterfly, which flies up into the summer sky, over the trees and far away, a symbol of evanescent beauty and a living proof that nature's imagination is richer than our own."

More than a decade later, Brian Greene's high-profile book The Elegant Universe *likewise points to string theory's aesthetic connections. Near the end of the book, moreover, Greene reflects on this theory's role in cosmology, highlighting with candor and modesty both the promise and the possible limitations of superstrings for the unification of physics. (The "M" in "string/M-theory" hints at Mystery.)*

Cosmology has the ability to grab hold of us at a deep, visceral level because an understanding of how things began feels—at least to some—like the closest we may ever come to understanding *why* they began. That is not to say that modern science provides a connection between the question of how and the question of why—it doesn't—and it may well be that no such scientific connection is ever found. But the study of cosmology does hold the promise of giving us our most complete understanding of the arena of the why—the birth of the universe—and this at least allows for a scientifically informed view of the frame within which the questions are asked. Sometimes attaining the deepest familiarity with a question is our best substitute for actually having an answer.

In the context of searching for the ultimate theory, these lofty reflections on cosmology give way to far more concrete considerations. The way things in the universe appear to us today . . . depends upon the fundamental laws of physics, to be sure, but it may also depend on aspects of cosmological evolution, from the far left-hand side of the [total universal] time line, that potentially lie outside the scope of even the deepest theory.

It's not hard to imagine how this might be. Think of what happens, for example, when you toss a ball in the air. The laws of gravity govern the ball's subsequent motion, but we can't predict where the ball will land exclusively from those laws. We must also know the velocity of the ball—its speed and direction—as it left your hand. That is, we must know the *initial conditions* of the ball's motion. Similarly, there are features of the universe that also have a historical contingency—the reason why a star formed here or a planet there depends upon a complicated chain of events that, at least in principle, we can imagine tracing back to some feature of how the universe was when it all began. But it is possible that even more basic features of the universe, perhaps even the properties of the fundamental matter and force particles, also have a direct dependence on historical evolution—evolution that itself is contingent upon the initial conditions of the universe. . . .

We don't know what the initial conditions of the universe were, or even the ideas, concepts, and language that should be used to describe them. We believe that the outrageous initial state of *infinite* energy, density, and temperature that arises in the standard and inflationary cosmological models is a signal that these theories have broken down rather than a correct description of the physical conditions that actually existed. String theory offers an improvement by showing how such infinite extremes might be avoided; nevertheless, no one has any insight on the question of how things actually did begin. In fact, our ignorance persists on an even higher plane: We don't know whether the question of determining the initial conditions is one that is even sensible to ask or whether—like asking general relativity to give insight into how hard you happened to toss a ball in the air—it is a question that lies forever beyond the grasp of any theory. Valiant attempts by physicists such as Hawking and James Hartle . . . have tried to bring the question of cosmological initial conditions within the umbrella of physical theory, but all such attempts remain inconclusive. In the context of string/M-theory, our cosmological understanding is, at present, just too primitive to determine whether our candidate "theory of everything" truly lives up to its name and determines its own cosmological initial conditions, thereby elevating them to the status of physical law. This is a prime question for future research.

SOURCE: Freeman Dyson, *Infinite in All Directions*, New York: Harper & Row, 1988; Brian Greene, *The Elegant Universe: Superstrings, Hidden Dimensions, and the Quest for the Ultimate Theory*, New York: Norton, 1999.

83

Cosmic Dust-Bunnies

John S. Lewis

John S. Lewis (b. 1941) is Professor of Planetary Sciences at the University of Arizona-Tucson. He matches his skills as a scientist with traits of a fine story teller: a keen, sometimes irreverent sense of metaphor and analogy; a way of entangling the reader in the web of his narrative; and a knack for knowing when it is either necessary or desirable to leave the mysterious unexplained. Although the book drawn on here is ultimately about other planetary systems, the following excepts from a chapter entitled "Genesis and Evolution" function as a minibiography of our solar system, with a brief final focus on the necessary conditions for our own lives within it.

We begin with a glimpse of the process of star formation, with emphasis on the stardust that becomes the stuff of planets.

When we look around us in our galactic neighborhood, at the spiral arm in which the sun presently resides, we see several . . . star-forming regions of various ages and various stages of progress. The Orion Nebula is a present-day hotbed of star formation, with brilliant, highly active young stars seen still embedded in the dense gas and dust clouds from which they formed. The Scorpio-Centaurus cluster is older, the gas and dust largely swallowed by forming stars or blown away by the active young stars. The mass of this cluster has been so much reduced by gas expulsion that it can no longer hold itself together gravitationally. Its member stars are flowing out from its center like shrapnel from a bomb explosion, escaping from the "open" (gravitationally unbound) cluster of young stars and mixing into the population of older stars in our local spiral arm of the galaxy. Over the time of a single revolution of the galaxy (two hundred million years), as stars born together drift apart and migrate through other spiral arms, all visible connections between these stars will be lost and they will be thoroughly mixed in among the teeming billions of stars in the galaxy. Indeed, whenever we can identify a star-forming region, we can measure its expansion age, which is the time since these stars were all very close together in a compact, dense stellar nursery. . . . That age ranges from a million to a few tens of millions of years for many different clusters.

The youngest clusters are enormously complex and dynamic. When we look at regions such as the Orion Nebula, where new stars are presently forming, we see an astonishing menagerie of strange phenomena. The young stars embedded in the nebula all have a distinctly higher luminosity than mature stars of the same color and temperature, with ultraviolet radiation making up a large percentage of the total. In many cases we can see direct evidence of an enormous, flat, warm, dusty gas disk in the equatorial plane of the star. Many also have clearly developed jets of enormously hot gas, traveling hundreds of kilometers per second, streaming out of both poles of the star like gushers in an oil field, spraying gas thousands of astronomical units into space. These hyperactive, hyperluminous stars are called T-Tauri stars, after the name of the first-discovered star of that type.

The next stage of Lewis's discussion zooms in on the processes that lead to planet formation.

Now that we know something of the nature and life histories of stars, we can look at the origin of stars with the origin of their planets in mind. It will be helpful to see how, at least according to present theories, the prestellar nebulae that spawn stars also give rise to dust disks, and dust disks give rise to planets.

The first question really must be, "Why do irregular, spheroidal interstellar cloud fragments shape themselves into flat disks?" The answer lies in the fact that these cloud fragments come together from different sources and directions at different speeds. It is exceedingly unlikely that the net result of this random growth-and-collapse process will have no net rotational motion. But the pieces of the cloudlet, because of their different histories, have not yet shared their energy and momentum with each other: in effect, different parts of the cloud move independently, being as yet unaware of what the net rotation of the cloud will be. As the cloud shrinks, collisions between parts of the cloud become more frequent, and eventually the parts all "agree" on where the rotation axis is, which direction the cloud is rotating in, and how fast it is rotating. Very slowly rotating clouds may collapse almost unimpeded, leading quickly to very massive stars that age extremely rapidly. These are pre-supernova stars. The most rapidly rotating cloudlets tend to split into two or more parts, each orbiting around their common center of mass. These separate parts may then evolve independently, giving rise to a gravitationally bound cluster of two, three, four, or more stars. In the wide middle ground of moderate rotation speeds, medium-size single stars—or close double-star systems—can be made.

Each of these cloudlets, once isolated, continues to radiate heat. Loss of heat from the interior of the cloud lowers its internal pressure (which supports the outer part of the cloud), and the cloud collapses further. This generates more heat, which must in turn be lost to permit further collapse. But each cloudlet is rotating around a well-defined axis. Collapse toward the spin axis is hindered by the rotational motion of the cloud: any collapse that occurs causes the cloud to spin faster, which further opposes collapse. On the other hand, collapse parallel to the spin axis, toward the cloud's equator, has no such difficulty. As a result the cloud quickly collapses, as observed, into a flattened disk in the equatorial plane of the system. The densest central part of the disk collapses eventually into a star (or a close binary pair). The dust in the disk settles down onto the equatorial plane, forming a very thin layer where its concentration is so high that the dust readily clumps together. This agglomeration is the cosmic equivalent of what happens under your bed when you're not looking: airborne dust settles to the floor, where slow air currents stir the dust. Dust particles, upon gentle collision, stick together to make fluffy domestic (or cosmic) dust-bunnies.

Lewis describes the varying kinds of agglomeration of material at different distances from the star (and hence different chemical composition of eventual planets), as well as the thermal evolution that accompanies these varying processes. His discussion of geochemical evolution then focuses in on organic compounds, "among them the building blocks of life," and the "vehicles" by which they may have been distributed in our own solar system.

There are many clues to life's role in the planetary drama, and many possible routes to its origin. Organic matter, which is essential to life as we know it, is incredibly widespread in nature. Giant interstellar clouds, which spawn stars and planets, are rich in organic molecules. Radio astronomers have found dozens of organic compounds, some with over a dozen atoms, in these so-called giant molecular clouds. Many of these molecules are highly reactive: at higher temperatures and pressures, and especially in the presence of liquid water, they spontaneously combine to make even more complex molecules and polymers.

Interestingly, many of the same molecules have been found in comets. . . .

Comets originate from the outer solar system, where temperatures during the nebular disk phase were not high enough to destroy the organic matter inherited from interstellar clouds. Some of the ices in comets may have been evaporated and recondensed; other comets may be made of solid materials that have remained unchanged since before the solar nebula itself was formed. Indeed, many comets may be bodies ejected along with the Kuiper belt "iceteroids" from the region in which Uranus and Neptune swept up local solids, long after the demise of the nebular disk.

Closer to the sun, and under the influence of disturbing chemical processes such as lightning, shock waves, and ultraviolet light from the T-Tauri phase of the sun, organic matter may have been synthesized in large quantities within the nebula. The outer half of the asteroid belt today bears the signature of abundant tarry organic polymers. The same materials make up as much as 6 percent of the mass of the black, volatile-rich meteorite class called the carbonaceous chondrites. . . .

But organic matter frozen in comet nuclei or distant asteroids is of little biological interest. Can this organic matter be made available in locations, such as the surface of the earth, where life can actually arise? Does the ubiquitous presence of organic matter in interstellar space guarantee that every young planet will be seeded, via comet impacts, with the building blocks of life?

The most obvious way to introduce organic matter onto a sterile young planet is to send a comet full of organic molecules crashing into it. A dozen or so kilometer-size comets pass through the inner solar system every year, each with a probability of one in a billion of striking a terrestrial planet. One comet this size strikes earth about once in every hundred million years. But this approach to importing organic matter is a woeful failure: the comet is so fragile, and the impact so violent, that the result of the impact is a fiercely hot (about one hundred thousand degrees) explosion fireball that completely destroys every molecule in it. But all is not lost: as pointed out by Chris Chyba of the University of Arizona, the dust shed by the comet is a different matter entirely. Tiny particles of fluffy dust released by the evaporation of cometary ices create a huge dust cloud surrounding the icy nucleus and stretching along its orbit. Whether the intact nucleus strikes a planet or not, the dust from its tail very likely will. And these fluffy little particles act like tiny parachutes, slowing down gently at high altitudes without ever getting hot enough to vaporize or burn up their precious organic cargo. . . .

Organic matter, whatever its origin, can do marvelous things in water. The simple building blocks of sugars, amino acids, and organic bases, once made by reactions in liquid water, are then available for incorporation into much larger molecules, much in the way that a one-room brick factory makes twenty-story brick buildings possible.

All this organic chemistry does not occur in splendid isolation: on earth, for example, it occurred on the surface of a chemically complex rocky planet. On the surface of such a planet, ferrous iron minerals, clay minerals, sulfides, phosphates and a host of other inorganic species may be present. Ferrous iron (an iron atom with two electrons removed, making an ion with a double positive charge, Fe^{++}), for example, which is almost ubiquitous in rocky solar system bodies, has the interesting property that it can protect organic matter from oxidation by low concentrations of oxygen. Many clay minerals can not only attract and adsorb (bind to their surfaces) many simple organic molecules, but in some cases they can also help these simple molecules polymerize into complex products of biological interest. Phosphate minerals dissolve to some extent in water and can bind to sugars. Sugars can bond directly together to make long chains called polysaccharides, of which starch is a familiar example. Sugars can also bind, in concentrated solutions, to organic bases. When all three of these components are present, units called nucleosides (each consisting of one organic base molecule bonded to one sugar molecule) can be linked together by phosphates into extremely long polymer chains called nucleotides. One such nucleotide is DNA, the molecule that carries genetic information. Another is RNA. Organic acids with long hydrocarbon chains in them (fatty acids) cooperate to make lipid bilayers, which are familiar to us as cell walls. Amino acids can link together to make short chains called peptides, or long chains called proteins. Many of these peptides and proteins, often those containing metal ions weathered out of rocks, serve as efficient catalysts that greatly accelerate the rates of important biochemical reactions. These organic catalysts are called enzymes.

Self-replication begins on the inorganic level. The surfaces of some types of clays can not only serve as a template for the assembly of complex organic polymers, they can also release the product and make another in the same template. Even more astonishing, if sheets of atoms are cleaved off the surface of such a clay, the separated clay sheets may capture ions out of solution to grow new layers that are exact copies of themselves. Even flaws in the pattern can be replicated, demonstrating the ability of clays to "mutate" as well as replicate. It may well be that organic matter was first replicated by and on inorganic clays, and that organic matter learned how to replicate from these clays. From the primordial red clay, life was first shaped. The Hebrew word for red clay is *adamah*.

But this magic does not happen everywhere. If the temperature is too low, the solutions freeze solid, the dissolved molecules are immobilized, and reactions cease. If the temperature is too high, the delicate organic molecules are destroyed. In short, if there is no liquid water, nothing of interest happens. If there is oxygen present, the organic matter oxidizes to carbon dioxide, nitrogen, and water. And if there is no source of high-quality energy, none of the building blocks are made.

Life is not for every planet—at least, life as we know it is not suited to exist everywhere. Planets that follow very eccentric orbits may alternatingly freeze and roast their passengers; planets that don't rotate may have almost no stable life zone between the dayside furnace and the nightside deep freeze; planets that closely approach other planets may develop chaotic orbits that cause wild temperature excursions or even collisions; planets orbiting multiple stars may not survive long enough to develop life. It's a tough universe out there.

SOURCE: John S. Lewis, *Worlds Without End: The Exploration of Planets Known and Unknown*, Reading, Mass.: Perseus/Helix Books, 1998.

84

Mystery at the End of the Universe

Paul Davies

Paul Davies (b. 1946) is one of the most prolific and admired among recent writers on physics, astronomy, and cosmology. The title of his book The Mind of God *is taken from the final sentence in Stephen Hawking's* A Brief History of Time, *which states that if we could understand "why it is that we and the universe exist . . . then we would truly know the mind of God."*

In fact, Davies's book is more about mind than about God: His main thesis is "that mind—i.e., conscious awareness of the world—is not a meaningless and incidental quirk of nature, but an absolutely fundamental facet of reality." His final chapter addresses issues that most of us inevitably ponder in the face an awesome cosmos: How do we explain it? How do we conceive of our presence and role within it? Recognizing that these are religious questions, Davies opens the chapter with an epigraph from Fred Hoyle: "I have always thought it curious that, while most scientists claim to eschew religion, it actually dominates their thoughts more than it does the clergy."

In his famous book *A Brief History of Time* Stephen Hawking begins by recounting a story about a woman who interrupts a lecture on the universe to proclaim that she knows better. The world, she declares, is really a flat plate resting on the back of a giant turtle. When asked by the lecturer what the turtle rests on, she replied, "It's turtles all the way down."

The story symbolizes the essential problem that faces all who search for ultimate answers to the mystery of physical existence. We would like to explain the world in terms of something more fundamental, perhaps a set of causes, which in turn rest upon some laws or physical principles, but then we seek some explanation for this more fundamental level too, and so on. Where can such a chain of reasoning end? It is hard to be satisfied with an infinite regress. "No tower of turtles!" proclaims John Wheeler. "No structure, no plan of organization, no framework of ideas underlaid by another structure or level of ideas, underlaid by yet another level, and yet another, *ad infinitum*, down to bottomless blackness."

What is the alternative? Is there a "superturtle" that stands at the base of the tower, itself unsupported? Can this superturtle somehow "support itself"? Such a belief has a long history. . . . Spinoza argued that the world could not have been otherwise, that God had no choice. Spinoza's universe is supported by the superturtle of pure logical necessity. Even those who believe in the contingency of the world often appeal to the same reasoning, by arguing that the world is explained by God, and that God is logically necessary. . . . Problems accompany these attempts to explain contingency in terms of necessity [but] the problems are no less severe for those who would abolish God and argue for some Theory of Everything that will explain the universe and will also be unique on the grounds of logical necessity.

It may seem as if the only alternatives are an infinite tower of turtles or the existence of an ultimate superturtle, the explanation for which lies within itself. But there is a third possibility: a closed loop. There is a delightful little book called *Vicious Circles and Infinity* which features a photograph of a ring of people (rather than turtles) each sitting on the lap of the person behind, and in turn supporting the one in front. This closed loop of mutual support symbolizes John Wheeler's conception of the universe. "Physics gives rise to observer-participancy; observer-participancy gives rise to information; information gives rise to physics." This rather cryptic statement is rooted in the ideas of quantum physics, where the observer and the observed world are closely interwoven: hence "observer participancy." . . . So, rather than appeal to timeless transcendent laws to bring the universe into being, Wheeler prefers the image of a "self-excited circuit," wherein the physical universe bootstraps itself into existence, laws and all. . . . Neat though such "loopy" systems may be, they inevitably fall short of a complete explanation of things, for one can still ask "Why *that* loop?" or even "Why does *any* loop exist at all?" Even a closed loop of mutually-supportive turtles invites the question "Why turtles?"

All three of the above arrangements are founded on the assumption of human rationality: that it is legitimate to seek "explanations" for things, and that we truly understand something only when it is "explained." Yet it has to be admitted that our concept of rational explanation probably derives from our observations of the world and our evolutionary inheritance. Is it clear that this provides adequate guidance when we are tangling with ultimate questions? Might it not be the case that the reason for existence has no explanation in the usual sense? This does not mean that the universe is absurd or meaningless, only that an understanding of its existence and properties lies outside the usual categories of rational human thought. [The] application of human reasoning in its most refined and formalized sense—to mathematics—is nevertheless full of paradox and uncertainty. Gödel's theorem warns us that the axiomatic method of making logical deductions from given assumptions cannot in general provide a system which is both provably complete and consistent. There will always be truth that lies beyond, that cannot be reached from a finite collection of axioms. . . .

It seems to me that, as long as we insist on identifying "understanding" with "rational explanation" of the sort familiar in science, we will inevitably end up with turtle trouble: either an infinite regress, or a mysterious self-explaining superturtle, or an unexplained ring of turtles. There will always be mystery at the end of the universe.

From this acknowledgement of the inevitability of mystery, Davies turns to discuss the practice or simply the attitude of mysticism (though he warns that this should not be confused with "the occult, the paranormal, and other fringe beliefs"). He takes rhetorical pains not to appear to suggest that scientific thinking be abandoned without a struggle. Science and logic may let us down but "only" (!) when we are "dealing with ultimate questions."

In fact, many of the world's finest thinkers, including some notable scientists such as Einstein, Pauli, Schrödinger, Heisenberg, Eddington, and Jeans, have also espoused mysticism. My own feeling is that the scientific method should be pursued as far as it possibly can. Mysticism is no substitute for scientific inquiry and logical reasoning so long as this approach can be consistently applied. It is only in dealing with ultimate questions that science and logic may fail us. I am not saying that science and logic are likely to provide the wrong answers, but they may be incapable of addressing the sort of "why" (as opposed to "how") questions we want to ask.

Davies seems to suggest a complementarity, if not a convergence, between mysticism and science. He cites numerous examples of scientists, among them Fred Hoyle, Richard Feynman, and Roger Penrose, for whom experiencing a breakthrough of scientific insight is akin to mystical experience. Facing questions of the "beyond" may also lead us back to the concept of infinity.

In our quest for ultimate answers it is hard not to be drawn, in one way or another, to the infinite. Whether it is an infinite tower of turtles, an infinity of parallel worlds, an infinite set of mathematical propositions, or an infinite Creator, physical existence surely cannot be rooted in anything finite. Western religions have a long tradition of identifying God with the Infinite, whereas Eastern philosophy seeks to eliminate the differences between the One and the Many, and to identify the Void and the Infinite—zero and infinity. . . .

The belief that infinity was paradoxical and self-contradictory persisted until the nineteenth century. At this stage the mathematician Georg Cantor, while investigating problems of trigonometry, finally succeeded in providing a rigorous logical demonstration of the self-consistency of the actually infinite. Cantor had a rough ride with his peers, and was dismissed by some eminent mathematicians as a mad-man. In fact, he did suffer mental illness. But eventually the rules for the consistent manipulation of infinite numbers, though often strange and counterintuitive, came to be accepted. Indeed, much of twentieth-century mathematics is founded on the concept of the infinite (or infinitesimal).

If infinity can be grasped and manipulated using rational thought, does this open the way to an understanding of the ultimate explanation of things without the need for mysticism? No, it doesn't. To see why, we must take a look at the concept of infinity more closely.

One of the surprises of Cantor's work is that there is not just one infinity but a multiplicity of them. For example, the set of all integers and the set of all fractions are both infinite sets. One feels intuitively that there are more fractions than integers, but his is not so. On the other hand, the set of all decimals is bigger than the set of all fractions, or all integers. One can ask: is there a "biggest" infinity? Well, how about combining all infinite sets together into one superduperset? The class of all possible sets has been called Cantor's Absolute. There is one snag. This entity is not itself a set, for if it were it would by definition include itself. But self-referential sets run smack into Russell's paradox.

And here we encounter once more the Gödelian limits to rational thought—the mystery at the end of the universe. We cannot know Cantor's Absolute, or any other Absolute, by rational means, for any Absolute, being a Unity and hence complete within itself, must include itself. As [Rudy] Rucker remarks in connection with the Mindscape—the class of all sets of ideas—"If the Mindscape is a One, then it is a member of itself, and thus can only be known through a flash of mystical vision. No rational thought is a member of itself, so no rational thought could tie the Mindscape into a One."

Davies ends his book by turning attention back on the finite creature pondering questions regarding the infinite and the ultimate. The epigraph to his fi-

nal section is a single sentence by Freeman Dyson: "I do not feel like an alien in this universe." The heading of the section is a three-word question borrowed from the author of Psalm 8 in the Bible:

WHAT IS MAN?

Does the frank admission of hopelessness discussed in the previous section mean that all metaphysical reasoning is valueless? Should we adopt the approach of the pragmatic atheist who is content to take the universe as given, and get on with cataloguing its properties? There is no doubt that many scientists are opposed temperamentally to any form of metaphysical, let alone mystical arguments. They are scornful of the notion that there might exist a God, or even an impersonal creative principle or ground of being that would underpin reality and render its contingent aspects less starkly arbitrary. Personally I do not share their scorn. Although many metaphysical and theistic theories seem contrived or childish, they are not obviously more absurd than the belief that the universe exists, and exists in the form it does, reasonlessly. It seems at least worth trying to construct a metaphysical theory that reduces some of the arbitrariness of the world. But in the end a rational explanation for the world in the sense of a closed and complete system of logical truths is almost certainly impossible. We are barred from ultimate knowledge, from ultimate explanation, by the very rules of reasoning that prompt us to seek such an explanation in the first place. If we wish to progress beyond, we have to embrace a different concept of "understanding" from that of rational explanation. Possibly the mystical path is a way to such an understanding. I have never had a mystical experience myself, but I keep an open mind about the value of such experiences. Maybe they provide the only route beyond the limits to which science and philosophy can take us, the only possible path to the Ultimate. . . .

Through science, we human beings are able to grasp at least some of nature's secrets. We have cracked part of the cosmic code. Why this should be, just why *Homo sapiens* should carry the spark of rationality that provides the key to the universe, is a deep enigma. We, who are children of the universe—animated stardust—can nevertheless reflect on the nature of that same universe, even to the extent of glimpsing the rules on which it runs. How we have become linked into this cosmic dimension is a mystery. Yet the linkage cannot be denied.

What does it mean? What is Man that we might be party to such privilege? I cannot believe that our existence in this universe is a mere quirk of fate, an accident of history, an incidental blip in the great cosmic drama. Our involvement is too intimate. The physical species *Homo* may count for nothing, but the existence of mind in some organism on some planet in the universe is surely a fact of fundamental significance. Through conscious beings the universe has generated self-awareness. This can be no trivial detail, no minor byproduct of mindless, purposeless forces. We are truly meant to be here.

SOURCE: Paul Davies, *The Mind of God: Science and the Search for Ultimate Meaning*, London: Simon & Schuster, 1992.

85

Do the Heavens Declare?

Owen Gingerich

*Owen Gingerich (b. 1930) is a Senior Astronomer at the Smithsonian Astro-
physical Observatory and Professor of Astronomy and of the History of Science at Harvard. As a scientist with a profound grasp of historical
scholarship he has but few peers.*

*The sermon presented here—in its summary of longstanding issues such as
cosmic teleology and "fine tuning," in its citation of poetry that reflects our
awe at the wonders of the cosmos, in its retrospective historical glances, and
in its brief meditation on our critical but tenuous stewardship of "our own
nest" within the universe—makes a fitting final chapter for* The Book of the
Cosmos.

Do the heavens declare the glory of God? Haydn certainly thought so when
he wrote his oratorio *The Creation*. According to his biographer, H. C. Lan-
don Robbins, Haydn was enormously impressed by his visit to William Herschel's forty-foot telescope, then the largest in the world, and this was the
direct inspiration for "The Heavens Are Telling," now the basis for the hymn
that begins:

> The spacious firmament on high,
> With all the blue, ethereal sky,
> And spangled heavens, a shining frame,
> Their great Original proclaim.

Among my eclectic group of friends there is a wide spectrum of opinion as to
whether this in fact is what the heavens declare. Physicist Steven Weinberg

wrote, back in 1977, that "the more the universe seems comprehensible, the more it seems pointless." He has complained that "that phrase has dogged me ever since," and a few years ago he backed off a bit, saying, "I didn't mean that science *teaches* us that the universe is pointless, but rather, that the universe itself suggests no point."

On the other side of the question, I have friends among the evangelicals who love to quote St. Paul's letter to the Romans, "For the invisible things of him from the creation of the world are clearly seen, being understood by the things that are made, so that they are without excuse."

I've read with interest what various fellow astronomers have had to say about the invisible things from the creation of the universe. I don't even count the impulsive and widely hyped comment when the detailed satellite results on the microwave background radiation became available—that wonderful fossil from the early stages of the universe—to the effect that if you are religious, it's like seeing the face of God. What I have in mind is something even more invisible, the mighty explosion popularly known as the big bang. "There is no way to express that explosion" writes the poet Robinson Jeffers,

> . . . All that exists
> Roars into flame, the tortured fragments rush away from
>   each other into all the sky, new universes
> Jewel the black breast of night; and far off the outer nebulae
>   like charging spearmen again
> Invade emptiness.

It's an amazing picture, of pure and incredibly energetic light being transformed into matter, and leaving its vestiges behind. And it's more than a quarter of century since some astrophysicists noticed how uncanny it seemed that, at that inexpressible explosion, there was just the right amount of energy for a life-bearing cosmos to emerge. With less energy the universe would have long ago collapsed back on itself, leaving insufficient time for the cooking of the elements deep inside stars—heavy elements necessary for life—and for the slow evolutionary development of life itself. With more energy, the cosmic broth would have thinned out too quickly for gravity to bring together the stars and galaxies. Like Goldilocks and the little bear's dish, this porridge was just right! And so they proposed the so-called anthropic principle: that the universe has been designed so that intelligent life is possible. In some profound sense, the heavens did seem to be telling the glory of God.

Let's review several other curious details of the cosmos, without which we would not be here today. I have already mentioned the extraordinarily fine balance between the kinetic energy of the original explosion and the gravitational attraction of the parts that tend to pull the universe back together again. Without this balance we could not have a large and old universe filled with structure. But why have such a vast cosmos? Can we find anywhere a starker example of profligate wastefulness? Yet here is a modern paradox: the block the builders rejected has become the cornerstone. Antiquity now seems a necessity for the ultimate emergence of life, and with age comes size. We now understand the role of billions of years, possible only in a cosmos of vast dimensions. We see the long, drawn out chemical evolution in the life and death of generations of stars. Slowly there emerges a chemically enriched universe with a small but extremely relevant fraction of heavier elements necessary for the complexity of life—iron for hemoglobin, to name just one.

One of the first scientists to consider how the chemical environment itself made life possible was the Harvard chemist L. J. Henderson. In 1913, after the work of Darwin, which emphasized the fitness *of organisms* for their various environments, Henderson wrote *The Fitness of the Environment*, which pointed out that the organisms themselves would not exist except for certain properties of matter. He argued for the uniqueness of carbon as the chemical basis of life, and everything we have learned since then reinforces his argument. But today it is possible to go still further and to probe the origin of carbon itself, through its synthesis in the nuclear reactions deep inside evolving stars. Carbon is the fourth most common atom in our galaxy, after hydrogen, helium, and oxygen, but it isn't very abundant. A carbon nucleus can be made by merging three helium nuclei, but a triple collision is tolerably rare. It would be easier if two helium nuclei would stick together to form beryllium, but beryllium is not very stable. Nevertheless, sometimes before the two helium nuclei can come unstuck, a third helium nucleus strikes home, and a carbon nucleus results.

And here the internal details of the carbon nucleus become interesting: it turns out that there is precisely the right resonance within the carbon to help this process along. As a technical term, resonance is not exactly a household word. You've no doubt heard that great opera singers such as Enrico Caruso could shatter a wine glass by singing just the right note with enough volume. In principle I know it could be done, because at our Science Center at Harvard we shatter half a dozen glasses each year with intense sound waves. It's necessary to tune the audio generator to just the right note where the glass begins to vibrate—the specific resonance for that particular goblet—and then to turn up the volume until the glass vibrates so violently that it literally explodes.

The specific resonances within atomic nuclei are something like that, except in this case the particular energy enables the parts to stick together rather than to fly apart. In the carbon atom, the resonance just happens to match the combined energy of the beryllium atom and a colliding helium nucleus. Without it, there would be relatively few carbon atoms. Similarly, the internal details of the oxygen nucleus play a critical role. Oxygen can be formed by combining helium and carbon nuclei, but the corresponding resonance level in the oxygen nucleus is half a percent too low for the combination to stay together easily. Had the resonance level in the carbon been 4 percent lower, there would be essentially no carbon. Had that level in the oxygen been only half a percent higher, virtually all of the carbon would have been converted to oxygen. Without that carbon abundance, none of us would be here now.

I am told that Fred Hoyle, who together with William Fowler first noticed the remarkable arrangement of carbon and oxygen nuclear resonances, has said that nothing has shaken his atheism as much as this discovery. Occasionally Fred Hoyle and I have sat down to discuss one or another astronomical or historical point, but I have never had the nerve to query him directly: "I say, Sir Fred, did finding the nuclear resonance structure of carbon and oxygen really shake your atheism?" However, the answer came rather clearly in the November 1981 issue of *Engineering and Science*, the Cal Tech alumni magazine, where Hoyle wrote: "Would you not say to yourself, 'Some super-calculating intellect must have designed the properties of the carbon atom, otherwise the chance of my finding such an atom through the blind forces of nature would be utterly minuscule?' Of course you would ... A common sense interpretation of the facts suggests that a superintellect has monkeyed with physics, as well as with chemistry and biology, and that there are no blind forces worth speaking about in nature. The numbers one calculates from the facts seem to me so overwhelming as to put this conclusion almost beyond question."

Indeed, some of these circumstances seem so impressive that those scientists who wish to *deny* the role of design have had to take into account its ubiquitous signs. Briefly stated, they have turned the argument on its head. Rather than accepting that we are here because of a deliberate supernatural design, they claim that the universe simply must be this way *because* we are here; had the universe been otherwise, we would not be here to observe ourselves. Simply because we're here, the universe must be full of details that in fact make intelligent life possible. In other words, no matter how many wonderful details of apparent design in the physical universe I can find—and the universe is indeed full of them—the detractors can simply say that since we are contemplating them, these details could be no other way, and that's that.

So, despite Romans 1:20, I doubt that anyone can convert convinced skeptics to a theistic world view by arguments from design. Even William Paley, with his famous watch and his conclusion that it pointed to the existence of a watchmaker, said that "My opinion of astronomy has always been that it is not the best medium through which to prove the agency of an intelligent Creator; but that, this being proved, it shows, beyond all other sciences, the magnificence of his operations."

For me, it is not a matter of proofs and demonstrations, but of making sense of the astonishing cosmic order that the sciences repeatedly reveal. Fred Hoyle and I differ on lots of questions, but on this we agree: a common sense and satisfying interpretation of our world suggests the designing hand of a superintelligence. Impressive as the evidences of design in the astrophysical world may be, however, I personally find even more remarkable those from the biological realm. As Walt Whitman proclaimed, "A leaf of grass is no less than the journey-work of the stars." I would go still farther and assert that stellar evolution is child's play compared to the complexity of DNA in grass or mice. Whitman goes on, musing that,

> the tree-toad is a chef-d'œuvre for the highest,
> And the running blackberry would adorn the parlors of heaven,
> And the narrowest hinge in my hand puts to scorn all machinery,
> And the cow crunching with depress'd head surpasses any statue,
> And a mouse is miracle enough to stagger sextillions of infidels.
>
> (*Song of Myself*, stanza 31)

In other words, the heavens do declare the glory of God, but only to the prepared mind. Is that good enough? In this hard boiled age of skepticism we look beyond the "evidences of design," as Paley put it in the title of his book; we want just the facts—and then we want proof.

Or do we? Actually, very little of what we believe hinges on proof, and very much depends on what makes sense, what hangs together in a broader framework of coherence. As a historian of science, I could devote an hour to explaining how Galileo, back at the beginning of the seventeenth century, had no proof for the motion of the earth. Galileo tried desperately to find a proof, but the best he could come up with, based on the existence of ocean tides, turns out to be completely phony. Nevertheless, his arguments for the Copernican system eventually won the day because he was able to bring together so many different phenomena that began to make more sense in a universe in which the earth spun and revolved annually around the sun.

Meanwhile, north of the Alps, Johannes Kepler also accepted the physical reality of the Copernican arrangement, and demonstrated that each planet has an elliptical orbit with the sun at one focus. He could not prove the truth of the earth's motion, but without the assumption of a fixed central sun, his orbital system didn't make much sense.

Then along came Isaac Newton, who, with some simple but potent assumptions concerning the nature of matter and motion and with some powerful new mathematics, was able to build up a marvelous system of both explanation and prediction. Still, Newton had no *proof* of the earth's motion. What he had was an elaborate picture of how the physical world worked, so thorough and so probable that most people had no difficulty in accepting it as truth. Everyone knew that if the earth was *really* moving around the sun, the positions of nearby stars should show an annual displacement, but the failure to find it did not discredit the heliocentric theory. Thus, when the discovery of annual stellar parallax finally took place around 1840, people could hardly get excited about this purported proof of the motion of the earth. . . .

In science, then, as in life generally, we do our best to create a picture that makes sense even when we don't have all the pieces of the puzzle in hand. The same principle applies to religious faith. There are incredible, unsolved problems, whether they are the immense challenge of evil in this world, or the minor problem of how to understand the Christmas star, which defies any literal astronomy. Some time ago I got a call from *Life* magazine, asking if a reporter could interview me for half an hour about prayer. *Prayer!* What I don't understand about prayer could fill a library, but that doesn't keep me from praying. For me, it was easy to become a Christian because I grew up in a Christian family, among Christian friends, in a loving Christian surrounding. I *stay* a Christian not because I understand what prayer is really about, or because I take every piece of the Bible literally, but because it gives me a coherent view of our relation to a greater universe. When I read Genesis 1, I don't see there a veiled account of the big bang theory. I see a brilliantly concise statement of God's creativity, the superintelligence that brought the laws of nature and the physical structures of the universe into being, and within that a quintessential statement that humankind is created in God's image, which to me means that we are God's creatures sharing God's consciousness, conscience, and creativity.

When I think about the arguments that self-reflective intelligence is inevitable, I always pause to consider how rash it is to declare that high intelligence provides a selective advantage toward survivability. Watching the swiftness with which we are consuming the earth's resources and polluting our own nest, I find it hard to envision a future for humankind. Does our intelligence give us the wisdom to save ourselves from our own technological cleverness?

And then, when I turn back to the life of Jesus Christ, and what he said about salvation, I can't help but think that there are many facets to salvation, and perhaps one of them has to do with the saving of the human race. Christ spoke of servanthood and sacrificial love, demonstrated in his life and death. Somehow it seems to me that unless we begin to get that message, in our selfishness we are doomed as a species.

I am not a theologian, nor have I solved the riddles of existence, of time, of sin and forgiveness, of prayer. I cannot prove that God exists, or that God's claim on our lives is what makes life ultimately meaningful. But do the heavens declare the glory of God? I think so. The universe is so full of such wonderful things that I can hardly think otherwise. Holding this view makes my understanding of the world richer and more coherent. But I can't prove it. I can only conclude with Pascal: The heart has its reasons that reason cannot know.

Let me finish with the prayer that Kepler placed near the end of his *Harmonice mundi*:

> I give Thee thanks, O Lord creator, because I have delighted in thy handiwork and I have exulted in the works of thy hands. Behold: now I have completed the work of my profession, having used as much of the ability as Thou hast given me. If anything unworthy of thy designs has been put forth by me, inspire me also so that I may set it right. If I have been allured into brashness by the wonderful beauty of thy works, or if I have loved my own glory among men, while advancing in work destined for thy glory, gently and mercifully pardon me; and finally, deign graciously to cause that these demonstrations may lead to thy glory and the salvation of souls and nowhere be an obstacle to that. Amen.

SOURCE: Owen Gingerich, "Do the Heavens Declare the Glory of God?" (a sermon delivered 5 December 1993 at Sage Chapel, Cornell University, and made available to *The Book of the Cosmos* by the kindness of the author).

Glossary

The following is intended as lexicographical first aid. For more extensive and technically rigorous treatment, please consult the books by Gribbin, Lightman, and Coles mentioned in "Further Reading" (p. 537).

absolute space. The Newtonian concept of endless, homogeneous, and isotropic space: It stretches in all directions; it exists independent of the bodies it contains; and relative *to* it all motion is conceived to take place.

absolute time. The Newtonian concept of beginningless and endless duration: it flows steadily onwards; it is independent of events; and relative *to* it all events take place.

anthropocentrism. The view that human beings are, literally or figuratively, central to the **universe**. Often mistakenly seen as entailed by **geocentrism**, versions of which in fact present the **universe** as anthropoperipheral (see Chapter 15).

anthropic principle. A set of claims (with varying strong and weak formulations) integral with the recognition that the **universe** must be conceived to be such that observers like us could have come, did come, and perhaps must have come to exist within it (see Chapter 73).

atoms. In ancient Greek usage (later adapted to modern scientific theory), elementary particles that are fundamental in the sense that they cannot be further divided (from *a toma*, "not divisible").

Big Bang. The term coined by Fred Hoyle (see Chapter 67) in derision of the idea—now accepted as the Standard Model (see Chapter 72)—that the universe began with the explosion of a primordial particle of unimaginable smallness and density.

black hole. A super dense, super massive collapsed **star** whose overbearing gravitational behavior precludes the emission of light or other radiation and so places the star and its vicinity beyond our observational horizon (see Chapter 75).

book. A physical device embodying a text (a "fabric") that communicates messages using images and/or symbols arranged in an orderly manner—usually produced by an "author" (perhaps with some help) for the enjoyment or edification of himself or herself, and possibly also of "readers." The device frequently takes the form of bound pages capable of being held in the hand, although many interpreters have seen other things, such as the **universe** ("the book of nature" or "the book of the cosmos"), as roughly matching this definition.

boundlessness. (Predicated of a given **space** of however few or many dimensions.) The condition of presenting no limit or barrier to an object moving within, through, or across the space. Now distinguished from **infinitude**. (Typically illustrated by an ant hiking across the surface of a sphere: He or she will never bump into anything. The surface is boundless, though not infinite.)

Copernicanism. The set of beliefs, associated with Nicholas Copernicus, to the effect that the system of the observable **planets**, including the **earth** now conceived as a **planet**, revolves about the sun. Often loosely extended to any rethinking of a system that involves the "relocation" of that which was previously conceived to be at the center (see Chapter 43).

cosmogony. A theory or story about the beginning or origin of the **universe**. (Hence, a branch of **cosmology**.)

cosmology. Historically: narrative, meditative, or interpretive discourse concerned with questions such as the beginning, substance, structure, aesthetics, **purpose**, and destiny of the **universe** as a whole. Academically: often restricted to the (still enormous) material or physical dimension of these questions and studied as a branch of astrophysics.

cosmos. A name for the **universe**; originally a Greek word (*kosmos*) bespeaking order and beauty.

creation. A term, when applied to the **universe**, that entails a view of the latter as something originating or deriving from the intention and agency of a creator.

curvature. The deviation of any extended geometrically describable entity (e.g., line, plane, **space**, space-time) from the "norm" of Euclidean rectilinearity (from straightness, flatness, etc.).

earth/s. Earth is the name of the sphere, the "terrestrial ball," upon which humans dwell. However, once this sphere came to be seen as a **planet**, the **planets** could be conceived of as earths.

empyrean. A term used theologically and poetically to denote Heaven, the realm of God, to distinguish it from the heavens (the astronomical sky).

entropy. A degree of disorder in a system. Sometimes used as shorthand for the second law of thermodynamics, which states that this degree of disorder in a given system only increases with the passage of **time**. (Exemplified by the practical observation that you will never get warm by huddling over a heater that is colder than you are.)

eti. Extraterrestrial intelligence.

etl. Extraterrestrial life.

expansion. In cosmology, the increasing measure of the cosmic scale—the ever-greater distance, among large-scale heavenly bodies, of everything from everything else—as evidenced by galactic red shift (see Chapter 64). Distinguished from **inflation**.

flatness. Describes a **universe** with a rate of expansion critically poised between a closed and an open **geometry**—between recollapse and infinite expansion. The sheer unlikelihood yet apparent actuality of our universe's exhibiting this critical behavior is the "flatness problem," which the theory of **inflation** aims to address (see Chapter 79).

galaxy. Derived from the Greek term denoting the Milky Way. With the recognition that the Milky Way is one of many such agglomerations of **stars**, the word became generic and could be used in the plural.

geocentrism. The theory that **earth** (because of its heaviness relative to other substances) is located at the center of the **universe**—or at least at the center of the planetary system (associated with the teaching especially of Aristotle and of Ptolemy). Often mistakenly seen as entailing a view that earth's human inhabitants are, literally or figuratively, central to the **universe**. See **anthropocentrism**.

geometry. Literally, the measuring of the **earth**. The branch of mathematics that, until after the mid-nineteenth century, was synonymous with the principles and definitions of Euclid, according to which lines, planes, and spaces are described rectilinearly. But then Riemann proposed a revised geometry that could be adapted to describe non-rectilear, curved **space** (see Chapter 57).

Great Year. The exceedingly long period of **time** from one conjunction of all the **planets** until a recurrence of the same.

heliocentrism. The theory that the **universe**—or, more minimally, the system of the observable **planets**, including the **earth**—revolves about the sun (Greek *helios*). See **Copernicanism**.

hermeneutics. (Etymologically connected to Hermes, the god whose job was the delivery of messages.) The science or methodology of interpreting texts—including, potentially, the "text" or the **book** of the **cosmos**.

imagination. The human capacity to form images. Historically, viewed within both religious and scientific circles as potentially vain and idolatrous to the extent that it can produce "idols"—distorted or distorting notions about reality—yet also potentially beneficial and creative to the extent that it enhances our intellectual and aesthetic grasp of things.

infinitude. (Predicated of a given **space**, temporal framework, set, or object, such as a **universe**.) Extension in at least one dimension and/or direction without end. Now distinguished from **boundlessness**.

inflation. An innovation within the Standard Model according to which the early **universe** underwent a brief "era" of extremely rapid, exponential expansion (see Chapter 79). Distinguished from mere **expansion**.

magnitude. *Absolute* magnitude is an expression of a star's intrinsic luminosity. Historically, however, *apparent* magnitude is a measure of how bright a **star** appears to a human observer, with larger numbers indicating lesser brightness—1 indicating roughly the brightest star and 6 the dimmest one visible to the naked eye under optimal conditions.

mind. Not only an individual's capacity of reflection and understanding, but also a more encompassing consciousness or intentionality that some cosmologists see as present in, or evidenced by, the **universe** (see Chapter 84).

moon/s. Within **geocentrism,** *the* moon was viewed as the **planet** closest to the **earth.** Within **heliocentrism,** the same moon is earth's satellite, but other **planets** may have their own moons (see Galileo, Chapter 24).

nebula/nebulae. Literally, a cloud, or clouds. Astronomically, the nebulae appeared as bright clouds but many were telescopically resolvable into myriads of **stars**—hence, galaxies. However, spectroscopy revealed still others to be clouds indeed, of gaseous composition (see Chapter 52).

New Astronomy. Depending on historical context, principally the new astronomy of Kepler in the seventeenth century, or the new spectroscopically aided astronomy of the nineteenth century.

parallax. The apparently varying movement or position of any object resulting from the varying location of an observer (as when one views the same object with one eye and then another). Parallax permits the calculation of distance, including astronomical distance, by means of triangulation.

philosophy. Literally, a love of wisdom. Traditionally conceived to encompass the search for or study of knowledge across different disciplines. Hence, *natural* philosophy—for a number of centuries the rough equivalent of what we call **science**—was the search for knowledge within the realm of the natural or the physical.

planet. Literally, a "wanderer." Traditionally distinguished not from **stars** absolutely but only from stars that do not wander, the "fixed" stars. The **earth** itself was recognized as a planet—and hence as a star (see Chapter 22)—only with the advent of **heliocentrism.**

Platonic year. See **Great Year.**

poetry, poetics. From a Greek root denoting creation. Poetry and poetics, like **book** and **hermeneutics**, usually apply literally to texts, to arrangements of words, but can by extension be applied to the **universe** and to its structure or its principles of composition (see Chapters 19 and 57).

purpose. The function, goal, or "end" (Greek *telos*) for which something exists or was created, or for which an action is carried out. Cosmologically, the question is whether the **universe** has a purpose or a "point" (see for example Chapters 72, 73, 84), and, if so, whether human beings can know something about that purpose, or are connected with it. This discussion falls within an area of study known as *teleology*.

relativity. In physics and cosmology, relativity is shorthand for Einstein's theories about the relationship between **space** and **time** (special relativity; see Chapters 58 and 59) and about the **geometry** of spacetime as an account of gravitation (general relativity; see Chapters 58 and 60).

rotation/revolution. Although (to the chagrin of translators) these two terms have historically not been distinguished very rigorously, "rotation" generally denotes the turning of a body about its own axis (e.g., "the earth rotates once every twenty-four hours"), while "revolution" denotes the circling or ellipsing of a body about a more-or-less central point or focus (e.g., "the earth revolves about the sun annually").

science. Etymologically, science means simply know-how, knowledge, or a branch of knowledge. In English, its meaning has narrowed to denote what used to be called natural **philosophy**. See also **scientist**.

scientist. A term first proposed in 1834, by analogy with "artist," to designate a student of **science**—"of the knowledge of the material world" (OED).

singularity. Hypothetically, a point or place (presumably its own place) of infinite density, perhaps "located" within a **black hole** or at the beginning of our **universe**, at time = 0.

space. One of the most radical and difficult concepts of cosmology. The Atomists (see Chapter 4) postulated space or emptiness as a precondition of physical movement. Etymologically, space is related to the Latin noun for hope (*spes*) and the German verb for saving money (*sparen*), both of which carry the core meaning of "room to move," be it in one's life or one's budget. With the advent of general **relativity**, space is no longer conceived of as merely neutral, empty, or **absolute space**, but as something with definite properties (see chapter 60).

spectroscopy. The study of the nature of physical objects such as **stars** by means of an analysis of their *spectra*—of the radiation (particularly light) that they emit (see Chapter 52).

sphere. Historically, an orb or "ball" that may be either solid or hollow. The **planets** have for millennia been considered spheres. But in pre-Copernican theory they were often thought to be carried along in their courses by hollow crystalline spheres turning silently (or with a music that human ears cannot discern), one sphere turning within the other. "Sphere" could also denote the empty or partly occupied territory within the hollow shell of the sphere itself. Hence, **earth** is located in the "sublunary sphere" (the area bounded by the sphere of the moon).

Standard Model. See **Big Bang.**

star/s. Traditionally, any radiant heavenly bodies, the two main categories being **planets** ("wandering stars," including the **moon** and the **sun**) and fixed stars—those whose movements seem not to vary relative to each other. **Star** now usually has the more restricted sense of self-luminous sun-like heavenly body and is distinguished from **planet**.

steady state. A cosmological model pioneered by Herman Bondi, Thomas Gold, and Fred Hoyle. For about twenty years in the mid-twentieth century it was a contender with the **Big Bang** hypothesis. Steady state postulated that, to an observer, the **universe** would always over **time** appear the same (see Chapter 67).

sun/s. Copernicus and Kepler both considered the sun to be utterly unique; but by the late seventeenth century, poets and visionaries were already referring to the **stars** as "other suns perhaps, / With their attendant moons" (Milton). Thus **sun**, along with **earth, moon, galaxy,** and **universe**—with the loss of belief in these things' uniqueness—came to be used in the plural.

time. "Everybody knows what time is, but nobody can *explain* what it is" (John Gribbin, *Companion to the Cosmos*, p. 404). **Relativity** denies that time is absolute and binds it together with **space** in spacetime. Expositors of the Standard Model see time as having a beginning, sort of (it can't of course be a beginning *in* time). Moses Maimonides, along with many other theists, views time as something created (see Chapter 14). It is also typically seen as being, or having, an arrow—one that shoots in one direction only, though why this should be so is an unsolved problem. In any case, it is probably not coherent, humanly speaking, to think of time as having a *brief* history.

universe/s. St. Augustine taught that the universe, being a creation, constituted all things turning toward the One (*uni-versus*). **Universe** is now used as a synonym for **cosmos** and denotes all there is, or at least all there physically is, regardless of whether this All is conceived of as a creation or "simply one of those things which happen from time to time" (Edward Tryon; see Chapter 79). If there are **worlds** beyond the horizon of our universe, then these can be conceived of as other universes (see Chapters 77 and 78).

vacuum. An emptiness, a vacuity, a true **space**. Aristotle, and apparently most natural philosophers up to and including Descartes and Leibniz, did not believe there was such a thing (because Nature abhors a vacuum). Relativity theory defines vacuum as a state of minimum (not absolutely null) energy.

world/s. Historically, **world** is a synonym mainly for **universe** rather than for **earth**. Accordingly, the earth is located in the world. With the blurring of this distinction, "other worlds" can now mean "other earths" or "other universes," depending on context.

Further Reading

Readers with a taste for more of what they have found in this anthology are best to begin by chasing up the sources I have cited at the end of each chapter (though some of these aren't too easy to find!). Part of the impetus of this work has been my belief that you're always best to get ideas and information from their source, or as close to the source as you can manage. So my chief suggestion is simply to lay hands on a copy of Maimonides, Kepler, Herschel, Hubble, Hawking, Ferguson, Gingerich, or some other writer from *The Book of the Cosmos* who has touched your imagination, and explore his or her writing more intimately than an anthology selection permits. There's no substitute for a long afternoon spent in the diaries of Maria Mitchell, or in *The Realm of the Nebulae* (Hubble), or in the autobiography of Cecilia Payne-Gaposchkin. Or pick up Michael J. Crowe's reader-friendly and excerpt-rich books *Theories of the World from Antiquity to the Copernican Revolution* and *Modern Theories of the Universe from Herschel to Hubble*. These also provide not-too-intimidating exercises that even nonscientists can perform, and have excellent bibliographies as well.

First-hand readings can then be supplemented with works such as John Gribbin's highly useful *Companion to the Cosmos*, or John D. North's magisterial *The Norton* (a.k.a. *The Fontana*) *History of Astronomy and Cosmology*, or Norriss S. Hetherington's *Encyclopedia of Cosmology*. I would also dip into Alan P. Lightman's entertaining and informative *Origins: the lives and worlds of modern cosmologists*, which comprises mainly firsthand interviews. A further helpful reference is *The Icon Critical Dictionary of the New Cosmology*, edited by Peter Coles.

Beyond these I would recommend accessible if occasionally demanding historical studies that approach a period or issue in cosmology from a particular angle: Alexandre Koyré's *From the Closed World to the Infinite Universe*; Thomas Kuhn's *The Copernican Revolution*; Hans Blumenberg's *The*

Genesis of the Copernican World; or Barrow and Tipler's massive study of *The Anthropic Cosmological Principle*. Two books that have stretched my understanding of space and its measurement are Max Jammer's *Concepts of Space: the history of theories of space in physics* and Albert Van Helden's *Measuring the Universe: cosmic dimensions from Aristarchus to Halley*. For a study of cosmology and literature one could hardly do better than to start with Marjorie Hope Nicolson's *Science and Imagination*. Or, for a sweep of "other worlds" discussions, turn to Steven J. Dick, *Plurality of Worlds: the origins of the extraterrestrial life debate from Democritus to Kant*; Michael Crowe's *The Extraterrestrial Life Debate, 1750–1900: the idea of a plurality of worlds from Kant to Lowell*; and Dick again, *Life on Other Worlds: the 20th-century extraterrestrial life debate*.

To start exploring more contributions to cosmology by women writers, please consult two works I unfortunately could not benefit from in compiling this anthology: *A Biographical Dictionary of Women in Science*, edited by Marilyn Bailey Ogilvie and Joy Harvey (forthcoming, New York: Routledge, 2000); and Pamela Gossin's *Beneath the Stars: A Literary History of Astronomy, Women, and Poetics, 1590–1990* (in progress).

And that's enough for a start!

Copyright
Acknowledgments

Index